D1480918

66778

HE
9784.5   Eddy, Paul
F82   Destination disaster
P264
1976

DATE DUE			
AUG 23 1995			
DEC 8 1995			
APR 22 1996			

LEARNING RESOURCES CENTER
MONTGOMERY COUNTY COMMUNITY COLLEGE
BLUE BELL, PENNSYLVANIA

DESTINATION DISASTER

Additional reporting and research by
Patrick Forman and Elaine Handler

The authors are grateful to Jacques Lannier, now Chef
d'Escadron of the Gendarmerie, Angers, France, for
permission to reproduce in Chapter 1 extracts from his
written recollections of the events of March 3, 1974.

DESTINATION DISASTER

From the Tri-Motor to the DC-10: The Risk of Flying

by Paul Eddy, Elaine Potter, and Bruce Page

Quadrangle/The New York Times Book Co.

Random lines of poetry from *Hymn of Breaking Strain* by Rudyard
Kipling, copyright 1935, and Kipling's *The Secret of the Machines*,
copyright 1911, appearing in the book *Rudyard Kipling's Verse:
Definitive Edition* are reprinted by permission of Mrs. George
Bambridge and Doubleday & Company, Inc. Lines from Werner
Heisenberg's, *The Meaning of Beauty in the Exact Sciences* are
reprinted from *Across the Frontiers* by permission of Harper and
Row. The lines from Simone Weil's essay *On Bankruptcy* are
reprinted from *Selected Essays 1934–1943,* translated by Richard
Rees 1962, by permission of Oxford University Press. The excerpt
from *The General Theory of Employment, Interest and Money*
by John Maynard Keynes is reprinted by permission of Harcourt
Brace Jovanovich, Inc. The copyright of *Musée des Beaux Arts*
from W. H. Auden's *Collected Shorter Poems 1927–1957* is held by
Random House, Inc. who gave permission for the extract to be
reprinted.

Copyright © 1976 by Times Newspapers, Ltd.
All rights reserved, including the right to reproduce this book or
portions thereof in any form. For information, address: Quadrangle/
The New York Times Book Co., Inc., 10 East 53 Street, New York,
New York 10022. Manufactured in the United States of America.
Published simultaneously in Canada by Fitzhenry & Whiteside, Ltd.,
Toronto.
Published in Great Britain by Granada Publishing Ltd.
Book design: Beth Tondreau

Library of Congress Cataloging in Publication Data

Eddy, Paul, 1944-
 Destination Disaster.

 Includes bibliographical references and index.
 1. Aeronautics—Accidents—1974. I. Potter,
Elaine, joint author. II. Page, Bruce, joint
author. III. Title.
HE9784.5.F82P264 1976   387.7′33′49   75-36263
ISBN 0-8129-0619-5

Second printing, November 1976

66778

STRAND   B.A.   6.25   1976/77

To the 346 passengers and crew of Ship 29

CONTENTS

PHOTOGRAPHS

PREFACE

THIS IS THE story, as complete as we can make it, of the first crash of a fully loaded jumbo jet, its causes and its consequences. It is also an inquiry into the nature and epic history of the civil aviation industry. The two stories are inexorably linked, for the world's worst air disaster was no accident in the true sense of the word: It resulted from the enormous commercial and political stresses which surround the aviation business and it was foreseen and actually forewarned.

The crash and, for a while, its postmortem provoked great public interest —and, in the United States, congressional interest—but that concern waned long before the investigation into the causes had produced any really meaningful results. Essentially, it has been left to a few American lawyers representing the relatives of those who died to pursue the truth, and their activities will cease once they decide that sufficient compensation, in monetary terms, has been won.

There are overwhelming reasons why that should not be so. The Paris disaster, and the events that led to it, demonstrated that while the commercial air transportation system in the Western world is remarkably safe, it is nowhere nearly as safe as it could be. Our story illuminates some of the lessons that should be learned, but we cannot pretend that any journalistic inquiry is adequate. Nor is it sufficient to leave the reexamination of aviation safety standards—as presently seems to be the inclination—to the industry itself and to those who are supposed to regulate it: It was precisely that policy which won awful retribution in the Forest of Ermenonville on March 3, 1974.

The only body which has the necessary power to inquire deeply into the nature and standards of the commercial aviation business is the Congress of the United States because the vast majority of airliners in service in the Western world are American-built. Undoubtedly, a full-scale congressional investigation would be time-consuming and expensive, it would require the assistance of a great many technical experts, and it would probably go largely unrewarded in terms of newspaper headlines. It would, however, save lives.

The matter is of no small importance to all of those who fly, however infrequently. The 346 people who died with appalling violence in the Paris disaster were a mixture of experienced travelers and first-time flyers, businessmen and holiday-takers, the eminent and the obscure, young and old of twenty-one nationalities. Most of them had never intended to take that flight and actually held tickets for other ones. They were assembled together, and therefore picked out for death, in the very last moments before takeoff through chance and circumstance. There, but for fortune. . . .

DESTINATION DISASTER

We can pull and haul and push and lift and drive,
We can print and plough and weave and heat and light,
We can run and race and swim and fly and dive,
We can see and hear and count and read and write. . . .

But remember, please, the Law by which we live,
We are not built to comprehend a lie.
We can neither love nor pity nor forgive—
If you make a slip in handling us, you die!
—Rudyard Kipling, *The Secret of the Machines*

A Legal and Moral Problem

About suffering they were never wrong,
The Old Masters: how well they understood
Its human position; how it takes place
While someone else is eating or opening a window
  or just walking dully along; . . .

In Breughel's *Icarus*, for instance: how everything turns away
Quite leisurely from the disaster; the ploughman may
Have heard the splash, the forsaken cry,
But for him it was not an important failure; . . .
          —W. H. AUDEN,
          "Musée des Beaux Arts"

SUNDAY, MARCH 3, 1974, was a duty day for Captain Jacques Lannier, commander of the Senlis District of the French Gendarmerie Nationale. Lannier was slightly annoyed at having to work on the first really fine day of spring, with the air crisp and clear, and the celebrated daffodils of the Forest of Ermenonville shining among the trees, pushing aside the last hardened crusts of snow. Senlis is a pleasant outer suburb some thirty miles northeast of Paris. The forest, covering twenty-five square miles of gently rolling hills, crisscrossed with hiker's trails and bumpy tracks, is one of the city's chief weekend resorts.

Still, it did not look like it would be a heavy day. There was a motor endurance rally that Lannier regarded as noisy and ridiculous but which, apart from some questions of traffic control, was not burdensome. There was also the village fete of Bois Hourdy—an affair of decorated floats, carnival stalls, and concerts, much more to the captain's personal taste and usually no trouble at all. Lannier decided to take his family to lunch in the Officers' Club at Senlis which had pleasant surroundings, not-bad food, and a reliable telephone switchboard in case anything should come up.

Just as they were settling themselves in the dining room, some time around 12:45, their waiter said that Captain Lannier was wanted on the telephone. It was Lannier's personal assistant, speaking from the control room at the Senlis Gendarmerie.

3

"Captain, I have just been informed that an aircraft has crashed between Le Plessis Belleville and Ermenonville."

"Is it a plane from the Aero Club?" asked Lannier. There was a flying club in the parish of Plessis Belleville, and his men were used to dealing with minor accidents of gliders and light aircraft. Rarely was it anything more than a matter of a few broken bones.

The message had come from one of the road-safety patrols on the Nord freeway, and it was not clear what sort of plane might be involved. Lannier agreed that the Nanteuil and Senlis squad cars should investigate. No need to do more. He exchanged a "bon appetit" with his assistant—the natural envoi of two Frenchmen speaking together just before lunchtime—and returned to his seat.

He had scarcely sat down when he was called back to the phone.

"Orly Airport informs us that the aircraft is a DC-10 of Turkish Airlines with 185 people on board. In addition, a violent explosion has just occurred in the forest near Mortefontaine."

Lannier, perhaps, did not quite believe it. As he says, he comes of a generation to whom propeller-driven DC-4s and DC-6s used to seem large. He had never heard of the DC-10 and had never really appreciated that in the jumbo-jet era a single airplane can carry three or four hundred passengers. And, of course, he had no way of knowing that the first report from Orly had underestimated by half the number of people on board the DC-10. Not that he would have acted differently. His intuition had already told him that this was a major disaster.

"Have all the squads on radio alert. Get all the available staff to converge towards Route Longue and D 126. Warn the hospitals, ambulances, and emergency centers. Inform Beauvais.* I am coming over. . . ."

Route Longue/D 126 is the intersection of two small roads that cross the Forest of Ermenonville roughly at right angles to each other. Once the wreck was located, Lannier wanted to have all his force concentrated in one handy spot.

A swift, blunt word of explanation for his family, then a dash home to get into uniform: a policeman at a disaster site is only half-effective without his identifying dress.

At police headquarters Lannier threw maps, binoculars, briefcase, telephone directory, and first-aid kit into one of the cars equipped with a shortwave radio. No time to waste finding a driver: switching on the revolving flasher, sounding the two-tone emergency horn, Lannier sent the radio car weaving through the Sunday drivers toward Ermenonville.

IT SEEMS ALMOST unbelievable that a 300-ton airplane, carrying 346 men, women, and children and their belongings, can plunge into the closely settled soil of Europe and disappear. Yet, it is a fact that wrecked aircraft

* Departmental HQ of the Gendarmerie.

are often difficult to find. The DC-10, to be sure, was too large to escape detection permanently, and this was a clear, sparkling day with the sun at its height. But there were only thirty-odd police and ambulance men immediately available to search a thickly wooded area roughly ten miles long by ten miles wide: each man's horizon was often limited to fifty yards or less. Apart from these official searchers, set in motion by the radio network, most of the people inside the search area were thinking in terms of lunch and a glass of wine.

Lannier and his men blundering about the woods knew that it is in the first hour after an accident that lives can be saved: when blood flow can be staunched and blankets can counteract shock. They had no way of knowing that this time first aid was of no possible relevance.

"Scene of the accident located," said Lannier's radio. *"It is very close to Chaalis Lodge."*

The lodge is on Route Longue at a point where several forest tracks join up. Lannier changed the convergence point to Chaalis and asked for more details. *"On the right of the lodge"* was all the Senlis squad-car radio could add.

Two minutes later, Lannier was at the track junction. There was no one in sight, and all the tracks had their wildlife barriers down and locked to stop tourists' cars from penetrating the forest. Clearly no squad car had been there, because policemen in a hurry do not close barriers behind them. Taking a key from the glove box, Lannier selected one of the low metal gates, almost at random, unlocked it, and swung his car through the gap.

After two kilometers, he felt uneasily that he was on the wrong path. He met a man and woman walking quietly among the trees. No, they had told him, they had seen nothing.

Back to the lodge. The squad-car driver came onto the radio: He was lost, had been going round in circles, but he thought Chaalis could not be more than a kilometer from where he was. Lannier was soon circling himself, thrashing his car along rough tracks cut out by forest tractors, hoping his car would keep moving.

Suddenly, at the junction of two tracks, he saw a red truck, the jeep from the Senlis fire station, and the blue Renault squad car. The patrolman came up as Lannier braked. It was 1:15.

"Bodies all over the place, like. There are no survivors." Lannier was mildly surprised. The patrolman normally spoke quite polished French and the use of the slang "like" suggested unusual stress. The man handed Lannier a baggage registration ticket. It read: "Turkish Airlines. Paris Flight TK 981. To: LON."

On the other side of the card was a small, clear picture of a three-engined jetliner, and the Turkish words: TURK HAVA YOLLARI. Lannier stopped the search, got on the radio and told the patrolman to start a log.

He half-ran, half-walked along a track, with wreckage littered on either side, that lead to what an hour ago had been a pleasant shallow valley called the Grove of Dammartin. There he confronted the appalling scene of the worst air crash the world has ever seen. It was, as he wrote later, a "nightmare," and for once that tired noun was aptly used.

> On my left, over a distance of four hundred or five hundred meters, the trees were hacked and mangled, most of them charred but not burnt. Pieces of metal, brightly colored electric wires, and clothes were littered all over the ground. In front of me, in the valley, the trees were even more severely hacked and the wreckage even greater. There were fragments of bodies and pieces of flesh that were hardly recognizable. In front of me, not far from where I stood, there were two hands clasping each other, a man's hand tightly holding a woman's hand, two hands which withstood disintegration. . . .

For the rest of that terrible day, as he and his men toiled in the sunshine to clear the Grove of Dammartin, the image of the hands haunted Lannier's mind: two human beings, ambushed by death, reaching to each other for a last, brief assurance of existence.

THE WORLD NEVER lacks for horrifying images: a bar wrecked by a terrorist's bomb; a village napalmed by military error; a coal mine flooded; a child's body by the roadside.

Lethal events can hardly be ranked in some exact order of horror. Nonetheless, there is a special affront to the senses in the scene of an airliner's destruction. Typically, there will be a considerable number of people involved in each accident, so the witnesses are confronted with a dense concentration of bodies. The impact is likely to have been extremely violent, so that the mutilations are grotesque and extreme. The dead will have been, for the most part, healthy and prosperous people and their disfigured corpses are usually surrounded by evidence of the duties and pleasures they were enjoying until the moment of disaster: books, items of jewelry, holiday clothes, briefcases, typewriters, and cameras.

The airliner in flight, maintaining its human occupants in warmth and ease against the gasping cold of the stratosphere, is a classic symbol of the control that modern humanity can achieve over the forces of the physical universe. Naturally, therefore, the wrecked airliner is an equally potent symbol of the fragility of that control and the drastic consequences that follow upon its failure.

The people who build and operate airplanes have a particular distaste for any plainly realistic account of a crash site. When the authors of this book helped make a television film about the crash of the DC-10,* the executives of the McDonnell Douglas Corporation, who built the airplane, were greatly

---

* "The Avoidable Accident," Thames TV (London), WNET/Channel 13 (New York), and *The Sunday Times* (London), 1975.

angered by the fact that the film included still photographs of the scene at Dammartin, calling it "emotionalism."

One may wonder whether an accident that takes 346 lives could be seen in terms other than emotional ones. But, perhaps the aviation industry has a point. Most of us are more frightened than we need to be of flying in commercial airplanes: people who will casually get into an automobile with a half-drunk driver (which, statistically, is a really dangerous thing to do) suffer prickles of sweat whenever their airplane dips a wing approaching an airport. Flying can, in some circumstances, be horrendously dangerous. But, in general, it is not, and the reason for the unease which afflicts most of us in airport departure lounges is that flying is an experience that is contradictory to the physical senses on which our emotions are based. Such fears cannot readily be tamed by intellectual conviction. David Hume noted in his *Treatise of Human Nature,* as a proof of human irrationalism, that a man suspended over a chasm in an iron cage will feel discomfort if he looks down through the bars beneath his feet—and no amount of knowledge about the strength of the cage and the certainty of its suspension will really eliminate that fear.

Essentially, aviation executives are saying that they do not like to see their business endangered by the evocation of fears which are rooted in the irrational depths of human personality, and this is a respectable enough position when one is reporting the average airplane accident. But the crash of the DC-10 in March 1974 was in no sense an average accident. Obviously, its scale was unique. But the events which led up to it were, if not unique, at least highly unusual.

Many airplane crashes are accidents in the most ordinary meaning of that word. They are implausible conjunctions of violent weather, piloting mistakes, and mechanical malfunction. Most of them, these days, occur immediately after takeoff or just before landing.

But the world's worst air crash, to date, occurred when a new and powerful airplane fell 12,000 feet out of a clear sky—and did so as a result of failures which not only could have been foreseen, but were specifically predicted.

To explain how and why this happened, it is necessary to go deep into the epic history and nature of the civil aviation industry, to examine the records of airlines and airplane manufacturers, and to analyze clearly the impact of mechanical, financial, and legal forces that bear upon the design of an airliner. But as one traces the story through engineering design shops, airline boardrooms, factory assembly lines, airliner flight decks, lawyers' offices, and the lobbies of politicians, it is possible sometimes to forget the flesh-and-blood reality with which this story properly begins and ends.

The best witness to that reality is Jacques Lannier, whose calm but devastating words should reverberate through the remainder of this book. Our account of the situation in the Grove of Dammartin relies chiefly upon the personal account which Captain Lannier wrote from his official notes and logs. He discovered that simply writing down his experiences with the DC-10 made him feel less oppressed by them.

March 3, 1974, was the first of seventeen consecutive days which he spent in the forest, organizing, together with firemen, soldiers, and Red Cross workers, the collection and removal of 22 tons of human flesh and bone.

Airplanes that crash near takeoff or just before landing (statistically, the most hazardous phases of every flight) are usually flying at not much more than 150 miles per hour, so that large sections of the structure retain their general shape and most of the victims' bodies are still intact in their seats. In this case, the airplane had struck at nearly 500 miles per hour, and even though the angle of impact was shallow, the kinetic energy was so great that it instantaneously demolished the whole airframe. Apart from the engines, almost indestructible masses of hardened steel and nickel, there were few pieces of wreckage more than a couple of feet long.

Fire frequently consumes most of the remains of air-crash victims, but in this case the ferocity of impact precluded any such cremation. The DC-10's wing tanks were carrying about 100 tons of fuel: as the tanks disintegrated this liquid was transformed into a huge cloud of minute droplets which hovered for a fraction of a second before vanishing in a thunderous fireball. Trees (and here and there flesh) were scorched and blackened, but the flame was too transient to fire the forest still damp from winter. When the first policemen arrived on the scene, there were only a few minor fires smoldering.

There was a scar in the forest about five hundred yards long and up to a hundred yards wide. It bisected a small oblong valley which itself was criss-crossed by tracks and, at about halfway, a rocky gully. All over this area there was scattered a gruesome mixture of human remains: trunks, legs, heads, arms, and pieces of viscera, hanging from trees, half-buried in the ground, and entangled everywhere with life jackets, papers, clothing, endless loops of wire, and minor personal effects. Looking down at his feet after his first *coup d'oeil* of the shattered landscape, Lannier saw somebody's tooth-brush. He noticed that the forest was silent. The spring birdsong was utterly extinguished.

There were about fifty policemen available within the first hour or so. Two men began photographing the wreckage and the bodies, and the others, reinforced as the afternoon went on by soldiers, firemen, and more police, began clearing the site. Their rough-and-ready rule was to pick up fragments and put them into plastic bags (when those ran out they used old fertilizer bags: there was nothing else immediately available), while larger, recognizable pieces were to be roughly assembled on stretchers.

At the beginning, though, it was almost impossible to move far in any direction without treading on some scrap of a human being entangled in the vegetation. It was necessary simply to ignore things that slipped and yielded under foot. Lannier once caught sight of a brain, resting whole and unmarked on a bed of moss. Just beyond it, running toward the head of the little valley, was one of the first press photographers. The man was not looking down: Lannier threw up a warning arm, but it was too late. The photographer's foot stamped the soft mass into pulp and he ran on, uncomprehending. A fireman brushed something aside with his foot: looking down, Lannier saw "a piece of backbone, about twenty-four inches long, admirably well cut away—cut away, and not torn away—as if by some monstrous butcher; the ribs had been sheared away with the same precision."

Some of the clearing teams, particularly the older and more phlegmatic policemen, recovered swiftly from the first shock of the sight and began doggedly combing the undergrowth. Others, however, seemed to become irrational and disoriented by the sheer scale of the horror under their eyes. (Many young Red Cross volunteers, hurrying to help save lives, as they originally supposed, simply fainted.)

The plan was to try to match each trunk with a head and limbs that seemed likely to belong to it, relying chiefly on hair color and skin pigmentation. But it turned out to be difficult most of the time to go beyond identification of sex, and even that was frequently impossible. Many workers began to collect anatomical categories, rather than attempting to reassemble bodies. On one stretcher Lannier found eight heads.

They had no idea how many dead they were dealing with, nor how the catastrophe had come about. In midafternoon came a shattering piece of news: Several bodies and aircraft parts had been found twelve kilometers away near the village of St. Pathus. Clearly they came from the DC-10. Was it now necessary to start searching for other bodies along the line from St. Pathus to Dammartin? *

Beyond this was another question: Were all the bodies in the grove necessarily from the airplane? Suppose some of them were hikers or picnickers? How many people, for God's sake, had been on board the aircraft in the first place?

Lannier soon found himself grilled by journalists who had arrived on the scene: "Captain, the possibility that this was an accident is receding, isn't it?" "Captain, the Turkish Government is convinced that it was sabotage; you, too, of course?" "Captain, why was the baggage of the passengers who boarded at Orly not searched?" "Captain, how many people have been killed?"

While this was going on, about 2:30 P.M., Lannier saw a little party approaching him: two or three men and a uniformed flight attendant with tears streaming down her face. These were the Turkish Airlines representatives who had driven out from Orly. A short man approached Lannier, trying to draw him aside from the journalists and displaying a small piece of paper in his half-cupped palm. Twisting his head, Lannier read it.

---

* The crash site was identified by the press as within the parish of Ermenonville, a well-known village because Jean Jacques Rousseau died there. Actually, the Dammartin Grove is in the less distinguished parish of Fontaine Chaalis.

"Three hundred and thirty-two passengers, eleven crew, and one baby. Are you sure?"

The man gestured evasively, fading back into the crowd, refusing to answer any more questions.

It was actually to be several weeks before the number was precisely agreed, and then it was at 346, not 344.* This was the first reliable indication that the workers at the site received.

By now, important persons were arriving to inspect the scene: the Prefect and Subprefect for Seine et Marne, the Mayor of Senlis, the Turkish Ambassador, the Deputy Public Prosecutor, police and army generals, and then the Minister of the Interior, Jacques Chirac.

Elements of the absurd, as well as the gruesome, began to enter into the situation. Lannier found himself engaged in an insistent dialogue with the wife of one of the local mayors, who asked over and over again whether he was sure that everyone was dead.

"Isn't there a single casualty anywhere? Give me your personal assurance that all the occupants of this aircraft have died!"

"How can I give you such an undertaking? It seems impossible that any human being could have survived the force of impact, but one never knows, you can go and have a look for yourself. . . ."

And the Prefect, it seemed, was not very pleased with the arrangements for temporary shelter for the bodies once collected. Lannier had arranged with the Mayor of Senlis that they should be taken in trucks to the Church of St. Peter in Senlis, no longer in use as a church and preserved as a historic monument. Neither of them could think of any other empty building large enough.

"Why was this church chosen?" asked the Prefect. "What about the morgue of the hospital at Senlis?"

Lannier stared at him. "The morgue at Senlis has room to accommodate ten bodies. I need room for three hundred!" (None of the Protestant, Buddhist, Moslem, or simply areligious families ever complained about the victims being taken to what had originally been a Catholic church.)

The Prefect decided that there were too many people present. "Have everyone who has no business here sent away. The press may stay. I shall address them."

Lannier issued "draconian instructions" but the crowd was "made up of policemen, rescue workers, stretcher bearers, legitimate officials, semiofficials, and local government personalities, each of whom could give just cause for being in this area."

However, the crowd did clear briefly when the Prefect began his press conference because just as he stood up a menacing crackle filled the still air.

---

* A sudden strike on March 3 at British Airways had produced an overwhelming demand for seats on all other Paris–London flights that day. There was, understandably, considerable confusion at Orly Airport, and as a result, the Turkish Airlines passenger manifesto for Flight TK 981 was incomplete and inaccurate.

The top of a tall pine tree, which was probably struck by a metal fragment at the time of the impact, suddenly broke off and came crashing down, very close to the prefect. He did not move an inch, but the fall triggered off a beauty of a panic. . . . People ran in all directions, one of the journalists did a beautiful belly-flop right in the middle of the path, scrambling to avoid the crashing tree which fell . . . far from that spot.

There were arguments, also, about what to do with property. Even amid scenes of annihilation, property must be respected, and from among the wreckage, the bodily remains, and the undergrowth, piles of documents, jewelry, checks, and banknotes of almost every Western currency were being collected. The Prefect thought that they should go to the Senlis salvage center. Lannier, backed by his superiors, insisted that valuables must be locked up at the police station. (They noticed that most of the watches had stopped at 12:42.) Meanwhile, it was decided that the fire brigade would look after the great piles of torn and bloodstained clothing that were beginning to accumulate.

Toward the end of the afternoon, Lannier was told that at the western end of the site (his command post was at the east, where the plane first hit) there were so many bodies coming to light that more collection teams were required. He set off on a tour of his grisly kingdom, taking a civil defense officer with him.

A dislocated doll brought into his mind a picture of a little girl going through a departure lounge. Nearby his companion stepped on a fishing rod with a football boot beside it. (Several footballs were found, still inflated: they were being carried by English rugby footballers who were homeward-bound after watching a match the previous day between France and England.)

Halfway along the great gash in the forest, they reached the rocky gully crossing the path of the airplane. This, it turned out, was the center of this earthquake of horror. Lannier guessed that at least one hundred people had died within 150 square yards. Their bodies, bisected by seat belts or shattered by impact with disintegrating airframe parts, were draped thickly among the trees of the little woodland path. In the midst of the shambles lay the great bulk of a jet engine, looking relatively undamaged.

Not one body was complete. "Although a few heads were still attached to chests, a great many bodies were limbless, bellies were ripped open, and their contents emptied; I noticed a woman's chest had joined a man's pelvis. . . ." Lannier saw an even more heart-rending version of the image which had arrested his mind when he had first arrived at the scene: a woman's hand firmly, desperately clutching the severed arm of a small child. Dully, Lannier wondered if the child might have been the owner of the doll he had seen earlier. Some other part of his brain was noticing that the casualties were of many races: one body, burst open at the foot of a tree, was clearly Asian.

"Everywhere, the scene was nightmarish; the Forest of Ermenonville had been turned into a battlefield, it was Verdun after the bloodiest battle."

One of Lannier's policemen was pulling at his sleeve. "Captain, you are wanted," he said. The man refused to let go of Lannier's arm, tugging remorselessly. "Although in his job he was used to seeing death, like many of us, he was completely overwhelmed. He had lost all sense of direction. . . ." On the way back to Operations Control, with the disoriented policeman still clinging to him, Lannier saw a fireman perched in a tree, trying to dislodge a headless bust from a branch. He pulled, and suddenly all the innards poured out on top of him. The man clambered down the tree and began vomiting.

And now sightseers—perhaps, more accurately, voyeurs—were beginning to crowd the scene. "How often," wrote Lannier

> have I not been sickened by the crowd of spectators gathering round an accident . . . crowding together, pushing each other to get a better look. . . . In the case of the DC-10, it was worse than anything we had ever seen before. I shall never know exactly how many thousands, or tens of thousands, of men, women, and children came only for the opportunity of seeing the ragged remains of a crashed plane and the mutilated bodies of people who had been hurtled all over the place.

He saw one family of mother, father, and a three-year-old boy pushing eagerly forward, complaining that they could not see properly because the police cordon held them back. They had come out in such haste that the woman was still in her carpet slippers, and the little boy had a bib under his chin. Neither she nor her husband could see, so they lifted the child onto their shoulders, and asked him to relate what he could see. Nearby, another woman was making comparisons about the size of the penis on a male pelvis which had been separated from its trunk and limbs.

One of the journalists reported seeing a young man bend down, pick up a hand, stuff it into his pocket, and hasten away into the crowd.

YET ALL THIS was taking place within a self-contained world of horror. The only participants were the dead, the officials, the news-gatherers, and the voyeurs. Outside the forest, the life of Europe continued unconcerned. Glinting in the clear blue sky, the airliners continued to climb one after another out of Orly, to wheel away for London, Amsterdam, or New York. Like the "expensive delicate ship" in Breughel's *Icarus,* that "must have seen something amazing, a boy falling out of the sky," they "had somewhere to get to and sailed calmly on." *

And indeed airports and airlines cannot cease their business because somewhere there has been an accident. This adds another measure of poignancy to the circumstances of an air crash: Whenever a flight goes down, there will be people waiting at its destination to meet friends and relatives. At midday, in any part of the week, the concourse of any big airport building is thronged

* W. H. Auden, "Musée des Beaux Arts," quoted at the beginning of this chapter.

with people beginning and ending journeys, and all this bustle continues, indifferent to those few of the passing thousands who have been picked out for catastrophe.

Flight 981 was due into London Airport just after midday local time. Leonard and Barbara Collen arrived at Terminal No. 2 in good time to meet their twenty-year-old daughter Esther, returning from Ankara. Esther had telephoned during the week to announce that she was engaged to a young Turkish engineer. "I'm coming home Sunday," she said, "and we'll go out and buy the wedding dress."

The Collen family drove to the airport from London's northern suburbs in two cars: They had their son, their daughter-in-law, and their eight-year-old daughter Rebecca with them. As they entered the terminal building, Flight 981 was shown with its arrival time as 12:20. Almost immediately it changed to 12:25 and then to 12:30.

Barbara Collen and the children strolled around the concourse for a few minutes. Her husband stayed by the indicator board, and when they came back to him, the arrival time had vanished altogether. In its place were the words: "Make inquiries at the BEA desk." British European Airways (since merged into British Airways) were handling agents for Turkish Airlines' relatively few flights into London.

"Something's gone wrong," said Leonard Collen. He was already convinced that there had been a crash and the look on his face convinced his wife. The attendants at the inquiry desk could say nothing useful to the people who now began to crowd around them, except that Flight 981 was not within the London radar area. At about 12:50 local time (one hour behind French time)—by which time Captain Lannier and his men had already assured themselves there were no survivors—an official arrived at the inquiry desk.

"There's been an accident," he said. "Would you come along with us?" A hundred or so people were waiting for news of Flight 981, and they trailed along into one of the airport's private lounges.

There was no announcement, no further details. But almost at once a large supply of free drinks was brought in. Some people started to weep. Mrs. Collen turned to her son, and said: "Get the baby [Rebecca] out of here quickly. Take her home."

Her recollection of the afternoon is one of almost complete emotional paralysis, and her husband's is the same. "I froze right up," he said later. "I was just frozen solid, like a block of ice I went. And I couldn't move. And the officers in charge, the airline people, you know, they insisted that they knew no details at all."

As the afternoon wore on, alcohol and tension worked on the people— English, Turkish, Japanese, a few Americans—waited in the crowded lounge. There was not much doubt in anyone's mind about what had happened, but still there was no formal word.

National traditions differ in expression of grief. The English and the Japanese, in their different ways, tried to maintain a stoic immobility. But some of the people present came from simple Turkish families, where there is no inhibition against physical manifestation of grief: from time to time people would wail, rock in their seats, and hurl themselves to the floor. Beside Barbara Collen, a man flung himself against the painted brickwork of the wall, cutting his head open, making the blood flow.

Then, at about 3:30 local time, a color television set was brought into the room and switched on. "They admitted nothing at all," said Leonard Collen. "They switched it on and showed the . . . news flashes that showed you pictures of the actual disaster scene."

"We couldn't even speak to one another," Barbara Collen recalls.

> Our throats were so tight we couldn't hardly breathe, let alone speak. And we saw this television switched on, and we saw damage, and the . . . the things hanging on the trees, and the . . . the piece that cut right through the forest and had taken all the bodies through. I mean, I could see that nothing . . . nothing could have survived. It wasn't even a big piece of engine left. And they covered it pretty well, you know . . . it didn't leave us in any doubt that there was no survivors. . . .
>
> My husband said: "Shall we go home?" I said no, let's wait for the passenger list, in case she's changed her mind. . . . We waited until we got the Istanbul list and of course the name Winifred Collen came through, because that's the name she traveled under, and not Esther. And we just looked at one another and said: "That's it."
>
> And we came out of the terminal building and up to the car park and we were chased all the way along by reporters and they said: "Is your daughter dead?" Because they knew all about her, funnily enough. Because when we were sitting down in that [lounge] that afternoon, Leonard and I couldn't talk to one another. We were too choked up. And a middle-aged man came and sat beside me, and he was very, very kind, and he spoke very quietly, and he said: "Who are you waiting for?"
>
> I said: "I'm waiting for my daughter." And he started to speak to me, and it was such a relief to talk to somebody. I couldn't speak to Len, I couldn't speak to the . . . priest or anything. And I told him all about Esther and I turned round to him and I said: "Who are you waiting for?" He said: "Oh, I'm not waiting for anybody, I'm the press. . . ." One of the officials came straight up. He said: "Press?" and he said: "Yes," and they just got hold of him and threw him out. . . . I wasn't angry with him. Actually it was a great relief to talk to him.

Extract from a memorandum, written June 27, 1972, by F. D. Applegate, Director of Product Engineering, Convair Division of General Dynamics Corporation (subcontractors to McDonnell Douglas Corporation for detail design and construction of the DC-10 fuselage):

> The potential for long-term Convair liability on the DC-10 has been causing me increasing concern for several reasons. . . . The airplane demonstrated an inherent susceptibility to catastrophic failure when exposed to explosive decompression of the cargo compartment in 1970 ground tests. . . .

It seems to me inevitable that, in the twenty years ahead of us, DC-10 cargo doors will come open and cargo compartments will experience decompression for other reasons and I would expect this to usually result in the loss of the airplane. . . . It is recommended that overtures be made at the highest management level to persuade Douglas to immediately make a decision to incorporate changes in the DC-10 which will correct the fundamental cabin floor catastrophic failure mode. . . .

Extract from an interview with John H. Shaffer, head of the U.S. Federal Aviation Administration (FAA) 1968–73, during which period the DC-10 was certificated by the FAA for commercial use, and during which it suffered a nonfatal accident that almost exactly prefigured the 1974 crash of Flight 981. Shaffer explains why he did not see any need for fundamental design changes in the DC-10:

I just happen to be one of those people that likes to be—or is—optimistic about the future. We thought we had it fixed—believed we had it fixed and certainly would not have been sitting around, just fat, dumb and happy, thinking that it had all been done. . . . We obviously didn't want any tragedy or any accident. So it was just one of those circumstances. We may have been lulled into a false sense of security. . . .

Extract from the affidavit of Donald W. Madole, of counsel, in support of the plaintiffs' motion for summary judgment on liability in Hope et al. v. McDonnell Douglas et al., U.S. District Court, Central District of California, March 10, 1975:

. . . It was determined that the direct and immediate cause of the crash was the inadvertent opening of the aft bulk cargo door at approximately 13,000 feet altitude, which resulted in a pressure differential over the cabin floor sufficient to cause its collapse. Controls for the No. 2 engine, elevators, mechanically controlled trim, and the rudder are routed under the floor of the DC-10 and their placement was the responsibility of McDonnell Douglas. . . . The collapse of the after cabin floor resulted in a loss of flight control capability of the subject DC-10. . . .

From the testimony of Marchal H. Caldwell, Chief Fuselage Design Engineer for the DC-10, McDonnell Douglas Corporation, in Hope v. Mc-Donnell Douglas:

Q. Do you believe the cargo door came off in flight?
A. I do believe that happened.
Q. What information did you have on that score?
A. The fact that the cargo door was found remote from the main damage site. . . .
Q. Did that suggest to you that the crash was initiated by that aft cargo door blowing out?
A. It only suggested that the aft cargo door had blown out. It did not necessarily suggest that the crash was caused by that.
Q. Did you ultimately decide that's what caused it?

A. The French SGAC * had issued that as the probable cause.
Q. Do you agree with it?
A. Yes, sir.

Later, from the same testimony:

Q. When you heard the cockpit voice recorder tape played, did there come a time when you heard the sound of the explosion?
A. Yes, sir.
Q. After the sound of the explosion were there any sounds of aural warning, bells, or buzzers?
A. Yes, sir.
Q. Was there any conversation by the flight crew members?
A. Yes, sir.
Q. I take it, then, that the fact that there was conversation after the explosion would indicate that they were alive and they were talking to each other at that time?
A. Yes, sir, that is correct. I would have that impression.
Q. Do you recall from the reading of the transcript whether they had any conversation about the flight controls?
A. I do not recall any specific comments. The only one that stands out in my mind is one of the flight crew, and I don't know whether it was the pilot or the co-pilot, said: "Help me pull" . . .
Q. Did you at a time after hearing the explosion ever hear the sound of the overspeed warning klaxon?
A. Yes, sir. Just prior to impact.

The worst accident in the history of civil aviation was not necessarily the most terrible event that overtook the human race in 1974. There was war in Indochina, famine in sub-Saharan Africa, disease and violence in a dozen countries, and a steady accidental slaughter on the roads and in the homes and factories of the developed Western world. But it was, for all that, something especially horrible—because it was something that need not have happened at all.

THREADING OUR WAY through the story of the DC-10, we shall come again to the Applegate Memorandum, as we call it. We shall cite it in full and try to explain its content as well as the circumstances that caused it to be written.

The title of this chapter is drawn from the documentary response made to Applegate's memorandum by his immediate superior within the Convair organization, J. B. Hurt. Hurt was Program Manager for the DC-10 Support Program, and the man largely responsible for Convair's relationship with McDonnell Douglas, as subcontractors for detail design and manufacture of the DC-10 fuselage. We will also cite Hurt's memorandum in full at an appropriate point. Here, we wish only to quote one especially vital pas-

* Investigating body of the French Ministry of Civil Aviation.

sage. Hurt says that he does not take issue with any of the facts outlined by Applegate; but, in effect, he questions whether the responsibility should properly lie with the Convair division of General Dynamics.

> We have an interesting legal and moral problem, and I feel that any direct conversation on the subject with Douglas [division of McDonnell Douglas] should be based on the assumption that, as a result, Convair may subsequently find itself in a position where it must assume all or a significant portion of the costs that are involved.

Putting the issue simply, Hurt, in his memorandum, recognizes the plain human duty to warn others of danger to life and limb. This, in itself, is an uncomplicated response: Any of us, seeing a man or woman about to cross, unknowingly, a bridge which is likely to collapse, or climb into an airplane whose mainspar we know to be cracked, would simply give warning. Hurt, however, is pointing to another responsibility which, in law at least, is laid upon the officers of a corporation. They have a duty to safeguard the corporate interest, and if the duty to give warning can legitimately be placed elsewhere, they may be entitled, or expected, to remain mute. As our story will show, that was the course that Convair chose to adopt. We do not feel, on the evidence available to us, that it is possible to say whether the real responsibility for not disclosing the possibility of danger lies in the end with Douglas or with Convair or with other actors in the story. Indeed, one of the inferences which we think can be drawn from this narrative is that the complexities of modern business bureaucracies are such that it may be very difficult to tell where exactly blame or praise lies.

We do not step back from the fact that there are people in the story whose actions we, at any rate, believe should be criticized. In particular, it seems to us that John Shaffer was quite wrong to suspend the rigors of the Federal Aviation Administration's normal practice after a near disaster in 1972 had demonstrated a need for regulatory supervision of safety improvements in the DC-10 design. Equally, it seems to us that Jackson McGowen and John C. Brizendine, as officers of the McDonnell Douglas Corporation, were responsible for the introduction of the DC-10 into the "stream of commerce," and they were under an obligation to fulfill promises about its safety. This they failed to do.

But insofar as we can understand the situation, it appears that all these people acted, not out of malice, but because at particular moments they were subjected to the demands of special loyalties—to the aviation industry in general, or to corporate parts of it in particular—when they performed acts which we now consider to be foolish or unacceptable. To see now, in the afterlight of the disaster, what should have been done is easier than to know how one would oneself have behaved in the same circumstances, under the same pressure.

Nations, certainly, would not survive if they could not, at times, provide special moral frames within which ordinarily sane and decent people will cheat, lie, and even kill. "If we did for ourselves what we are doing for Italy," said Count Cavour, "what shocking rascals we should be."

Corporations, especially the large and complex ones with which we have to live, now appear to possess some of the qualities of nation states—including, perhaps, an alarming capacity to insulate their members from the moral consequences of their actions. It may be an inevitable tendency: It is nonetheless one which needs to be watched, understood, and controlled. The story of the DC-10, and of the disaster of March 3, 1974, may be understood in terms of men who did things "for Italy" that they would never have done for themselves.

2

The Numbers of Safety

The whole problem is confined within these limits, viz. To
make a surface support a given weight by the application of
power to the resistance of air.

—SIR GEORGE CAYLEY,
On Aerial Navigation (1809)

AT ONE MINUTE past 3:00 P.M. on July 30, 1971, Pan American Flight
845 slid away from its departure gate at San Francisco Airport. This was a
Boeing 747, carrying 199 passengers and 11 crew, fueled for a flight to
Japan. In the crew room a few minutes earlier Captain Calvin Dyer and First
Officer Paul Oakes had calculated takeoff speeds for a departure from
Runway 28L.

As the terminal building receded, Oakes tuned his radio to the airport's
recorded information service and learned that Runway 28L was closed. At
that moment, a net of sinister misunderstanding began to close around
Flight 845.

Taking off in a big jet transport naturally bears little relation to the
heroic days of flying, when the pilot, trundling across a grass field, hauled
back on the stick as soon as the wind on his face seemed strong enough.
Each takeoff now needs to be a precisely calculated operation, in which
aircraft weight, air temperature, atmospheric pressure, and wind conditions
are accounted for.

The 747 needs about a mile-and-three-quarters of runway to accelerate
to its takeoff speed. Runway space is critical, but curiously enough scarcely
any airports provide markers with runway lengths displayed on them. The
pilot at takeoff sees a wide pavement stretching into the distance ahead of
him. But nothing he looks at through his windscreen tells him how long
it is. The assumption is that every dispatcher clearing aircraft for takeoff at
a busy airport knows exactly the serviceable length of any runway. At San
Francisco on that particular day, the assumption was not correct.

There is an established ritual to each takeoff. In a 747, the captain handles the flight controls and lines the nosewheel up on the runway "threshold"
line. The flight engineer handles the throttles, bringing the engines up to
takeoff power (which is kept as low as possible for each occasion, to save

19

fuel and increase engine life). The first officer watches his air-speed indicator and calls out to the captain as the plane goes through each of the three critical speeds which rule every takeoff. "V-1" is commitment speed—after which the plane is going too fast to stop before hitting the end of the runway and *must* take off. "V-R" is "rotation speed," at which the captain lifts the nosewheel off the ground. And "V-2" is the speed at which the airplane can remain airborne and controllable, even if an engine should fail.

At the cost of using greater power, takeoff speed can be reduced by pushing the wing flaps further out and down. With the extra power to overcome air resistance, the heavily flapped wings will lift the aircraft sooner, and so less distance is needed for acceleration. On the assumption that they would have 10,600 feet available on Runway 28L, Captain Dyer and his crew had elected to use only ten degrees of flap. They would not reach V-2 until 171 knots.* When the automatic information system told them that 28L was closed, Dyer and Oakes asked for another runway, and after some argument they were sent to 01R. "Start at the painted line," the Pan Am dispatcher told them, "and you will have 9,500 feet plus clearway ahead of you."

In fact, the line of 01R had been moved to keep jet blast off a nearby freeway. There was, as a result, only 8,500 feet available—not really enough for a 747 takeoff with the flaps set at only ten degrees.

Investigation by the National Transportation Safety Board (NTSB)† showed, as one might expect, that the system for alerting flight dispatchers to changes in runway status needed considerable tightening up. No one, however, could quite explain the way that Captain Dyer and his crew handled the takeoff. They were told that on 01R in the prevailing conditions they would need twenty, rather than ten degrees of flap, and would need "wet settings" on the engines—that is, the maximum power output, which is produced by using water injection, and is only allowed for two-and-a-half minutes. Such extra power and extra flap extension would be needed for the shortened runway. Amazingly, however, neither Captain Dyer nor First Officer Oakes—nor indeed the second officer, nor either of the two flight engineers—thought to recalculate the takeoff speeds. Oakes left the "bug" markers on his air-speed indicator for the V-1, V-R, and V-2 speeds that had been previously calculated, although with the new flap and power settings they were appreciably higher than they should have been. The fact that Captain Dyer had been flying for Pan American for thirty-two years, and had been one of the first captains selected for 747 service, shows merely that in flying no one is immune from primitive error.

At 3:28 P.M. the engines opened up to full "wet" power, and 316 tons of airframe, fuel, cargo, and human bodies started to gather speed. Given the flap settings and the power settings, the nosewheel should have been

---

* Aircraft speeds are quoted in knots—nautical miles per hour, which are longer than statute miles: 171 knots equals 194 miles per hour.
† Report NTSB/AAR/72-17, May 24, 1972.

lifted at 157 knots (V-R), and if it had been the plane would have just escaped trouble. But First Officer Oakes, watching his "bugs," was waiting to see 164 knots come up. Accelerating through the extra seven knots took about four seconds—or, more dramatically, nearly a quarter-mile of runway.

Oakes actually made his V-R call somewhere around 160 or 161 knots, because by then the end of the runway was "coming up at a very rapid speed." * As the air speed passed 165 knots, the men on the flight deck felt the plane lift and shudder. Behind them in the passenger cabin the floor burst open, and the air filled with debris.

Runway 01R at San Francisco ends on the bay shoreline, and built out into the water is a long pier with handrails and gantries, carrying the approachlights for aircraft coming in the other way to land. Flight 845 simply charged through some three hundred feet of the pier structure. The impact alone might have destroyed a smaller aircraft, either by airframe fracture, loss of flying speed, or some combination of the two. The damage to the 747 was comparable, perhaps, to a hit from an antiaircraft shell.

"Missiles" of angle iron ripped through the hull. Undercarriage bogies, wing flaps, and fuselage bulkheads shattered on impact with the light gantries. The cabin floor above the wheel wells sprang up two feet in front of the passengers' eyes and split open, spraying metal fragments. Luckily, this section of the cabin above the wing was sparsely populated, but one seventeen-foot shaft of a two-by-two-inch angle iron came up under seat 46G, cut through the leg of a Japanese passenger in 47F, crushed the arm of his neighbor in 47G, passed through the cabin roof, and lodged inside the vertical fin. Another piece cut its way through the three righthand lavatories and passed out through the rear of the fuselage.

Large parts of the 747's undercarriage had been carried away or driven up into the hull, and wreckage was embedded in the tailplane and elevator system. But damage to the architecture of the hull and flying surfaces is not the only result of missile penetration. An airplane is packed with delicate systems, as a human body inside its envelope of skin is packed with veins, nerves, and arteries. Almost anywhere—but especially in the vulnerable underbelly where wings, wheels, and fuselage join together—a missile can sever vital channels of energy and control.

Virtually all jet transports are controlled through hydraulic systems: that is, by high-pressure fluid carried through the structure in seamless, flexible tubes. Extensive rupture in any part of a hydraulic system will drain the whole system of its power. As Captain Dyer's 747 plowed through the light gantries at San Francisco, three hydraulic systems were ruptured in the airplane's belly.

Early jet aircraft generally used two duplicated hydraulic systems. With the change to all-power controls in the second generation of jets, most design teams installed three. A 747, however, carries four hydraulic systems, any one of which can fly the aircraft. In Captain Dyer's plane, three systems were

* NTSB report.

knocked out, and a shard of metal passed within four inches of rupturing the fourth. By that margin, though, the aircraft and the lives of 210 people were saved.

Having survived this potentially fatal impact, the 747 circled over the sea for an hour and forty minutes while the crew dumped fuel, and seaborne observers tried to estimate whether there was enough undercarriage left to avoid the further risk of a belly landing. (Two passengers were doctors, and applied first aid to the seriously injured passengers.) Awaited by a gathering of fire engines, ambulances, and rescue engineers, the battered aircraft came in to land at Runway 28L, from which the crew had hoped to take off in the first place. It was a rough landing, because the elevators were not working properly and the devices which are supposed to reverse engine thrust and so slow the plane down failed on three engines, causing the plane to veer off the runway. Further, there were some serious injuries among the passengers when the escape chutes failed to work properly.

But even so, it was a remarkable piece of aeronautical survival. It was a demonstration of the fact that a big airplane can be more resistant to the blunders, accidents, and miscalculations that are inseparable from the process of flight. Properly applied, size, power, and redundancy, which is in the end something rather more complex than just the provision of extra systems, are the essential principles of aerial safety.

THERE HAVE ALWAYS been choices of method available to human beings trying to fly, and of course, it was first achieved with lighter-than-air devices like balloons and airships. But to the more clear-eyed investigators, it was readily apparent that the fixed-wing rigid airplane would be the most efficient general-purpose system. Before the Battle of Waterloo Sir George Cayley, who had made a stable and reliable model glider ("It was very beautiful to see this noble white bird sail majestically from the top of a hill to any given point of the plain below it . . ."), was certain that once such machines could be provided with a suitable power source, then "we should be able to transport ourselves and our families, and their goods and chattles, more securely by air than by water, and with a velocity of from twenty to two hundred miles per hour." * Cayley correctly anticipated that this would begin "a new era in society," though he was sorry to think (in a considerable underestimate) that "a hundred necks" would have to be broken before "all the sources of accidents can be ascertained and guarded against."

It is sobering for us to recall, amid the technological panoply of the twentieth century, that most of the principles of mechanical flight were worked out by a Yorkshire country gentleman who was born before the Declaration of Independence was composed. The essential requirement,

* On Aerial Navigation: *Journal of Natural Philosophy, Chemistry and the Arts,* November 1809.

Cayley saw, was not simply great power, but a high ratio of power to weight. Given a suitable source of power, the rules governing the behavior of wings and streamlined bodies should not be difficult to establish—and with certain exceptions in the region of Mach 1, the speed of sound, this has proved to be the case. Each advance in aviation has been a complex process, frequently rooted in disciplines apparently far removed from the science of airplane design. But the movement from one stage to another has always been decided by the availability of power.

Cayley, who eventually developed a glider big enough to transport his coachman across a small valley (the man gave notice instantly upon landing), was able to show by careful calculation that the expansion steam engine, the wonder of his own age, would probably always be too heavy for aeronautical use. It would be more promising, he thought, to investigate the working of pistons through "the expansion of air by the sudden combustion of inflammable powders and fluids."

And it was a careful development of the internal-combustion engine that made the first powered flight possible. But before their Flyer took off just by Kill Devil Hill on December 17, 1903, the Wright Brothers had made two crucial innovations.

One was to solve the outstanding problem of airplane control: that of lateral stability. Cayley and later nineteenth-century glider exponents understood that a hinged rudder working in the same way as on a ship could control an airplane's direction. From this it was a short step to the elevator, controlling attitude: imagined as a horizontal rudder, a hinged surface at the rear of an airplane will, when angled upwards, raise the nose and when angled downwards, push the nose down. But although the problems of direction and attitude had therefore been solved, no one before Orville and Wilbur Wright knew how to control an airplane rolling around its lengthwise axis (and the belief that the answer lay in shifting the pilot's weight cost the lives of several glider pioneers).

The Wrights were aware from wind-tunnel experiments that bending the rear edge of a wing *down* will increase the lift on that wing, and that bending it *up* will produce the reverse effect, destroying lift. They devised a means of bending or twisting an airplane's wings to control rolling motion, and the principle survives in the modern aileron system—although hinged sections are now used—to avoid bending the whole wing.

The Wrights had thus completed the three-dimensional control system of the modern airplane. (As a matter of detail, their elevator was on the nose of the airplane, not the tail, but the principle was the same.) However, they could find no engine with a ratio of power to weight sufficient to maintain their machine in the air. With admirable self-confidence they returned to their workshop to produce another innovation: the first successful aero-engine. As used in the early flights, the engine weighed 11.2 pounds per horsepower, although by "stretching" the power output this ratio was improved to six pounds per horsepower. (The Wrights' whole eight-year research and development budget for engine and airframe came to $1,000, something of a contrast to figures which will occur later in this book.)

Large improvements were made to aero-engines in the years of World War I, and the classic of the period, the American Liberty engine, weighed only 2.1 pounds per horsepower. But total output stayed for another decade at four hundred-or-so horsepower, which was sufficient only to lift fairly primitive airframes, carrying modest supplies of fuel. Wood and fabric dominated the airframe structure, and there were few machines that could cross more than a narrow strip of water. Therefore, during the first thirty years or so of its existence, the airplane was little more than a military instrument or a specialist mail carrier.

The problem lay not so much in the availability of materials for building better airframes and better engines—aluminum and special steels began to appear in the 1880s—but rather in the chemistry of petroleum fuels. Greater power could only be obtained by compressing an ever larger original volume of air and "inflammable liquid" into a given space above each working piston. But the heat of compression will make some fractions in a mixture of petroleum hydrocarbons explode before others, and often this will happen before the piston has reached the correct point in its cycle. Premature ignition —pinging or knocking—will eventually destroy an engine, and during the 1920s it made an effective barrier to increases of power.

There was some improvement after 1921 when Thomas Midgley of the Dayton Engineering Company introduced tetraethyl lead to stabilize fuel behavior. But the truly decisive work was done by Vladimir Ipatieff, originally of the Czarist artillery corps, later of the Universal Oil Product Company in Chicago. Ipatieff was the founding genius in the business of manipulating hydrocarbon molecules, and he made "high octane" aviation spirit—and therefore aviation—into a commercial proposition. Generally, the unstable fractions of petroleum are those with more hydrogen and less carbon in their molecules. Before World War I Ipatieff had worked out the basic principles of breaking up and reordering the atoms of a hydrocarbon to produce octane, named for its eight carbon atoms, which behaves reliably in high-compression engines. Despite his background as an Imperial Army officer, Ipatieff's patriotism kept him in the Soviet Union until the intensifying Stalinist terror of the late twenties seriously endangered his life. When he eventually fled to the United States, Ipatieff was one of the most significant of all technological defectors. He was able to show how to make low-grade gasolines into fuels of any required octane rating.

The era of high-compression engines using 100-octane fuel began in the early thirties with units of 1000-hp and the prospect of 2000-hp to come. One classic engine of the period, the Rolls-Royce Merlin, was eventually made to produce more than three horsepower per pound. It was this generation of engines, and the sturdy, all-metal monoplanes built around them, which really inaugurated the "new era" that Cayley had foreseen. The Douglas DC-3, with two 1000-hp engines, was the first fully realized example.

According to the records of the Douglas Company, one experienced airline pilot, when shown the prototype from which it was developed, said emphatically that it was too big and would never fly. It was, after all, sixty feet long and weighed some ten tons.

As it turned out, the DC-3 flew well enough to be the first airplane that would enable an airline to make a profit without the U.S. Mail contracts that were used as a discreet subsidy in the early days of the industry. The new engines enabled the airplane to escape from its specialist role and become, for better or worse, one of the basic vehicles of Western society. The DC-3's four-engined successors, such as the Lockheed Constellation and the Douglas DC-6, could lift enough fuel to begin making intercontinental passages.

But at the upper end of its power range the high-compression piston engine became a hideously complicated device, using up much of its own power in the supercharger needed to force air into its hungry induction system, depending for its life upon intricate cooling systems and electrical equipment. By the end of World War II, some aviation companies had begun to suspect that the gas turbine would replace the piston engine. However, the commercial development of the turbine passed through some paradoxical stages before arriving at the present big jet era. Contrary to one standard illusion, modern technology does not advance with breathtaking speed along a predictable linear track. Progress goes hesitantly much of the time, sometimes encountering long fallow periods and often doubling back unpredictably upon its path.

THE "PURE" JET—a gas turbine engine which has no propeller and works by direct acceleration of a column of air—very rapidly outdid the most advanced piston engines in terms of brute power. Seven or eight years after the war, jets were sufficiently developed to give ten thousand pounds or more of thrust, and on a rough comparison that was 4000-hp, exceeding the 3500-hp of the twenty-eight-cylinder Pratt & Whitney Double Wasp, generally taken to represent the pinnacle of pure piston technology.* And once the problems of metallurgy and lubrication were dealt with, jets turned out to be beautifully reliable.

There is simply less to go wrong in a gas turbine than in a piston engine. Essentially there is only a fanlike compressor driving air into a combustion chamber, where it is mixed with burning fuel to make a high-pressure gas stream, which then forces itself through the blades of a turbine into the outside air. The turbine, which may be pictured as a fan working in the reverse sense, provides power to drive the compressor and keep the system feeding itself. The components of a gas turbine merely revolve. They do not fly back and forth like those of a piston engine. And instead of several

* Divide thrust (pounds) by 2.5 to get an approximation of shaft horsepower.

thousand violent explosions every minute, combustion occurs in a steady, continuous flow.

These intrinsic qualities attracted aeronautical engineers very early, and during the twenties several technical papers were published canvassing the use of gas turbines to drive aircraft propellers. The difficulty was to build a turbine which would withstand high temperatures and also be light in weight, and this kept gas turbines on the ground (chiefly in electrical power stations) during the interwar years. When the problems of heat and weight were finally solved, it was done as part of a process which dramatically recast the aeronautical role of the turbine.

Theoretical aerodynamic studies showed during the twenties that the propeller, however much power might be fed to it, would not be able to drive an aircraft past 500 miles per hour. At this velocity, with the aircraft approaching the speed of sound, the propeller blades (adding the speed of their rotation to the speed of forward motion) would already be operating at supersonic speeds—and losing most of their propulsive effect amid a tangle of shock waves. There was no solution. It could easily be shown that piling on more power would merely accelerate the drop in propulsive efficiency. In the United States the airplane might be evolving into a worthwhile instrument of commerce, but in Europe its purposes remained chiefly military, and in military aviation speed dominates all other considerations. In 1928 it occurred to the young RAF officer Frank Whittle—and shortly afterwards to Pabst von Ohain in Germany—that in the pursuit of speed a gas-turbine engine might dispense altogether with propellers. They applied to the problem a piece of high school Newtonian physics: for every action there is an equal and opposite reaction. Air drawn into a gas turbine engine, then compressed, violently heated and allowed to escape through a rearward-facing jet pipe would produce a large forward pressure inside the engine, a thrust readily transmittable to the airframe.

The politics of Europe made this aspect of gas turbine attractive, and the outbreak of war made it irresistible—with the result that by 1944 both Britain and Germany had jet fighters in service which could outpace any actual or possible piston-engined airplane.

Military preoccupations, in other words, shaped the development of the gas turbine toward the pure-jet, which encounters no theoretical speed limitation until it is going at several times the speed of sound. Naturally, the inauguration of the Cold War maintained and intensified this bias.

Unhappily, though, a jet is much *less* efficient than a propeller at most nonmilitary aircraft speeds. In the subsonic range, a propeller can move a large diameter airstream at a speed which is reasonably close to that of the airplane itself: in such circumstances, propulsive efficiency is maximized. A jet moves a thin, hot airstream at a speed which is vastly greater than that of a subsonic airplane. The result is a propulsive inefficiency which makes the jet an uneconomic fuel user. It also makes an airplane that is awkward to handle (rather like an automobile in permanent high gear) and it produces a great deal of noise. The "scream" of a jet is caused by the "shear" between its fast-moving exhaust and the still surrounding air—the noise, as usual, is a clue to wasted energy, and inevitably is worse when it is least acceptable, that is, at takeoff when the greatest amount of thrust must be used. In the band between 400 miles per hour and 1,000 miles per hour, the jet succeeds, not because it works well, but because the laws of aerodynamics prevent the propeller from working at all.

During the 1950s, it seemed that there might be two ways of applying gas turbines to civil aviation with a proper regard for propulsive efficiency. One was to make the airframe go faster than sound, thus getting closer to the region of the pure-jet's best efficiency. The disadvantage of that approach, as British and French taxpayers have learned from the Concorde development costs, is that achieving supersonic flight brings into the airframe all of the complexity that the turbine revolution previously eliminated from the power plant.

In those same years, the old idea of the turboprop, in which most of the turbine's power is used to drive a propeller, seemed an attractive solution. Operating in the propeller's best speed range of 350 to 400 miles per hour, turboprop airliners like the Vickers Viscount and Lockheed Electra displayed superb flying qualities, fuel economy, and mechanical reliability.* Although the early models of the Boeing 707 were said to fly like "civil bombers" and burned fuel heavily, their pure-jet engines gave them a 25 percent speed margin, which was decisive on transcontinental and transoceanic routes. In those days of cheap oil prices, they were irresistibly superior profit-makers, although the decision might easily have gone the other way had the pace of Arab political development been different.

These early jetliners, in fact, were flying close to Mach 0.9, on the lower border of the "transonic region," which is approaching the maximum speed of a simple light-alloy airframe. Air resistance at this speed remains moderate, given well-designed swept-back flying surfaces. Around Mach 1, the speed of sound, resistance increases drastically, and then declines again in the calm of the "supersonic region" around Mach 2. The trouble is that engines are needed to break through into supersonic and hypersonic flight, and when such speeds have been attained, a specialized airframe is required to deal with the frictional heat from the airstream.

The conflict between the pure-jet, which is inefficient at the ideal airframe speed, and the prop-jet, which cannot reach the best speed (Mach 0.9) of the subsonic airframe, has now been solved by the fan-jet, which reaches its best propulsive efficiency at around the speed of sound. But there are no

---

* The Electra's record was flawed by a freakish, but lethal engineering problem. Resonances of a kind not seen before weakened the engine mounts, and in two cases engines broke away in flight, causing crashes which at first were inexplicable. By the time the problem was found and cured, the Electra's reputation was seriously damaged.

bargains in the physical universe, and making the fan-jet—the power unit of the DC-10, the 747, and the Lockheed TriStar—into a working reality consumed large quantities of time and money.

Looked at crudely, it is a prop-jet which has, instead of a propeller, a "fan" resembling a many-bladed propeller enclosed in a duct or pipe. The problem with a high-speed propeller is that the normal atmosphere is too insubstantial to absorb usefully all the power that is being dispensed. The air simply escapes from behind the propeller blades—as power increases more and more of it collapses into shock-wave formation or escapes centrifugally at right angles to the intended direction of flight. But in a fan engine the duct cramps the air in behind the blades, preventing centrifugal escape. Because the air is held in a controlled flow, this kind of "propeller" retains its efficiency further up the speed range. The machine is also called a "bypass" engine because the greater part of the airflow bypasses the engine core. Only a small proportion is taken in, fueled and used to drive the turbine. The overall result is a large, relatively slow column of cold air, with a thin, hot, fast exhaust inside it. Because the fast air is wrapped up, as it were, a fan-jet is much quieter than a pure-jet.

The layman's natural question is why the fan-jet was not developed sooner, and most of the answer is that it did not fit readily into the mainstream of advanced engine development. In both America and Europe this has continued for many years to follow the original bias of gas-turbine technology toward the military excitement of supersonic and hypersonic flight. And very large proportions of all engine development programs are done on military contracts. No true fan-jet (as distinct from low bypass engines) emerged until the U.S. Defense Department commissioned the enormous C5A Galaxy subsonic freighter, which required the General Electric TF39 fan-jet to drive it.

The fan-jet's descent from its true ancestor, the prop-jet, was in any case physically laborious. Temperatures inside the turbine of a pure-jet are severe enough, but at least the hot gas is going through rapidly (wasting, indeed, large parts of its energy on the atmosphere). But in order to drive a thirty-blade fan some seven feet in diameter, a much larger proportion of the gas stream's energy must be extracted, which means slowing up the stream and sharply increasing turbine temperature. Even in the early pure-jets, turbine blades ran cherry red. In a big fan-jet, they must run white-hot, under stresses of about 1000g (or say, 14 tons per square inch).

It was a long time before engine builders could be reasonably confident that the problems of the high bypass engine were soluble. Ernest Eltis, a Rolls-Royce engineer, still recalls the anxious debates of the 1950s when the firm's management came to the conclusion that finance would not be available to develop *both* an engine for the supersonic transport market *and* fan-jets for slower but probably larger aircraft.

The engineers were asked to say which it was to be. In those days, the prospective fuel efficiency of the fan-jet did not seem quite so powerful an argument, and the lesson of the 707 era appeared to be that the faster airplane would always win the revenue battle. But the Rolls-Royce engineers eventually declared against supersonic engines, starting themselves thereby along the road leading to the RB-211 engine of the Lockheed TriStar. And despite the difficulties of technology and finance that Rolls-Royce subsequently encountered in the transonic region, the decision was undoubtedly sound, for the fan-jet has restored the company to the front rank of the world's engine manufacturers. (Subsequently, Rolls-Royce was obliged to become engaged in the supersonic engine business. In 1967 it took over its main British rival Bristol Siddeley and inherited the contract, which it shares with a French company, to build the Olympus engines for Concorde. The project has, of course, been an unhappy one and the supersonic jet appears, for the moment at least, to have reached a commercial dead end.)

Solving the fan-jet's problems required the revival of a technique used in German military jets of 1944–45, when turbine blades were largely made by hand. Confronted with heat failure, the German engineers led cold air into the turbine hub, and out through cooling channels inside each one of the turbine's hundreds of blades. To make air-cooled blades into a practical proposition for mass-produced civil engines required elaborate metal handling techniques: Blades are cast with great precision by a "lost-wax" process, and the minute systems of cooling channels are cut by electric-spark erosion. When the first big fan-jet went into airline service—the Pratt & Whitney JT9D in the Boeing 747—it was still burning out turbines at a rate which alarmed both the National Transportation Safety Board and the financial directors of the airlines.* But even so, the problems were in principle defeated, and the fan-jet brought to aviation another quantum jump in available power.

Whereas pure-jets and low bypass engines could produce 16,000- to 20,000-pound thrusts economically, the fan-jet could offer 45,000 pounds and more. It was the prospect of this startling power increase which produced the logic of the wide-body, transonic † airliner, the jumbo jets—Boeing 747, Lockheed TriStar, and DC-10. "Up to then," said a member of the Lockheed design team, "you built an airplane by looking at the power you could get and hanging an airframe onto it. This time we felt we could build the airplane as we wanted it, and the power would be there."

WHEN ONE OF these vast machines, like a cathedral in motion, rumbles along an airport runway, onlookers are frequently afflicted with that same unease that the Douglas sixty-foot monster evoked in 1933. How can anything so gross sustain itself safely in the scarcely tangible medium of air? But because flight is an experience paradoxical to our senses, physical in-

---

* See Chapter 9.
† Transonic means the speed region around Mach 1—from Mach 0.85 to Mach 1.2. On either side of it lie the subsonic and supersonic regions.

tuitions are a poor guide to the rules which govern it. In abstract principle, there are several reasons why bigger machines should generally be safer.

Safety is chiefly a contest against the assaults of the physical environment, and these have proportionately less effect upon a larger structure, other things being equal. To the little *Sirius* of 703 tons, the first ship to steam across the Atlantic, a sixty-foot wave was an awe-inspiring hazard, for it was one-third as high as *Sirius* was long. In 1859 engineers and seamen of orthodox experience felt, intuitively, that Isambard Kingdom Brunel was crazy when he proposed to build the *Great Eastern* to 18,915 tons. But in principle Brunel was right, if ahead of his time. To his ship, the height of a sixty-foot wave was less than one-tenth of her length.

Granted that the demands of structural strength and control have been met—*and given that the power is available to drive it*—the big ship or airplane has another reason, besides sheer scale, for its superiority. Generally, systems of safety and control increase their unit weight less rapidly than does the structure as a whole, and so the big machines can have more of them without sacrificing economic efficiency.

As the size of a ship increases, double-bottom frames, watertight door gears, and communications equipment form a lesser proportion of the total displacement, and the same effect applies to aircraft. A single hydraulic system for a jet transport weighs about a thousand pounds and in the 80-odd tons gross weight of the Boeing 727—the most widely used of all jetliners, especially on intercity routes—the two hydraulic systems make up about 2.5 percent of the total. But a single hydraulic system is not much more than 0.1 percent of the 300 tons of a 747. Therefore, a 747 can readily have four hydraulic systems, making in the San Francisco case the difference between survival and disaster.

There seems to come a moment in the history of any branch of technology when the physical problems have in essence been mastered. Statistical inquiry alone cannot identify the moment, although it should coincide with a period in which accidents become increasingly rare. It is a matter of historical judgment: of distinguishing between an era in which people move tentatively, afflicted by failures whose causes remain largely mysterious, and an era of technical confidence, when failure is only rarely a matter of physical inadequacy, and more frequently results from an inability to exercise social control over complex industrial processes.

Such a distinction may be perceived, for instance, in the history of marine navigation. Few people today, stepping aboard a ship, feel any of that disquiet that even experienced travelers feel on boarding an airplane. Yet the sea, within modern times, inspired greater fears and with good cause: during its wars with France the British Navy lost 10 ships to the enemy, but 344 to the sea. And the introduction of metal hulls and steam power at first produced no great improvement, for the state of the art was too primitive. Ships disintegrated at sea, ships foundered for unknown reasons, engines failed, boilers burst, and fires broke out.

Gradually, however, the rules of construction and stability were worked out, and the engines which at first gave only 100- or 150-horsepower were evolved into serious power units. In one sense, Brunel's opponents were right about the *Great Eastern:* strong as she was and shaped to confront the ocean, her engines were much too feeble. She had less than 1-horsepower per gross ton, which was no more than the little wooden steamers living on sufferance among the seas.

Fifty years later, when ships like the *Mauretania* and *Lusitania* came into service, Brunel's spacious conception was realized. The *Mauretania,* enjoying the power revolution based on the introduction of steam turbines, had more than 2-horsepower available for each of her 31,000 gross tons. It was thought that ships of such size and strength would be immune from virtually all of the dangers of the sea.* It was a reasonable aspiration, but it was not turned into reality until there had been an exemplary disaster.

The *Titanic* was built as the White Star Line's answer to Cunard's *Mauretania.* This was a period of fierce competition on the Atlantic sealanes, when any advance by one company compelled a swift response from its rivals. Possession of the Blue Riband for the fastest transatlantic passage was a matter of serious business rivalry.

Although larger than the *Mauretania* and built somewhat later, the *Titanic* was in certain respects less advanced, for only one-third of her power came from turbines. She was promoted, however, as the "unsinkable" ship, in a piece of overconfidence which has haunted designers of ships and airplanes ever since.

This supposed unsinkability was not predicated simply on the fact that she was then (46,000 tons) the world's biggest ship. Much progress had been made with the theory and practice of subdividing hulls, so that it should be possible to "seal off" any puncture by closing off watertight doors. The *Titanic* met all of the demands for subdivision laid down in the shipping regulations of the time. (She also met all the lifesaving regulations: that is, she had boats for about one-third of those on board.)

It was important for competitive reasons that she should make a speedy maiden voyage from Southampton to New York, and the more so because with her less advanced engines she could not match the *Mauretania*'s top speed. Her officers drove her on at twenty-one knots through an area known to be sown with icebergs, and just before midnight on April 12, 1912, she struck one which cut, smoothly, a three hundred-foot gash in her double bottom. She sank in less than three hours, taking 1,494 people with her, and leaving behind one of the most enduring of all images of technological hubris. She demonstrated that mastery of the physical environment is held only by constant vigilance and constant humility.

Investigation showed that although her subdivision conformed with regulations of the time, it was considerably less thorough than it could have been.

* Curiously enough, she had 68,000-hp from three turbines: roughly the same as the total power of a Boeing 747.

It was also found that the operation of the ship had been reckless. As a result, new rules were laid down for the construction of ships, for operation in iceberg areas, and for carrying of lifesaving equipment.

Aerial navigation in the 1970s has perhaps reached a stage analogous to that attained by sea navigation in the first decade before World War I. Improvements, of course, can and will be made in the technology of aircraft, just as improvements have been made in marine technology since 1912. And indeed, it may be assumed that no system will be flawless. There will always be freakish accidents, and there will always be the possibility of sabotage.

But after the sinking of the *Titanic,* there were no great peacetime disasters on the North Atlantic passenger run, and that fact probably owes more to constructional regulation, to the supervision of the sealancs, and to the moderation of economic conflict in the shipping industry, than to advances in the state of the art.*

There is some support for this general proposition in the introductory record of the 747. No new aircraft before the 747 flew as much as half a million hours without a fatal accident. Insurers had grown accustomed to an unlovely burst of accidents while the operators ironed out the "bugs" from each new type of airplane. The 747 fleet, however, exceeded two million hours before suffering a fatal accident—a record unlike that of any earlier aircraft.

Our inquiry into the crash of the DC-10 outside Paris in March 1974 provides evidence, from another direction, in support of the proposition. Our argument is that, like the sinking of the *Titanic,* it arose out of human failures in the legal and economic organization of an industry. The story can be traced back as far as the origins of the McDonnell Douglas Corporation, and the competitive circumstances of the industry in which it operates.

* An extraordinary series of errors led, in 1956, to the sinking of the *Andrea Doria* after being rammed by the *Stockholm.* However, all but 43 of the 1,706 people on board were rescued, due, in large part, to the efficiency of the U.S. Coast Guard. The *Andrea Doria,* although virtually on her beam ends, remained afloat for eleven hours. Further evidence for the role of regulation is the difference in fatal accidents caused by nationally registered and "flag of convenience" ships. Both use the same technology. But although convenience registrations are only 15 percent of world tonnage, they caused in 1972 44 percent of maritime fatalities.

3

"Build Comfort and Put Wings on It"

Douglas built this ship to last, but nobody expected
The bloody thing would fly and fly, no matter how they
 wrecked it—
While nations fall, and men retire, and jets go obsolete,
The Gooney Bird flies on and on, at eleven thousand feet.

Chorus: *They patched her up with masking tape, with paperclips
 and strings,*
And still she flies and never dies—Methuselah with wings.—
 —AIRMEN'S SONG (ANONYMOUS)
  *celebrating the durability of the Douglas DC-3
  (the "Gooney Bird").*

AEROSPACE ENTERPRISES ARE only just emerging from their heroic days. Many of them still bear heavily the imprint of the founding families, and McDonnell Douglas, on both sides of its ancestry, is the classic instance. The Douglas Aircraft Company and the McDonnell Company—respectively, the builders of the most successful piston-engined transport and the most successful heavy jet fighter, the Phantom Two—were both creations of patriarchal individualists. Both companies made their founders into immensely rich men. But in spite of all the technical and financial success that attends the beginnings of our story, a crucial part of it is the process by which the first monarchs of the air-transport industry were deposed. Their position was originally won after a complex battle between airlines and aircraft companies which in many respects resembled the battles fought thirty-five years later over the jumbo-jet market. The differences, however, are as instructive as the similarities.

In corporate mythologies usually much is made of the supposedly humble origins of the founding father, and this is the case with both Donald Douglas and James McDonnell. Although neither of them came from wealthy backgrounds, they were in fact both well connected with the solidly prosperous eastern middle class of the years before 1914. Donald Douglas, the son of a New York bank official, was accepted into the U.S. Naval Academy at

Annapolis in 1909 and later transferred to the Massachusetts Institute of Technology (M.I.T.). Without a doubt, he displayed energy, courage, and intellectual address in building up his aviation enterprises. But his was not in any sense an underprivileged start in life. After some experience teaching aeronautical engineering at M.I.T., and as a consultant to the U.S. Government, Douglas spent five years working in California for Glenn L. Martin, one of the very first generation of airplane builders, and there he gained his first substantial design experience. The concept of strategic bombing began to emerge in the closing phase of World War I, and Douglas designed for Martin a heavy bomber which was a considerable advance upon those going into service in Europe during 1918.

In 1920, after leaving Martin, Douglas set up on his own. He was aged twenty-eight and had $600 capital. His main financial backing came from a wealthy aviation enthusiast named David R. Davis, who put up $40,000 for an attempt to build a machine capable of flying nonstop from coast to coast. In his honor, the company was called Davis–Douglas Company.

The "Cloudster," a clean, sturdy single-engined biplane, did not manage a coast-to-coast flight. But it provided the company with the design basis for a very successful line of military aircraft: chiefly torpedo bombers for the U.S. Navy and light bomber-observation planes for the Army. In 1924 four Douglas biplanes flown by Army pilots made the first aerial circumnavigation of the world, spreading roughly two weeks' flying over a leisurely six-month schedule. (One crew lost its plane off Iceland, and Douglas shipped out a replacement.)

The company's pioneer vitality derived chiefly from this prosperous military connection. On winning his first naval contract, Douglas got a group of Los Angeles businessmen, including Harry Chandler, publisher of the Los Angeles *Times,* to guarantee promissory notes for $15,000. He then turned the Davis–Douglas Company into the first Douglas Aircraft Company. After that, expansion was financed by bank loans secured by government orders. In 1922 the Navy placed a follow-up order for thirty-eight torpedo bombers; in 1925 the Army ordered seventy-five Douglas 0-2 biplanes, and a couple of years later the U.S. Post Office ordered fifty modified 0-2s as mail planes.

In 1928 Donald Douglas organized a new Douglas Aircraft Company in Delaware to take over the assets of the original. There were 300,000 shares issued, of which Douglas himself received 200,000 as consideration for the assets of Douglas Aircraft of California. Perhaps the most important of these assets was a substantial body of experience in aircraft design and manufacture. Opportunities for the development of such experiences were about to increase dramatically.

By the turn of the decade, it was clear that there was a potentially enormous market for long-distance air travel within the United States. The long boom of the twenties had created unprecedented wealth in many parts of America, and not even the Great Depression could wipe it out. (Indeed, the fall in prices after 1931 *increased* the effective purchasing power of people who still had jobs and incomes.) Even the fastest trains took three days to make the coast-to-coast crossing: America, outside the eastern seaboard, was essentially a series of unconnected provinces.

Technically, aviation was ready to make major advances. The engineering of all-metal structures had been mastered, so that with retractable undercarriages and simple streamlining it was possible to design clean and efficient airframes. The moment had now arrived for something new, and in 1931 Jack Frye, vice-president of Transcontinental and Western Airlines (now TWA) suspected that one of his rivals might be about to produce it. Specifically, he had heard that United Airlines had invested $4 million for a fleet of sixty Boeing 247s: twin-engined, all-metal monoplanes to carry ten passengers each. When delivered, they would enable United to mount a schedule for twenty-hour crossings coast to coast. Frye's airline was then using the Ford Tri-Motor, which took thirty-six hours. Frye sent out letters to five firms, including Douglas, asking them to see if they could devise something to equal or surpass the machines United was about to receive. He suggested, rather modestly, a three-engined airplane which would cruise at 145 miles per hour with twelve passengers.

Frye's request intrigued Douglas and his design team, although their experiences with commercial aviation up to that point had not been especially happy. In 1925 Western Air Express had inaugurated a mail run from Salt Lake City to Los Angeles with Douglas aircraft, but had switched quite rapidly to larger Fokker aircraft. Half a dozen years later, however, the Douglas engineers under Arthur Raymond and "Dutch" Kindelburger were well aware that powerful new engines were becoming available, and that this could transform civil aircraft design. By way of research Raymond, who disliked flying, was persuaded to make a coast-to-coast journey in a Ford Tri-Motor.

The "Tin Goose," as it was called, possessed no beauty. It had a slabsided metal fuselage hung under a thick wing, three 300-hp Wright Whirlwind engines, and a fixed undercarriage. Compared to the wood-and-wire biplanes it had replaced, it was a strong and reliable machine, but Raymond found a great deal wrong with it.

"They gave us cotton wool to stuff in our ears, the 'Tin Goose' was so noisy," he wrote. "The thing vibrated so much it shook the eyeglasses right off your nose. In order to talk to the guy across the aisle you had to shout at the top of your lungs. The higher we went, to get over the mountains, the colder it got in the cabin. My feet nearly froze. . . . When the plane landed on a puddle-splotched runway, a spray of mud, sucked in by the cabin air vents, splattered everybody."

Recovering from his odyssey, Raymond declared that if the Douglas Company were to go into the commercial market, it would be necessary to "build comfort and put wings on it." Possibly it required a designer coming from the specialized world of naval aviation to see that commercial airplane designers had come to accept standards of amenity that canceled out many of the supposed advantages of speed. If a man of normal physique gained two days on a business trip from New York to Los Angeles, only to lose it in recovering from the debilitation inflicted on him, then there was no net gain.

Raymond set aside Frye's ideas about a three-engined airplane, because with two of the new 710-hp Wright Cyclone engines he could get 60 percent more power than the "Tin Goose" had with three Whirlwinds. He had his engineers start by building a full-scale mockup of their proposed passenger cabin—nowadays, a routine step, but then a radical departure.

The DC-1 (D for Douglas, C for commercial, 1 for first), when it appeared in 1933, looked very similar to the Boeing 247. Both were low-wing monoplanes with two engines tucked inside smooth cowlings, exhibiting the now-familiar airliner configuration of a small, streamlined windscreen just over the nose and a long row of passenger windows along each side. Both of them cruised at a little under 200 miles per hour. The 247 was not an entirely unsuccessful aircraft, but the DC-1 was the beginning of a series which was to display decisive commercial superiority. The difference with Arthur Raymond's design lay in the care lavished on its interior.

To begin with, the DC-1 could carry twelve passengers, two more than the Boeing 247. And the cabin interior was uncluttered, with the seats ranged along a clear, well-carpeted aisle. In the 247, the cabin space was penetrated by the mainspar, the heavy beam joining the two wings together through the fuselage. Raymond Kindelburger and John K. Northrop * had evolved a "cellular" wing with two spars which could be integrated in the fuselage structure and did not penetrate the cabin.

Each of the DC-1's well-upholstered seats had a reading lamp and a foot-rest. At the rear there was a galley with electric hotplates and a lavatory. The cabin was heated, had a ventilation system designed to let in air without letting in noise, and was soundproofed. To anyone who flies in an old Douglas airplane today—and the DC-3 is still, after forty years, the most numerous type on the civil register—it seems a noisy aircraft. But in the thirties, the DC series were miracles of silence.

There was only one DC-1, but the plane that went into production, the DC-2, was only slightly different.† In 1934 the Dutch Airline KLM entered a DC-2 in the England–Australia air race. The winning machine was a slim two-seater De Havilland Comet racer built especially for the event, but the DC-2 came second carrying passengers and mail, having covered eleven thousand miles in ninety hours elapsed time. The next year, a DC-2 set the U.S. west-to-east record for transport aircraft at twelve hours and forty-five minutes.

The Douglas Company lost money on building 138 DC-2s, but the losses were repaid many times over by its immediate descendant, the DC-3, which first flew at the end of 1935. It was the result of an approach, a few months

---

* Later the founder of the Northrop Corporation.
† The DC-1 was sold to Britain, then to Spain, and was destroyed while flying supplies for the Loyalist side in the Civil War.

earlier, from Cyrus R. Smith, president of American Airlines, who said that the DC-2 fuselage was too narrow to accommodate a comfortable sleeping berth. There were by that time 1000-hp engines available from both Wright and Pratt & Whitney, so a slightly wider aircraft could be built without sacrificing performance.

Curiously, Donald Douglas himself was not very ebullient about the DC-3 project when American Airlines first mooted it. And later he seems to have attributed its success as much to good luck as to good design. "It is certainly the best-loved plane we have ever produced," he said later. "But the circumstances that made it great just happened. They were not of our making." *
Possibly, Douglas was thinking chiefly of the plane's military apotheosis in World War II, but that was only made possible by the extraordinary success of the DC-3 as a civil transport in the prewar years. The extra size and power seemed to give the passengers (twenty or so) exactly what they wanted: by 1940 there were nearly three hundred DC-3s in service, and Douglas aircraft dominated the trunk networks that were completing the task of knitting metropolitan America into a single nation. A whole generation of pilots and engineers were brought up on Douglas aircraft and Douglas methods; meanwhile, Douglas's rivals, such as Boeing and Lockheed, were forced to sink large sums into new, and hopefully more advanced aircraft, in the hope of finding some way of getting back into the market.

The DC-3 was not quite the obvious choice as the main instrument of Allied aerial logistics in World War II. Its basic design, after all, was almost ten years old when America entered the war. There were more advanced aircraft available, such as the Fairchild Packet or the Budd C-93 (built of stainless steel), and some of these machines were better laid out as bulk military transports than the DC-3. Cyrus Smith of American Airlines became Chief of Staff of Air Transport Command and decided that the new Curtiss-Wright C-46 Commando ought to be superior to the airplane he had ordered from Douglas in 1935.

But the DC-3's age was probably its decisive advantage. Thousands of hours of airline operation had "de-bugged" the DC-3 more thoroughly than any plane had been de-bugged before and more than any new machine could possibly be in the stress of war. The C-46 suffered from leaky fuselage joints, hydraulic failures, fuel blockages, and carburetor icing: in twenty-two months' operations there were thirty-one recorded cases of C-46s catching fire or blowing up in the air. As it turned out, the air transport commands in both America and Britain relied basically upon the DC-3. Dwight Eisenhower named it as one of the five critical pieces of equipment which made worldwide war possible (the others being the bulldozer, the jeep, the two-and-one-half-ton truck, and the amphibian DUKW).

During the war more than ten thousand DC-3s were produced. The American government financed and built a new factory for Douglas at Long Beach, on land which was leased from Donald Douglas's Montana Land

* Arthur Pearcy, *The Dakota* (London: 1972), p. 12.

Company. (After the war the Long Beach factory was sold to the company, and it is now the California headquarters for the amalgamated McDonnell Douglas Corporation.) The DC-3 was the largest single component in Douglas's astonishing 1944 sales figure of $1,061,407,485. The company that Donald Douglas had started fourteen years earlier with six employees had become, for the moment, the fourth largest business in the United States.

Immediately after the war, naturally, there was a sharp falling off from these phenomenal levels of activity. Nonetheless, the first postwar decade appeared to establish Douglas as the invincibly dominant power in commercial aviation. War-surplus DC-3s were sold to airlines throughout the Western world (Aeroflot, in the Soviet Union, organized its postwar expansion around Russian-built DC-3s) and along with those aircraft went war-surplus pilots and war-surplus mechanics who knew how to fly and maintain Douglas airplanes. The DC-3 could still meet most demands for a general-purpose commercial transport, and for those whose needs were more advanced, Douglas began selling bigger aircraft which made an easy step up for anyone who could understand a DC-3.

Douglas held this dominant position virtually without challenge until at least 1955. And then, within four years, it was utterly wiped out. By the middle sixties Douglas had not only lost its position as the world's greatest supplier of transport aircraft; it was a company lumbering towards bankruptcy. The agency of this startling reversal was the Boeing Airplane Company of Seattle, which consummated one of the great acts of commercial retribution just about thirty years after its own Model 247 was overwhelmed by the DC-1, DC-2, and DC-3.

BOEING WAS, FROM the start, a company standing slightly apart from the patriarchal traditions of the aviation industry. Its founder, William Boeing, was a lawyer and timber merchant. Although airplanes fascinated him, his lack of engineering education deterred him from trying to run the company as a personal concern. He concentrated his energies upon finance and administration—a sufficiently hair-raising task, given the switch-back economics of the early days—and built up a design team of notably high theoretical ability. The company was an emergent business bureaucracy, rather than an individualist empire, well before the fashion began to turn that way in the aviation industry. The stereotype version is that a corporate bureaucracy will be less adventurous than an individualist entrepreneur. But industrial history sometimes deals roughly with stereotypes, and this is the case when comparisons are made between Donald Douglas's highly personal operation and the managerial challengers from Seattle.

The Douglas design philosophy, according to the company's own account of itself,* is encapsulated in a couplet from Alexander Pope:

* *Flight Plan for Tomorrow* (The Douglas Aircraft Company, 1966), p. 76.

> Be not the first by whom the new are tried
> Nor yet the last to lay the old aside.

It is a well-seasoned and frequently profitable business principle. Broadly, the idea is to let other people lose their money fooling with new technologies, and then move in yourself once the problems have been sorted out. (Of course, an economy containing only firms which behaved according to this seemingly rational principle would not make much technical progress.)

The Boeing Corporation, as it is now called, is not without its enemies— apart from anything else, it incurs the suspicion of a good many U.S. citizens simply by being a member in good standing of the "military-industrial complex." No one, however, could say that its aeronautical engineers have been reluctant to appear as "the first by whom the new are tried." And essentially it was Boeing's greater technological daring—and the success which attended it—which destroyed Douglas as an independent company.

Conceivably, the victory of the DC series over the Boeing 247 was good for Boeing in the long term. One result was that the company spent the thirties, not in stamping out repeated examples of a known successful design, but in trying to extend the state of the art. In particular, the Boeing engineers worked on the problems of high-altitude flight with multiengined aircraft. Power was available in the thirties to lift aircraft out of the heavy, turbulent lower atmosphere (amid which the DC-3 cruised at eleven thousand feet) into thin, calm, high-altitude air where airframes would be more efficient and passengers would ride more smoothly. Also, at twenty thousand feet or so, airplanes would be able to go straight over major mountain ranges, instead of having to weave around them.

These improvements, however, could only be realized if high-altitude passengers could breathe a normal atmosphere inside the cabin, instead of having to wear oxygen masks.

The Lockheed Corporation of Burbank—also kept out of the major airline markets by the success of the DC-series, but emerging nonetheless as a technically powerful firm—was first to fly a pressurized airplane. This was the twin-engined XC-35 of 1937. But Boeing was at work on designs for larger aircraft, and in 1940 the Model 307 Stratoliner, the world's first pressurized airliner, went into service with TWA and Pan American. There was a little poetic license in its name, because at best it only got halfway to the stratosphere. But its four engines could lift thirty-three passengers to twenty thousand feet, which was a very considerable feat at the time.

Douglas, meanwhile, had been working on a four-engined successor to the DC-3, and at first the plans were for a pressurized machine, to be called the DC-4E. But after expending $3 million on development, Douglas abandoned the DC-4E. The explanation, given more or less as the Boeing Stratoliner was going into service, was that "progress must come by orderly evolution of sound, well-developed principles." *

* *Flight Plan for Tomorrow*, Douglas.

The DC-4, when it eventually appeared, *unpressurized,* in 1942, was certainly soundly engineered. These qualities no doubt influenced its success as a wartime transport with the U.S. Army and Navy, whose orders accounted for most of the 1,243 DC-4 "Skymasters" which were built. The airplane helped considerably in establishing Douglas's wartime and postwar prosperity. But it was not an advanced machine for its time, and neither was the twin-engined DC-5 (which, confusingly, was produced before the DC-4).

At this same period Boeing, having put the four-engined B-17 ("Flying Fortress") into service before America entered the war, was developing the astonishingly advanced B-29 "Superfortress," conceived as a pressurized bomber capable of 370 miles per hour at twenty-five thousand feet. Just then, such a performance would have been thought creditable in a single-seat fighter aircraft. It was made possible in the 45-ton B-29 by designing a long, slim wing which was alarmingly heavily loaded by the standards of the day, and which required unprecedented quantities of flap area to get the plane on and off the ground.

The B-29's early history was troubled—chiefly because of engine fires, for the Wright RR3350 engine was also extending the limits of the possible to produce 2200-hp. But its eventual success consolidated Boeing's reputation for designing high-performance, long-range load carriers: the B-29 could fly seven thousand miles. When, at the end of the war, the German work on swept wings and jet propulsion fell into Allied hands the Boeing designers, already accustomed to struggling with new technologies, were natural beneficiaries.

In 1946 the propeller-driven DC-6 finally brought Douglas into the era of pressurized flight. But in the following year, Boeing went ahead into the jet era with the XB-47 medium bomber.

The XB-47 was not exactly a well-loved aircraft. Notoriously, it suffered from the "coffin corner" effect often found in early jets, there was a combination of speed and altitude at which any acceleration would cause a high-speed shock stall and any decrease was likely to cause an aerodynamic stall due to loss of airspeed. In either case the plane fell out of the sky.

It was as difficult to land as it was to fly, because it had a simple bicycle undercarriage. Still, it fulfilled its military tasks, and it gave the Boeing design team experience of the classic multiengine jet layout, with swept wings and engine-pods hung under the wings on forward-sloping "pylons." In July 1954, Boeing flew a new four-engined jet transport, the Model 367-80, which was ordered by the Air Force as the KC-135 tanker. Having done a good deal of development work on the government account, Boeing remodeled this airplane to produce the 707 commercial airliner. The prototype flew on December 20, 1957, and obtained almost at once an order for sixty aircraft from Pan American.

In retrospect, it is not easy to see just how bold a decision that was both for the airplane builder and the airline. Two years earlier, the first jetliner, the British De Havilland Comet I, had been withdrawn from service after two instances of high-altitude disintegration. The causes were still imperfectly understood. (Sadly, the British industry had little experience with pressurized aircraft, and investigation showed that in the Comet design insufficient allowance had been made for the fatigue caused around the plane's large windowframes by "hoop stress." This phenomenon is caused by the fact that a pressurized fuselage contracts and expands slightly at various stages of flight, complicating the stress patterns in the airframe.) The Comet death toll added to the considerable frequency of military jet disasters and made passengers uneasy about the safety of jet airliners. Airline managements were uneasy about the fuel consumption of jet engines, and pilots were uneasy about learning to fly aircraft which behaved quite unlike propeller-driven transports. Boeing, which had not had a large civil fleet in operation since before the war, was still relearning the differences between military airplanes, which are coddled like racehorses between their relatively infrequent exercises, and commercial hacks which must ply their trade seven days a week regardless of weather and ill-kept runways.

And there was the formidable objection that the famous Douglas Company was laying its bets *against* the jet airliner. Douglas, having pursued a policy of steady consolidation, seemed impregnable as market leaders through most of the 1950s. In 1950, two-thirds of all the aircraft on scheduled services in the noncommunist world were Douglas-built. During the following five years, a good many DC-3s retired, but even so, Douglas still accounted for 50 percent in 1955. A DC-6 succeeded a DC-4 as the President's personal transport. And there was a new Douglas airliner on the way which would develop naturally from existing and well-liked Douglas aircraft, and which would use a new, apparently promising—but now forgotten—power plant called the "compound engine."

Much official opinion, both American and European, favored the compound engine in the fifties, and some people thought that engineers at Pratt & Whitney and Rolls-Royce were premature in their enthusiasm for gas turbines.

Piston-engine designers since 1910 had been looking for ways of using the combustion energy which normally goes to waste in the exhaust. One method was to make the hot gases drive a turbine to supply power to the supercharger. The idea of the "compound" engine was to apply such power direct to the propeller shaft. The Wright Company in America and the Napier Company in England were both important engine builders and both advocates of the system. Both companies were to lose their importance with its eclipse.

The Douglas DC-7 used the Wright Turbo-Compound 3350, which in "normal" configuration would have been able to produce 2700-hp. Three turbines driven off the exhaust raised the power output to 3700-hp without any increase in fuel consumption. The result was to increase aircraft range by about 20 percent, and the DC-7 could fly the Atlantic nonstop without restriction. When it went into service in 1956 it was regarded as the ultimate in airline practice in spite of the challenge of the turbo-props. That supremacy lasted until October 1958, when the first Boeing 707s came into service, and it was then obliterated.

The impact of the jet revolution upon Douglas's position can be brutally expressed in production figures. In 1956 the firm delivered 106 airplanes. In 1957 it delivered 123 DC-7s alone, plus 44 DC-6s. Even in 1958, the year of the 707's debut, Douglas produced more aircraft than Boeing, Convair, Lockheed, and Fairchild altogether.

Then in 1959 Douglas delivered *one* DC-6, *no* DC-7s and, at enormous cost, twenty-one examples of its belated candidate for the jet market, the Douglas DC-8. Boeing meanwhile delivered seventy-three of its 707s.

The mistiming of Douglas's attempt upon the jet market was curiously like the case of the DC-4E and the sudden retreat from pressurization. There had once been a clear prospect that Douglas would win the jetliner race. In 1952, two years before Boeing began to turn the KC-135 tanker into the 707, Douglas had announced the DC-8 as its jet contender. Airline executives were invited to the Santa Monica plant and shown round a full-scale mockup of an aircraft, which they were told would be ready for service in 1958, and would fly at 560 miles per hour. They were impressed, as was the influential magazine *Aviation Week,* which reported that while Boeing was haggling for government subsidies, Douglas had taken the bold decision to finance jet-transport development itself. The decision had produced a "clear-cut lead," and Douglas could be expected to dominate the jet era just as thoroughly as the piston era.

But in 1953, at a meeting celebrating the fiftieth anniversary of powered flight, Donald Douglas said his company was giving up its lead. "In our business," he said, "the race is not always to the swift, nor the first to start. I have always held the conviction that airplanes should make money as well as headlines. . . ." Work on the DC-8 ceased, so that the prospects of the company depended almost entirely upon the DC-7.

It was an epic miscalculation, but of a sort that was readily made at the beginning of the jet era. Lockheed also abandoned work on its pure-jet airliner in 1952, and concentrated upon the prop-jet Electra, which, after one year of high success in 1959, failed so badly that the company dropped out of the civil market for a decade. Again, General Dynamics suffered huge losses through delays in the development of its Convair 880/990 (Coronado) jetliner series. But Lockheed and General Dynamics were sufficiently diverse corporations to absorb the losses and turn to other activities. Douglas, deeply reliant upon the commercial transport market, had no choice but to remain in the race, however dreadful the odds had become. This meant, in 1959, getting the DC-8 into airline service whatever the cost.

In 1952 it had seemed likely that the development budget of the DC-8 would be $40 million spread over six years. But once Douglas lost its lead, the cost of catching up multiplied many times. In the event, the development costs of the DC-8 and its smaller, twin-engined companion, the DC-9, probably came to $500 million.

Douglas did make some progress, for the DC-8 went into service only one year behind the 707, and it got plenty of orders. An airline's profits in those hectic days were decided almost entirely by the rapidity with which it could get jets onto its long-distance routes, and the demand was much greater than Boeing could alone supply. By 1961 Douglas broke its run of losses with a pretax profit of $5 million. But the signs of recovery were misleading. The Boeing assault had inflicted wounds which were even deeper than they appeared to be.

The profitability of an airplane company depends upon how soon it can "climb on top of the learning curve" with each new model. The price received for the first few examples will bear little relationship, generally, to the cost of building them. Apart from research and development expenses, a production line is supremely inefficient when it has just been set up, with time and materials going to waste as workers are trained "on the job." But with a settled design and steady output, the construction costs of a single machine can be cut by four-fifths or more. And at this point the manufacturer begins to see profits and begins to think about derivative models with high sales potential but minimal start-up costs.

Boeing's 707 production line had been worked up on military contracts and was able to produce profitable airplanes almost at once. The resulting revenue enabled Boeing to strike two more deadly blows at its rivals in the jet-transport market. First, in 1960 Boeing introduced the Model 720, a Boeing 707 slightly scaled down for use outside the intercontinental routes. Three years later came the Model 727 medium-haul jetliner, with three engines clustered at the rear of the fuselage. Numerically, this is the most successful civil aircraft ever built—there were more DC-3s, of course, but most were originally military—and its advanced versions seem likely to go on selling almost indefinitely.*

Douglas, still trying to make a profitable DC-8, failed to respond with a new design. In five years from 1960, Boeing outsold Douglas in jets by more than two to one. Then in the mid-sixties Boeing went into the short-haul market with the tubby little 737, one engine under each wing, and introduced the enormous 747 in 1968. Ten years after the 707 made its first revenue flight, Boeing had taken over the jet market as thoroughly as Douglas had taken the propeller market. When the U.S. Government ran a competition for a supersonic transport program (now abandoned), Douglas had neither the money nor the time to compete seriously. The contract went to Boeing.

* By the end of 1975 Boeing had sold 1,244 Boeing 727s and delivered 1,178 of them. The 707 is still in production (sixteen are currently being built) and, in 1974, was sold to Communist China.

IT NOW CAN be seen that Douglas's independent existence was in doubt from the end of 1960, but the faith in Douglas of a generation of airmen was so potent that nobody would have believed it. During the previous two years the company had lost $108.6 million pretax, and had been forced to borrow heavily to avoid a cash crisis. The DC-8 was an excellent basic design, but early examples failed to meet their performance guarantees, and Douglas had to introduce a rework program, bringing the aircraft back to the factory and modifying them. Engineering confidence—and the learning curve—suffered heavily as a result.

Donald Douglas had in 1958 handed over the presidency of the company to his son, Donald Junior. But few people who worked in the company in those days thought that the essentially autocratic nature of the company genuinely changed; indeed, Donald Sr. remained chief executive of the company right to the end.

The Douglas Aircraft Company in the sixties was a jovial, but not an entirely happy ship. All major executives were expected to gather at midday for martinis in the Heritage Room, where most days the founder would present a guest—sometimes an airplane customer, but often enough a film producer or a favorite writer.* The midday court in the Heritage Room became the main decision-making forum for Douglas Aircraft, if forum is the appropriate word in a situation where one man's writ dominates almost completely.

During the early sixties, while Boeing expanded from the long-haul into the short- and medium-haul markets, Douglas could have countered with its twin-jet DC-9. Rather smaller than the Boeing 727, it resembled the French Sud-Aviation Caravelle and the British BAC 111. But Donald Douglas Sr., worried at the company's inability to get the DC-8 program flowing smoothly, delayed the production decision on the DC-9. As an alternative, the Douglas Company decided to act as sales agents for the Caravelle, which had been flying since 1959, in the hope of getting enough ordered to start a production line in California. It was a bizarre project, and Douglas didn't sell a single Caravelle.

Most of Douglas's senior executives were in despair at the delay of the DC-9 program—none more so than Jackson R. McGowen, a talented engineer and an even more talented salesman. McGowen had been with the company virtually since his graduation from Indiana Technical College in 1939. He had been chief project engineer for the DC-6 and DC-7, and in 1958, he took on, successfully, the task of rescuing some part of the long-haul market for the belated DC-8. On joining the board in 1961, he began to proselytize his own well-founded belief that the DC-9 was capable of being the company's best prospect since the DC-3. It promised to be one of the simplest jets yet built, needing only a two-man crew, but able to lift seventy passengers from a runway less than a mile long. Given a good start, the DC-9 might indeed

* The day the merger with McDonnell became official, Douglas became a "dry ship." It now requires a dispensation from the chairman before liquor can be served.

have remade Douglas's fortune. As it was, the vicissitudes of the little plane's career finished the company off as an independent concern.

The DC-9 missed the first wave of short-haul orders that were scooped up by the Caravelle and 111, and when McGowen at last persuaded the board to begin building it in 1963, it was a speculation in advance of orders. In 1965, with the first flight due, there were still only fifty-six orders: nothing like enough to cover costs. Then, abruptly, the mid-sixties airline boom arrived (between 1964 and 1967 the world's major scheduled airlines alone made profits of $1.5 billion) and everybody wanted planes. By the end of 1966 Douglas had obtained orders for 409 DC-9s, with options for another 104, and had obtained at the same time 138 orders and 40 options for improved-model DC-8s.

"Sell the airplanes," McGowen told his salesmen ebulliently, "and we'll find a way to build them." The years of the drought were over, it seemed, and in order to make sure of the fact Douglas offered some truly remarkable financial concessions to customer airlines. Normally, a manufacturer asks for 25 percent of the airplane price on order, and the balance on delivery. Douglas, however, elected to take promissory notes at fixed interest rates which rapidly fell below the standard interest rate. In other words, Douglas was lending money to the airlines more cheaply than it could borrow money itself. By the end of 1966 the airlines had taken up about $100 million of the cheap "credit," and Douglas was committed to extending another $220 million or so over the next two years.

McGowen, as head of the newly formed transport division "responsible for aircraft" had to produce the aircraft, and it may be doubted whether his talents as an industrial organizer were equivalent to his talents as a sales-team boss. He retained all his confidence, saying shortly before the whole Douglas card house collapsed that any troubles were due to "external factors," and there was "nothing to apologize for. We have a good staff." Most subsequent investigators have doubted this verdict, saying that the line management at Douglas was simply inadequate for the gigantic task imposed upon it when the corporate style changed suddenly from inertia to hyperkinesis. Douglas was suffering from shortages of equipment, skilled workers, and working capital. The crisis of 1959, when production almost came to a halt, had weakened the company's financial base and had also forced it to cut its skilled workforce. Now that skilled men and women were needed urgently, they could not be obtained, and there was little prospect of subcontracting more work out because the Vietnam War was taking over all the industry's spare resources.

The only solution seemed to be to take on unskilled workers and train them. But Douglas had no experience of such operations, and the learning rate of the new workers fell far below forecast, pushing up overtime and pushing up costs. Production schedules fell behind, and as the airlines threatened lawsuits over the delivery delays, yet more unskilled hands were hired, driving costs up further. By 1966 Douglas had been forced to hire three men for every two it should have needed: the 30,000 workforce had increased to eighty thousand. According to the predicted learning-curves, it should have been taking forty-eight thousand work-hours to assemble a DC-9. In reality, it was taking nearly eighty thousand hours.

At the same time, the subcontractors who were available were falling steadily behind their own dates. An airplane must be assembled in sequence, and a missing part can delay the whole line for days. Delays became so serious that Douglas engineers had to build wooden mockups of under-carriage legs just to keep the planes moving until the dummy legs could be replaced by real ones. And on top of all this were the crazy pressures of "customerization."

The salesmen had been told to sell, and they had reacted with remarkable imagination. Boeing was offering a complete "family" of aircraft, from 707 down to 737. To compete, Douglas had to offer "stretched" versions of both the DC-8 and DC-9 well before it had mastered the problems of producing the unstretched models. Following tradition, the airlines ruthlessly played sales teams off against each other, and at times it seemed that the DC-8 and DC-9 were being redesigned ad hoc in a hundred airline offices. Any prospective customer could choose from *one hundred* different galley configurations (though most of the galley subcontracts were running late); he could have one, two, or three fuel tanks; there were four different power ratings for the engines, and in the passenger cabin there were eight hundred different items that came in choices of color. "We have thirty-two shades of white paint," Jackson McGowen told one visitor to his plant.

Douglas managed to keep on delivering airplanes, not too far from schedule, and reported profits showed a slowly rising trend during the first half of the sixties. Indeed, on April 20, 1966, Donald Douglas Junior told shareholders at the annual meeting that the company was in one of the most successful phases of its history. But the reality, whether the management grasped it or not, was otherwise. The production lines, almost at that moment, were beginning to fail under the strain, and the delivery schedules were collapsing. On May 31, Douglas admitted in a quarterly report to shareholders that "earnings this year, if any, will be nominal." By September, it was clear that there would be a heavy loss.

On October 7, 1966, Douglas's seven banks received their first real indication of how serious the crisis was. The loss on the year would be about $70 million, and the company's debts might be twice as large as its assets. So far as the bankers could see, the Douglas board had no idea what to do about the crisis.

Three days later Douglas's line of credit was abruptly cut off. The company could not even meet the weekly payroll without emergency loans. A takeover, if necessary on humiliating terms, was the only possible alternative to bankruptcy.

4

The Spiritualist of St. Louis

We are here to grow souls. We are free to discover and create and make life on earth . . . into heaven or hell. It is at times a hell of a crucible . . . but it works. And at the end, the successful candidates joyously graduate to the next plane of existence.

—JAMES S. MC DONNELL, CHAIRMAN
McDonnell Douglas Corporation

JAMES SMITH McDONNELL displayed interest in the Douglas Aircraft Company some time before its financial debility became obvious. In 1963 he bought 200,000 Douglas shares and spent a weekend in Palm Springs with Donald Douglas Sr., discussing possible merger plans. It turned out that McDonnell's conception of a merger closely resembled a takeover with his own firm in charge. Douglas suggested another idea: maybe his firm should take over McDonnell. The patriarchs parted, as they had met, not the best of friends.

McDonnell Aircraft, dating from 1939, is considerably the younger of the two components that make up McDonnell Douglas. Its founder, however, is of an age to have been in uniform during the last year of World War I and to have attended M.I.T. not many years after Donald Douglas. His origins were perhaps more solidly prosperous than Douglas's, for his father was a considerable businessman, rather than a salaried official, who ran a grocery store in Arkansas and was active in the specialized business of financing cotton crops. He sent one son (Hunter, who became an architect) to Columbia, another (William, who became head of the First National Bank, St. Louis) to Vanderbilt in Tennessee, and his third son, James Smith McDonnell, to Princeton, where he enrolled in 1917.

"Mister Mac," as he likes to call himself (often in the third person, as in the request: "Let there be no moaning at the launch pad when 'Mister Mac' puts out to space"),* believes as firmly as any of the great entrepreneurs that man is the carver of his own destiny. Indeed, he illustrated the stock certificates of McDonnell Aircraft with a picture of a muscular gentleman sculpting his own form from the living rock.

* St. Louis *Post-Dispatch*, July 1, 1975.

James McDonnell is, of course, a man with genuine and formidable talents. But, like Donald Douglas, he came of a generation of young Americans who were not sent out to carve their destinies until they had been equipped with toolkits as good as any that money could obtain. And, like Douglas, the first material in which he carved successfully was the smooth rich-grained stuff of government contracts.

It would not be hard to make "Mister Mac" sound something like a hypocrite. A self-proclaimed acolyte of peace who pursues with relish the profits of military aviation—"the Banshee," as he once put it, "rang the multihundred million dollar gong, the Phantom Two the multibillion dollar gong"—is apt to encounter some cynicism, whatever he says about the efficacy of such machinery in "waging peace." And the cynicism is scarcely disarmed by the kind of answer he gives when asked to assess the results of "waging peace" in Vietnam: "There has been much military unpleasantness in the last decade. There is a big ache in my psyche where Vietnam is. I say now, let's lift our eyes to the future." *

The evidence suggests, though, that McDonnell's attitudes are genuine, owing little to the techniques of synthetic image building. For many years other Republican businessmen have been slightly scandalized to see McDonnell busying himself as a fund-raiser and general advocate for the United Nations Association. Few builders of war planes take front-office clichés about world peace seriously enough to support causes which are politically unpopular in their own circles. McDonnell does and thinks it a pity that other industrialists do not follow his lead by giving all their workers a paid holiday on United Nations Day.

An official of the National Aeronautics and Space Administration (NASA) once said, not without affection, that "Mister Mac," builder of space capsules, was "an anachronism," and that was an apt assessment. Rather than a twentieth-century manipulator of images, McDonnell appears to be a genuine Victorian survival. Obviously, he is richly endowed with the Victorian qualities of energy and business acumen. But he also seems to possess that crucial gift—first widespread among the English business classes, persisting longer among Americans—for conceiving that schemes of profit and interest to oneself will automatically benefit the human race in general. It is an optimistic, idealistic outlook, one which may readily "lift its eyes to the future," forgetting the grim wreckage left by a Phantom making a firing pass.

McDonnell has been influenced throughout his life by one of the characteristic passions of the late Victorians: The belief that scientific and mathematical methods, having explained so much of the physical world, should be extended to the investigation of spiritual and ethical phenomena. It should be possible to prove the existence of the soul by methods such as those used to describe magnetism or the kinetics of gases. It was not simple whim that caused so many McDonnell aircraft to bear names like Phantom, Demon,

* St. Louis *Post-Dispatch,* July 1, 1975

Banshee, and Goblin. Though perhaps only half-serious, it was another affirmation of an engineer's belief in supernatural dimensions.

Young McDonnell's particular favorite among the scientific spiritualists was F. H. Myers, author of *Human Personality and Its Survival After Bodily Death.* Today, perhaps, people like Myers and Frederick Sidgwick seem fusty, almost comic, figures, except to those active followers who labor with the mathematics of dice throwing and thought transference. (McDonnell's first ambition, by way of a career, was to travel to England to assist Myers in such work.) But the frame of mind was a potent one in its time, offering the seductive idea that systems of ethical behavior might be produced by scientific investigation—with the advantage that, once promulgated, they might expect the same universal acceptance as the laws of thermodynamics. Politically, it tended to manifest itself in idealistic schemes for new world orders of reason and light, such as those of Woodrow Wilson, another hero of James McDonnell's young days. Wilson's unsuccessful campaign to get American ratification for the League of Nations is one of McDonnell's earliest political memories. He still regards it as one of the great turning points of contemporary history, and from it traces his devotion to the United Nations.

Such materially progressive, religiously based scientific idealism is not an obviously contemptible philosophy, even if quantum physics and the "uncertainty principle" has made nonsense of its intellectual structure. McDonnell is a careful and conscientious employer, with labor stability rarely seen in an industry notorious for sweeping, ruthless layoffs. "Mister Mac" himself is the kind of man who will break all the corporate PR rules by skipping a national media interview in favor of a children's model-aircraft show and who gives every employee's firstborn child a pair of baby shoes.

Nonetheless, idealism of this kind can be associated with some alarming intellectual traits. The lofty moralist, having seen his pure solution nullified by the inadequacies of human nature, is apt to show little interest in compromises which may modify its failure. Indeed, he is often ready to contemplate with equanimity consequences which appall less highly motivated creatures. Woodrow Wilson, once he was balked of his idealistic and generous peace treaty, agreed complacently to a disastrously vengeful substitute that shocked and alarmed the greatest cynics of Versailles. Wilson's follower, grown to head one of the world's great engineering concerns, could perceive that in the 1960s the dream of universal world peace was not immediately obtainable. While he did not cease to advocate it, he was untroubled by the sound of Phantoms penetrating his office (ringing the multibillion-dollar gong) and taking off to devastate the countryside of Indochina.

McDonnell's idealism, in other words, is one which may be served by large statements of ultimate principle, regardless of whether they work out in terms of gritty, contemporary detail. "In this business," he once said, "getting things right can make the difference between life and death." * He was speaking of the Apollo and Gemini spacecraft McDonnell built, and was

* St. Louis *Post-Dispatch,* February 25, 1969.

celebrating the enormous team effort which had insured that all sixteen astronauts using those machines had returned to earth alive. "Mister Mac's" own personal commitment to getting the details of safety right are not in doubt. In directing the affairs of McDonnell Douglas, however, he may have assumed too readily the diffusion of that personal commitment throughout a complex and far from harmonious corporate structure.

YOUNG JAMES McDONNELL'S first proposals for a career were not well received by his businesslike father. As an alternate to psychic research, he proposed politics. Parental opinion was that James was too shy and other worldly for politics, and that psychic research, even if it could be made academically respectable, was far too other worldly in itself. Aeronautical engineering, as McDonnell recollects, was thought in those days to be "pretty much out of this world, too" but by comparison it appeared respectable. Having majored in physics at Princeton, he received the family blessing to go on to M.I.T., then the only place in America where formal aeronautical training could be had.

He was a methodical young man, celebrated at Princeton for having spent a whole week in his room considering the evidence before deciding in favor of the existence of God. He therefore set down a life-plan, under which he would work as an engineer with any and all of the pioneer firms who would let him do so—if necessary, unpaid—and then at the age of forty he would set up his own firm for the "designing, testing, selling and repair of aircraft and spare parts thereof."

For the next twenty years McDonnell experienced all the vagaries of American aviation between the wars. More than once he was asked to move on and find himself a new job, and an attempt to get ahead of his forty-year deadline for independence met with outright disaster.

It was once a master ambition of American aviation to produce a plane which would become the "flivver of the air," as ubiquitous as the Model T Ford. McDonnell raised cash to build a machine for one of the design contests intended to encourage this development, but while he was flying his "Doodlebug" to demonstrate its structural integrity to the judges, a piece of its tail broke off. Showing that there was nothing wrong with his own financial integrity, McDonnell ignored his parachute and tried hard to preserve everyone's investment by getting the plane down in one piece. But the "Doodlebug" was a total wreck, and McDonnell's spine required considerable cartilage repair.

In December 1938, a few years later, McDonnell left a good job as Chief Project Engineer (Landplanes) with Glenn Martin, and on July 6, 1939, he set up his own firm, the McDonnell Aircraft Company. He had $165,000 to work with, which was his own savings topped up with venture capital he had raised, mostly from St. Louis businessmen, together with $10,000 from Laurance Rockefeller, a Princeton contemporary. He had one employee, a secondhand typewriter, and a rented office at the St. Louis municipal airport. At the end of one year's operations, he had no orders, no income, and no profit.

It was, nonetheless, a propitious moment to start an aircraft firm. And McDonnell, showing that his father may have underestimated the son's political flair, had perceived that if rearmament were to lead to large government orders, then Washington would be compelled to look for Middle West contractors to avoid the charge of undue favor to the established aircraft firms of the East and West Coasts. Business began for McDonnell Aircraft with an order for $7,672 worth of parts for Stinson army observation planes, and nearly all income in the war years came from subcontracting work for longer established manufacturers. A large proportion of the 3,000 tons of wartime parts that McDonnell built was tailplanes for the military DC-3 program. But McDonnell was not one of the mushroom aluminum-cutting concerns that vanished along with the gigantic demand for World War II aircraft types. The returns from subcontracting were used to finance research into the next generation of aircraft.

The firm was regarded favorably in Washington. In 1941 it received a contract to develop the XP-67, a heavy twin-engined propeller-driven fighter, and there was a contract for a twin-engined bomber trainer. But the true foundations of McDonnell's success were laid with a $20,000 Army Air Force research contract in 1940. No metal was to be cut: The only task was to investigate the possibilities of jet propulsion, then becoming a practical possibility among European military designers. The little company's results were impressive enough to be remembered, and on New Year's Day 1943, James McDonnell was called to Washington and invited to tender for the building of a Navy jet fighter. The result was the first Phantom, the FD-1: not much like the famous Phantom II to look at, but (in 1946) the first jet to operate with the carrier fleet.

Like Douglas, McDonnell found the naval connection rewarding. In 1949 the F2H Banshee appeared as a carrier fighter and strike aircraft, in time to become one of the busiest aircraft of the Korean War ("immortalized," as company literature puts it, in James Michener's best seller, *The Bridges at Toko-ri*). Then there was the F-101 Voodoo long-range fleet fighter and reconnaissance plane, and the F3H Demon all-weather fighter.

The ringing of the "multi-billion dollar" gong began in 1953 with design work for Phantom Two. Originally it was to have been a relatively simple, single-engine single-seater, but more and more elaborate Navy requirements turned it into a heavy and complex twin-engine two-seater. With its humpy fuselage, drooped nose, slightly cranked wings and bent-down (anhedraled) tailplane,* it is a long way from being one of the world's most beautiful

---

* The anhedral tailplane was developed by McDonnell's engineers to cope with the potentially deadly consequences of "deep-stall." Early tests showed that during steep climbs the Phantom's swept-back wings could blanket the horizontal stabilizer, causing the plane to stall with its nose in a pitched up position. It was impossible to locate the horizontal tail lower on the fuselage because of the jet exhausts.

airplanes. It is, without doubt, one of the most effective. Seventeen years after its first flight it can still, in the hands of experienced pilots, match the most advanced opposition in aerial combat. It long ago transcended its role as a naval fighter alone, having served with eleven air forces, including the American, as a ground-attack plane, air-supremacy fighter, and—potentially —nuclear bomber.

According to the company literature, only 8 percent of the world's surface is geographically immune to strikes from Phantoms serving with the U.S. forces, which is certainly waging peace on the grandest possible scale. (In 1957, a year before the Phantom Two first flew, McDonnell sponsored an informal competition among employees to find a name for its new fighter. The overwhelming choice was Satan, but "Mister Mac" vetoed the selection saying: "No Sir! Satan's evil and we're not having anything to do with him." He suggested Mithras, after the Persian God of Light, defender of the Truth, but that was rejected on the grounds it sounded effeminate. For lack of anything better, Phantom was chosen two hours before the wife of the assistant secretary of the Navy was due to officiate at the plane's christening.) At one point the Phantom held the world absolute speed record (1606.3 miles per hour) plus all the available speed-to-height records.* Nearly five thousand Phantoms have been built, and they have scored some 250 aerial combat victories, easily the largest total of the jet propulsion era.

The Phantom Two was the first great triumph of one of the most formidable of American aerospace executives: David Lewis, a ruggedly handsome engineer, salesman, and administrator from Georgia Tech. Lewis did not design the Phantom alone; it was, inevitably, a team job. But he was head of Advanced Design when the plane was conceived and later, as head of marketing, he got the credit for having sold it to the U.S. Navy as their main fighter, edging out the Chance-Vought F-3U Crusader (later designated F8U-3). In doing so, Lewis virtually doubled the size of "Mister Mac's" company, and most people assumed that he was the man being trained to take over from the founder. That has not yet happened. In October 1970 Lewis, seemingly becoming restive at McDonnell's unwillingness to hand over final control, moved to the presidency of General Dynamics. But this, as things turn out, makes him more important in the development of the structure that finally gave birth to the DC-10. Without going too far ahead of our narrative, what this meant was that Lewis remained at McDonnell's side during the takeover and remodeling of Douglas, and during the birth of the DC-10 program. He then moved to General Dynamics in time to appear on the "other side" of the contract which GD had won to design and build the fuselage of the DC-10 and its various doors.†

* Eventually the Soviet MiG-23 ("Foxbat" in NATO code) took the record from the Phantom. But the Phantom's successor, the McDonnell Douglas F-15 Eagle, has now regained it for the United States.
† Mr. Lewis continues his spectacular career: In 1974 he won for General Dynamics the "arms contract of the century" to supply F-16 lightweight fighters to several European countries as well as the USAF.

Donald Douglas Sr. (previous page) began his company in 1920 with $600 of his own money and a rented workshop at the back of a Los Angeles barber shop. For thirteen years his company prospered, modestly, from military contracts until Douglas engineers revolutionized the air travel business by designing the first comfortable passenger airliner. The DC-1 and the DC-2 gave Douglas its breakthrough into the commercial airplane market: the DC-3, launched in 1935, gave it almost total dominance. Ten thousand civilian and military (Dakota) versions were built up until 1946 and hundreds are still flying. The little airplane became a legend and won for Douglas the affection of a whole generation of pilots and mechanics. No other Douglas plane has come close to rivaling the success of the DC-3—indeed, the DC-4E and the DC-5 which followed it both flopped. However, as late as 1958 Douglas remained the most successful commercial planemaker in the Western world but that dominance was wiped out by Douglas Sr.'s decision to delay development of what would have been America's first jet airliner, the DC-8. Douglas's attempts to recover from the consequences turned the production lines into a shambles. In 1967, all but bankrupt, Douglas was forced into a "marriage" with McDonnell Aircraft. The DC-10 was the first product of the new partnership.

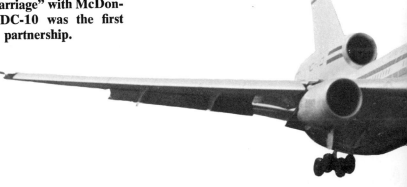

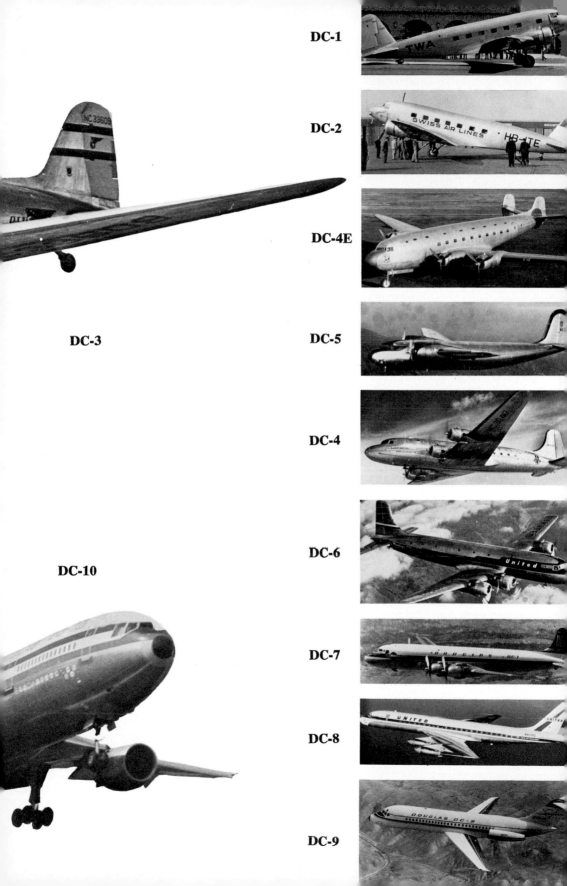

DC-1

DC-2

DC-4E

DC-3

DC-5

DC-4

DC-10

DC-6

DC-7

DC-8

DC-9

James Smith McDonnell in 1939

James "Old Mac" McDonnell started his company almost twenty years after Douglas Sr. and was once content to build parts of the tailplane for the DC-3. But a succession of fighter planes, and especially the remarkable Phantom Two, gave McDonnell the necessary wealth to buy his way into the commercial airplane business. He first proposed "marriage" to Douglas in 1963 but was rebuffed. Three years later he would not take no for an answer—"It had become more than just a business deal. . . ."

The Banshee, immortalized by the Korean War

Phantom One, produced in 1946

**Phantom Two (F-4)—
still in production after seventeen years**

**The F-15 Eagle,
destined to replace the Phantom**

Douglas Sr. gave up formal control of the Douglas Aircraft Company to his son in 1958 but, with his dog Wunderbar, remained an overshadowing influence, holding court almost every day in the Heritage Room. After the takeover Douglas Sr. lost that influence and Douglas Jr. was moved to St. Louis at a $50,000-a-year salary cut. Seven years later when "Old Mac" handed over the reins of the new company to his nephew Sanford, Douglas Jr.—who had hoped to get the top job—resigned. Today, the affairs of the McDonnell Douglas Corporation are controlled from the space-age HQ in St. Louis located just a couple of miles from the site of the austere office where "Old Mac" began to fulfill his life-plan thirty-five years ago.

Douglas Jr. and father in 1966, facing bankruptcy

McDonnell Douglas HQ

"Old Mac" and his nephew Sanford (right), now president of McDonnell Douglas

Brizendine, McGowen, and Lewis celebrate the delivery of the first DC-10 fuselage from General Dynamics

The job of "putting out the fires" at Douglas and developing the DC-10 was given to David Lewis. But although he pulled off a remarkable rescue act at Long Beach, Lewis did not stick around for the applause: instead he took over the presidency of General Dynamics—which builds the fuselages for the DC-10. Jackson McGowen took over Lewis's job until he, in turn, was replaced by John Brizendine.

John Brizendine   **David Lewis**

BY THE MIDDLE sixties McDonnell Aircraft was a remarkable success story. It was regarded with much favor by Wall Street, as proof that an aerospace company could be tightly run and aggressively profitable. Much traffic was made with tales about how "Mister Mac" issued his executives with egg timers to limit their long-distance phone calls, and used a slide rule to calculate the least quantity of whiskey required to mix an eggnogg to get an office party off the ground. It was nonetheless true that the vast majority of McDonnell's business came from military and space sales to the U.S. Government (in 1966 the Phantom II accounted for 90 percent), and McDonnell executives regarded this as a dubious situation.

Although James McDonnell profitably disposed of the 200,000 Douglas shares he bought in 1963, saying that the risks of civil plane making were "incommensurate," no one thought he really meant it. In 1966 Donald Douglas Sr. confided to his friend Roger Lewis, then head of General Dynamics, that he was certain "Mister Mac" would try again.*

It might have been better if McDonnell had succeeded at his first attempt, because by the time he did gain control, the situation at Douglas had become a great deal worse than disorganized. In the last few months of its independent existence, the Douglas company made what can only realistically be called highly misleading statements to obtain money, while some members of the New York Stock Exchange indulged in one of the most notorious cases of insider trading that came to light during the 1960s.

As president of the company, Donald Douglas Jr. must bear most of the blame for what happened. His real responsibility, however, should be looked at against the fact that Donald Sr. retained the role of chief executive officer. Donald Sr. continued to go to the Santa Monica offices with regularity, holding court in the midday sessions at the Heritage Room, and he was attended most of the time by his new wife (formerly his secretary), together with an enormous dog called Wunderbar. Donald Jr., according to his associates at the time, understood that the company's standards of management and financial control required urgently to be modernized. He tried, insofar as his influence served, to recruit new men with up-to-date ideas. Even a man of the flintiest character, however, might find it hard to reform a proud and well-established company while under the shadow of his own imposing father and of a stepmother. Donald Jr.'s character is usually assessed as amiable, and somewhat suggestible, rather than flintily self-assured: certainly his desire for change did not produce results in time to avert the approaching disaster. (One consistent passion in his life, it should be noted, has been a campaign for equal opportunity employment in the aerospace industry. In his role after the takeover as a vice-president of McDonnell Douglas, his Washington reputation as a serious believer in racial equality has helped

---

* Roger Lewis, who left GD to try to revitalize the U.S. railways, is nicknamed "Amtrak" Lewis, to distinguish him from his successor David Lewis, no relation.

the McDonnell division out of one or two awkward clinches.*Perhaps Donald Jr.'s chief corporate failing during the mid-sixties was a certain constitutional overoptimism. This manifested itself first in connection with the problems that Douglas was having in simply getting its airplanes built.

Giving evidence much later in a lawsuit brought by Eastern Airlines, demanding compensation for losses due to nondelivery of aircraft, Donald Jr. gave a vivid account of the state of production at Douglas just before the company collapsed:

> Late, very late 1965 and early 1966 we started running into extremely difficult problems. Our main problem was the tremendous impact that the tremendously rapidly accelerating Vietnam war was having. Our lead time on all materials was expanding almost on a week-by-week basis. If you could get a forging in eight weeks it would go to ten weeks and pretty soon 20 weeks and 30 weeks. . . . They would find they were running out of engines and find that they were running out of tires and brakes and everything. . . . Most especially in 1966 we had a terrible problem of hiring skilled people. . . .†

But at the time all this was happening, Donald Jr. gave little hint of it to shareholders and investors. On April 20, 1966, he told the company's annual meeting that Douglas Aircraft was "in one of the most satisfactory phases of its history." ‡ In June, by which time the chaos was even more horrifying, he found himself able to state that Douglas was "going along nicely," and in July when the company went to the securities market to raise $75 million through an issue of debentures, his publicly expressed optimism had been tempered only very slightly.

The Douglas debenture issue became one of the classic stock-exchange scandals of the 1960s, and the lawsuits resulting from it are not settled yet.

Douglas Aircraft did, of course, badly need money by 1966, and $75 million was the very least of it. There was simply not sufficient operating capital to finance the production of all the airplanes that Jackson McGowen's sales team had promised to the airlines. Bank debts were already uncomfortably large, so the money required could only come from actual investment.

Yet the major shareholders, such as the Douglases, father and son, and the other board members, were not eager to issue new shares. If only the company could stagger out of its immediate problems, its huge backlog of orders looked like producing several golden years of profits. Which of the existing shareholders could be expected to welcome the idea of spreading those profits over some larger number of equity shares?

---

* At one time McDonnell Douglas's contract with the USAF for F-15 fighters was said to be in jeopardy because of allegations that the company discriminated against black employees. Douglas Jr. is given the credit for convincing Washington that the company was truly an equal opportunity employer.
† *Eastern Airlines, Inc.* v. *McDonnell Douglas Corporation*, U.S. District Court, SD, Miami, civil case number 70-1129.
‡ Douglas Aircraft Company, Inc.: Annual Meeting, April 1966.

The company chose a classic solution: to raise new cash by issuing fixed-interest securities called, in this case, debentures, though they might as well have been called bonds (but with the right to convert into stock). Bondholders, unlike shareholders, must be paid their fixed return regardless of the profitability of company operations, but they are also unlike shareholders in that they do not get extra remuneration when business booms. The device is an attractive one because it maintains the "gearing" or "leverage" of the company. Putting it bluntly, the equity shareholders who are already on the inside have other people's money working for them at rates which, with luck, will be far less than they receive themselves. (There seems no doubt that Donald Jr. genuinely convinced himself that, given an injection of new cash, the company could look forward to a rosily prosperous future. In the early part of 1966 he borrowed a great deal of money to increase his personal equity holding in the business.) The law, however, specifies that great care must be taken in presenting the financial statements of a business for which money is being raised. The law also demands care over promises made about the way the money will be used. In some respects the Douglas debenture prospectus fell strikingly short of those requirements.

Douglas had failed to develop any really modern cost control and forecasting machinery, and up to 1966 this had not mattered too much. But in the bizarre conditions of that year, with costs escalating because of the Vietnam War, the primitive Douglas systems failed entirely, and all profit forecasts "went hog-wild," in the words of a plaintiff attorney some years later. On February 15, 1966, the company forecast a profit of $19.2 million for the fiscal year 1966.* By April 12, it was thought this might be $21.5 million. By May 31, it was down to $10.5 million, and seventeen days later that had turned to an expected $470,000 loss. After that point, the forecasts became more and more like guesses. There might be a loss of $26,000 or a profit of almost $1 million, or perhaps a loss of $2.27 million.† Not until much later in the year was anyone sufficiently pessimistic to guess at the actual loss of $75 million.

These projections, however, were all private and internal. When the debenture stock went on sale on July 12, the *public* forecast was that "net income, if any, will be nominal." ‡

Normally, debenture prospectuses concentrate on anatomizing the assets possessed by a company and say nothing about prospective profits. Profit projections are thought to be far too chancy to be inserted into such dour and sober documents, and even a slight experience of profit forecasting makes the wisdom of this course apparent.

Douglas Aircraft, however, perhaps needed to say something because only

* Douglas's fiscal year ran from November to November.
† Report of S. de J. Osborne of Lazard Freres, produced in *Harry H. Levy* v. *Douglas Aircraft Company, Inc.,* U.S. District Court, Southern District, New York (66 Civ. 3382-CBM).
‡ Douglas Aircraft Company: prospectus July 12, 1966.

three weeks earlier it had announced a second-quarter loss of $3.46 million. This was the first piece of dirty linen to show through in public, and it produced a notably plaintive letter from one of the Douglas board, George F. Getty II, addressed to Donald Jr. Mr. Getty's first knowledge of this loss came when he read about it in the newspapers, although he had been at a board meeting on June 1, the day after the unhappy second quarter closed. "You did not report anything," he complained to Donald Jr., "that would indicate to your directors that the anticipated financial results for May 1966 would show a rather spectacular loss."

Getty might have been even more upset if he had known that the loss for the second quarter ($7.51 million) was actually more than double the amount that Douglas had announced. The true figure, however, was not disclosed: In September 1975, after a marathon lawsuit launched by some of those who bought the debentures, the U.S. Federal District Court ruled "that Douglas' failure to disclose the pre-tax loss was either deliberate or so highly unreasonable as to constitute reckless conduct. Furthermore . . . the omission of the fact [from the prospectus] presented a significant and obvious danger of misleading purchasers of the debentures." *

That was not all. The prospectus promised that the money would be used chiefly to expand Douglas's inventory of aircraft parts—a legitimate objective, which urgently needed to be accomplished. In fact, almost the whole of the money was diverted immediately to pay off bank loans, and so far from building up a position for the future, all it did was give the lurching concern a temporary respite. In 1975 Judge Constance Baker Motley ruled that in saying it was going to build up inventories Douglas had been guilty of misrepresentation—investors would not have bought the debentures at the rate at which they were offered "if they had known the truth." *

But of course it was to be some time before the truth emerged. Donald Jr. was outgoing and confident when he went to New York in early July to talk to one hundred underwriters who were going to participate in selling the $75 million-worth of debentures. (The stock would become eligible for conversion into equity shareholding in 1990.) As one aerospace analyst, Lancaster Greene, recalls the conference, Douglas was "in fine fettle, highly optimistic about the company's prospects, and insistent that efficiency was only just around the corner." He blamed the corporation's problems primarily on a shortage of commercial airplane parts, caused by the Vietnam War. Douglas, like Jackson McGowen, believed that the Vietcong were about to knuckle under and was *encouraged* by the fact that his Pentagon friends

---

* *Harry H. Levy v. Douglas Aircraft Company Inc., et al.,* U.S. District Court, SD New York, September 30, 1975 (Civ. 3382).

* *Levy v. Douglas, et al.* (Civ. 3382). The court ruled that Douglas's break-even forecast was *not* reckless or fraudulent because at the time it was made it had "some basis." However, the court added the rider that the company failed to disclose "the assumptions upon which that prediction was based." Had it done so, "a reasonably prudent investor would not have concluded, given the assumptions, that it was highly probable that the forecast would be satisfied."

did not want to buy any more Douglas Skyhawk fighter bombers because they didn't want to end the war with surplus equipment.

The debenture issue went ahead because people who buy debentures are not concerned with short-term fluctuations in the price of common stock, and Douglas raised its $75 million. By August 31, with the prospectus less than a couple of months old, the forecast of "nominal, if any" earnings had been replaced by internal estimates of a $50 million loss, and that was itself a grossly optimistic position.

It has to be said that some highly privileged investors lost their confidence in Douglas even before the "nominal, if any" statement had been made public. On June 20, Douglas revealed that it was not going to make much profit in 1966 to the mighty stockbroking firm of Merrill Lynch, Pierce, Fenner & Smith, who were to be the leading underwriters for the debenture issue and whose name on the bill was of no small importance to lesser investors. Merrill Lynch was, of course, formally recommending the debentures and had underwritten them to the tune of $14.075 million. Yet, within hours of hearing the bad tidings from Douglas, employees of Merrill Lynch began passing them on to selected clients.

Inevitably, those clients began selling. The Madison Fund, a New York investment company, had bought 6,000 shares of Douglas stock (at around $90 a share) in early June. Fifteen minutes after hearing the bad news it instructed Merrill Lynch to sell the entire holding; Investors Management Company acted as an investment advisor to several mutual funds and had previously advised two of them to buy a total of 121,000 Douglas shares. Now it advised them to sell; Van Strum and Towne Incorporated had advised the Channing Growth Fund to buy Douglas stock as recently as June 20. On June 21, Van Strum's president overheard at lunch that Douglas was only going to "break even." The president called Merrill Lynch for confirmation and, although he received no definite information, the Fund disposed of its Douglas portfolio that afternoon.

In all, according to the Securities and Exchange Commission, some 154,500 shares were traded between June 21 and June 23.* The price held at around $90 because of optimistic newspaper reports about the future of the aircraft industry, but there was little doubt that the value of Douglas stock would slump when the bad news got out. Douglas realized what was going on and, to its credit, robbed the inside traders of their advantage by rushing out a press statement on June 24, revealing that it had lost money during the second quarter. The stock price dropped to $69.

By September Douglas Jr. himself was selling stock. Among those who were now aware that the company's stock had been wildly overvalued were

---

* In 1968 the SEC took action against Merrill Lynch. The company's New York Institutional Sales Office was suspended from trading for twenty-one days and its West Coast Underwriting Office for fifteen days. In addition, the SEC ordered that one employee, Archangelo Catapano, should be suspended without pay for sixty days and six other employees suspended without pay for twenty-one days. Three more Merrill Lynch employees were censured for their part in the affair.

Douglas Jr.'s personal bankers who earlier in the year had loaned him money to increase his shareholding. The loan had been secured by the stock: Under pressure from the bank, Douglas Jr. was obliged to sell 10,000 shares and repay the loan. To outsiders, aware of the transaction but not the reasons behind it, it looked as though Douglas Jr. had lost confidence in his ship. By October the share price was down to $30.

The Douglas Company, according to an official history published that year (1966), "was founded on precepts of Scottish thrift, honesty and candor which have been tempered by expediency." No doubt the words were written in good faith, but any investor who read them at the end of 1966 might have been allowed a wry and cynical smile.

FOR MOST OF its life, the Douglas Company had done the majority of its banking with East Coast concerns—it was felt, apparently, that bankers 3,000 miles away were less likely to interfere with the running of the business. But in 1963 a new financial vice-president named A. W. Leslie came in from Trans World Airlines, and he thought it might be a wise precaution to have some local financial allies. He recruited the Bank of America from San Francisco together with Security First National of Los Angeles. It was a prescient move, for when crisis became catastrophe in October 1966, it was the non-Californian banks (First National City, Morgan Guaranty, Chase Manhattan, Mellon National, and Continental Illinois) which wanted to pull out the rug immediately. And it was Leslie's two Western banks which persuaded them to hold on long enough for something, if not independence, to be saved.

Mild unease had been spreading among the bankers throughout the year. Douglas's financial officers, never very forthcoming, had less and less to say about the company's operations. Still, the shares had been selling for more than $100 each early in the year, and most brokers expected profits of around seven dollars a share. A New York analyst, Elliot Friend, who returned from a visit to Douglas with cautious profit predictions of three dollars a share, recalls that he was chided by his colleagues for being overconservative.

When the news broke, at the end of September, that Douglas had lost $50 million in the first three-quarters of the year, the shock was profound. On October 7, a consortium of bankers confronted the Douglas management—including Donald Sr. and Donald Jr.—and nothing they learned at the meeting made them feel better. The bankers were not too worried about what Douglas owed them, for only $25 million of a possible $100-million credit had been taken up. But the predicted loss for the year now appeared to be $70 million. It emerged that through its Canadian subsidiary Douglas owed $46 million to the Canadian Imperial Bank. And the financial staff was predicting that next year the company would have to borrow $300 million simply to survive.

The worst of it, from the bankers' viewpoint, was the apparent unconcern of the two Douglases. Their position was that nothing was seriously wrong, and if only the bankers would produce $200 million of credit, then the production lines would soon be spinning along profitably. It was three days after this meeting that Morgan Guaranty and Chemical Bank formally suspended Douglas's credit.

During the next three months the Douglas management, line employees, and customers lived from hand to mouth, and the company's shareholders—some of whom had bought at $110 a share—saw their investment collapse. Donald Sr. and Donald Jr. looked in almost every conceivable place for assistance. Lockheed was asked to put money in but replied that the shares were not worth ten dollars, let alone the forty dollars that the Douglas management was asking. Even a group of men who arrived in Donald Jr.'s office, "offering a large sum of money indicated to be Las Vegas funds" had to be entertained.* Ford and Chrysler, Howard Hughes and Laurance Rockefeller were approached, but without result. The crude fact was that inside a year Douglas had gone from being a blue-chip investment to a problem that nobody wanted—or anyway, that nobody wanted with the founder and his son still in charge.

On November 11, the Douglas board, worried no longer about diluting the insiders' equity in future profits, gave authorization for an issue of $50-million worth of common shares. Merrill Lynch, already being sued over the summer's debenture issue, had to tell the board that the SEC simply would not allow another public issue of any kind. Douglas was "unbankable."

The big investors and "deal men" of the American financial world were well aware by 1966 that difficulties in the world of aerospace companies tend to be large and intractable ones. Most big companies in big trouble are brought sooner or later to the attention of Colonel Henry Crown, a rugged entrepreneur from Chicago, whose personal fortune of $300 million enables him to move boldly and profitably in and out of companies which need assistance. It was Crown, who—for a price—kept General Dynamics afloat after the misfortunes of its Convair 880/990 jetliners threatened to sink the company totally.

Through an intermediary, Nathan Cummings of Consolidated Foods Corporation, the Douglases made an approach to Crown. "It's another GD situation all over again," said Cummings. "If you're right," replied Crown, "that's the best reason for me *not* to be interested."

The only condition on which the Californian banks could persuade their Eastern colleagues to keep Douglas alive with day-to-day injections of money was that Lloyd Austin of Security First National "convey to the Douglases the extreme magnitude of the company's problems," and get them to look for a strong takeover partner. In mid-November father and son went to New York to find financial advisers who could help them put a deal together. Merrill Lynch clearly faced the embarrassment of being co-defendants with Douglas over the debenture-issue lawsuits, and so the Douglases turned

---

* Osborne report, cited previously.

to the investment house of Lazard Freres, formerly of New Orleans, now among the great powers of New York and London.

André Meyer, of Lazard's New York, is said by many people to be the greatest of all modern "deal men." For a fee that eventually totaled $1 million, he agreed to see what Lazard's could do, and he put one of his senior partners, Stanley de Jongh Osborne, onto the job.

Since leaving the Harvard Business School in 1927, Osborne has been one of the most regularly successful company doctors in America. He began work on the Douglas problem over Thanksgiving weekend, and by the time he moved out to California on November 28, he had several serious takeover prospects in mind. Inevitably, one of the most promising was McDonnell: during the weeks of the crisis, about the only firm buyers of Douglas shares had been "Mister Mac" and associates from St. Louis. (James McDonnell's elder brother William, having moved independently to St. Louis, was now head of the First National Bank, and naturally became his brother's chief financial henchman on the impending Douglas deal.)

The Douglases still inclined to the view that, but for the jumpiness of the bankers, there would be no need to talk in terms of takeover. Osborne had a team of six with him in California, and after they had spent three days going through the Douglas plant and the Douglas books, he got the board together and gave them the bluntest of possible messages.

"Gentlemen," Osborne said, "you're bankrupt."

"They just couldn't believe it," recalled one observer of the scene. "Don Douglas Sr. died seventeen deaths. . . . His own baby was going down the tubes and he hadn't realized it."

THERE WAS NO time for any delicacy about takeover negotiations. Million-dollar fee or not, Lazard's were not prepared to vouch for Douglas in the time available for inspection. The suitors were told that they could come into the company, ask their own questions, and get their bids ready for presentation by New Year's Day.

Five candidates came up to scratch for the event: General Dynamics, Signal Oil, North American Aviation, Fairchild-Hiller—and McDonnell Aircraft. There was no doubt which was the least welcome to the existing Douglas management, for when there had been merger talk in 1963 Donald Douglas Sr. had explicitly doubted McDonnell's competence to enter the civil-aviation field:

> . . . A good merger is one that will be greeted enthusiastically by potential customers. We harbor serious doubts that this would be the case with respect to Douglas and McDonnell. In particular, some of our airline customers have expressed their concern that a Douglas–McDonnell merger would adversely affect Douglas' long-standing policies and competence in the commercial transport field. Only a few weeks ago one of our best DC-8 customers, with whom we were negotiating a sale, expressed this concern

in the strongest terms. When the customer was assured that no merger was in prospect, we got the order.*

The Lazard men believed that any successful takeover candidate would have to provide $100 million in cash right away—with possibly more to come —and, in effect, a whole new management team as well. Stanley Osborne's written assessment was that Douglas had gone "in a decade from the tightly, centrally controlled management of Douglas Sr., with a position of unquestioned leadership in the commercial aircraft industry to a decentralized, divisionalized company which over the years has lost much managerial talent and is left with inadequate leadership." The men running the aircraft division were "overly sales oriented," and the corporate staff was at best "mediocre."

Worst of all, Osborne found that all of the calculations that had been made of Douglas's financial needs took into account only the money needed to survive and continue the building of DC-8s and DC-9s. There was no provision for research and development on a DC-10—yet, as Osborne saw, "even if Douglas can be saved, its whole future depends on the next generation of aircraft." There was no possible Douglas rival to the Boeing 747, already near to entering service. And Douglas had no new candidate for the fast-growing medium-haul market. Having missed the right moment with the DC-8 and the DC-9, Douglas seemed to be on the point of repeating the same mistake again.

Of the bidders, General Dynamics and Fairchild-Hiller were eliminated for lack of immediate cash. Both North American and Signal Oil were seen as "friendly offers," which meant, bluntly, that the Douglas family liked them, but the Lazard team and the bankers did not, because they were less likely to act radically upon the existing management.

North American Aviation is a neighbor of Douglas in Los Angeles, and began life as a Douglas subdivision. Its chairman and president, J. R. Atwood, was a former Douglas engineer. There was no doubt that North American had the cash, and the expertise to help—unlike Signal Oil, which had the cash, but was only prepared to offer one executive in top management. The contest came down to North American versus McDonnell.

"Mister Mac" won by a remarkable display of coolness and nerve. When he entered the chart room at the Douglas headquarters to make his presentation, he was already the largest shareholder in the company, with 300,000 shares—amply demonstrating his determination to get into the commercial airplane business. (The founder and his son were by then reduced to nine thousand between them.) Discreetly, he did not emphasize this piece of leverage, but spent the time describing the cost-control skills his executives would bring to the joint company. He was accompanied by the impressive figure of David Lewis, who was promised as full-time leader of the rescue.

The key to his offer was that he proposed to buy one-and-a-half million Douglas shares *at once,* offering $43 each, which in negotiation he then im-

* Douglas Aircraft Company: Annual Meeting, April 17, 1963.

proved to $45.80. This meant an immediate injection of $68.7 million for Douglas—and McDonnell clearly had more to come if required. The risk for McDonnell was that the Justice Department might quite possibly veto the takeover on antitrust grounds—which would mean that McDonnell would have given Douglas a permanent, interest-free loan of $68 million plus. None of the other bidders were prepared to take a risk like that. North American, the other serious bidder, was insistent on waiting for Justice Department approval before turning its cash into shares. Roger "Amtrak" Lewis, who made the unsuccessful presentation for General Dynamics, thought that "Mister Mac" was unbeatable, because he seemed prepared to up the ante indefinitely. "It had become more than just a business deal for him," Lewis concluded.

The Douglas negotiating committee consisted of Donald Douglas Sr., three outside Douglas directors, and Osborne of Lazard's. At 2 P.M. on Friday, January 13, the full board came together, and the negotiators unanimously recommended acceptance of the McDonnell bid.

It is fair to guess that unanimity cost Donald Douglas some heaviness of heart. Within a very few days, he knew, he must lose all real influence over the company he had founded with $600 and built up into one of the world's most famous engineering enterprises. His son, who had been in charge—at least, nominally—while the actual debacle occurred, spared his father the humiliation of admitting defeat personally. At 3:30 P.M. Donald Jr. went to the phone, called "Mister Mac" in St. Louis, and told him he had won.

Stanley Osborne now rendered service to show that those who spend $1 million with Lazard Freres get real value for their money. He went to Washington to persuade the Justice Department and Attorney-General Ramsey Clark that the merger would be beneficial. On the face of it, this was no easy task.

Naturally, there was the important argument that a Douglas collapse would probably lay off eighty thousand people in California and create a ripple effect throughout the economy. Even so, the Justice Department still thought —perfectly accurately—that a merger between McDonnell and Douglas represented a huge new centralization of economic power in the aerospace industry. That is exactly the kind of situation that the Justice Department is supposed to prevent. From an antitrust lawyer's viewpoint, if Douglas required fresh money, as it obviously did, that money should come from outside the aerospace industry.

Osborne, who as a former aerospace adviser to President Kennedy, was well-connected in Democratic Washington, put forward the argument that his own inquiries into the chaos at Douglas showed that the company needed a good deal more than money. It needed a big injection of new management talent and that could only come from someone expert in the same kind of business.

When the matter came to Ramsey Clark himself, the Attorney-General manifested a good deal of unease. It seemed to Clark that any company which had a backlog of orders as big as Douglas's ($2.3 billion) had ought, somehow, to be able to stand on its own feet. "You seem to be saying there's a great sunset across the river," he complained, "and all they have to do is get there. Why don't they?"

"That river's so damn wide," replied Osborne, "that they just aren't going to get there on their own."

Not without reluctance, Clark decided to allow McDonnell to gallop to Douglas's rescue. With the takeover once consummated, Donald Douglas Sr. was made Founder-Consultant and Honorary Chairman of McDonnell Douglas. Real power, of course, moved decisively into the hands of James McDonnell, who became the unhonorary chairman of the new concern. After a $50,000 salary cut, Donald Jr. was removed to St. Louis to work on "corporate administration" under the beady eye of "Mister Mac" himself. And David Lewis was installed in California as head of the Douglas Division.

There he achieved one of the famous commercial miracles of aviation history. He got the DC-8 and DC-9 production lines flowing again. He paid the banks their money back, and he launched the company into its next generation of effort by getting the DC-10 program under way. But the way that he did it was not entirely the way the bankers and the Justice Department had come to expect.

5

The Great Airbus Stakes: A Game of Skill and Chance

Our knowledge of the factors which will govern the yield of an
investment some years hence is usually very slight, and often
negligible. . . . Enterprise only pretends to itself to be mainly
actuated by the statements in its own prospectus, however
candid and sincere. Only a little more than an expedition to the
South Pole is it based on an exact calculation of benefits to
come. . . .
—JOHN MAYNARD KEYNES,
The General Theory of Employment, Interest and Money

DURING THE LATTER 1960s, the leaders of the newly formed Mc-
Donnell Douglas Corporation used to say they were looking forward to a
party date some time in summer 1976. The idea, as they put it, was for a
selection of executives to gather in the rose garden at "Mister Mac's" house
outside St. Louis, where they would drink Old Fashioneds (apparently the
founder mixes a very effective Old Fashioned) and raise glasses to the Mc-
Donnell–Douglas connection. That was the earliest date at which they imagined it might be possible to call the affair a success.

It was too soon, for the problems of the aerospace industry have not
lessened during the intervening few years. But even then survival in the civil
airplane business required the undertaking of enormous gambles in which
even the date of the outcome, let alone its nature, must be outside the scope
of reasonable prediction.

Until he received Donald Douglas Jr.'s phone call in January 1967, informing him of victory in the takeover battle, James McDonnell had been
in a tolerably rational business. The U.S. Government may not be quite the
munificent customer it once was, but it is a highly convenient one for a manufacturer of novel and complicated machinery. The government pays for research and development as the work goes forward. And it disburses large
progress payments at the beginning of the production run, which is just when
the problems of cash flow are nastiest for plant managers wrestling with the
"learning curve." It would be wrong to say that, by contrast, running a com-

mercial airplane concern requires the instincts of a professional gambler. Las Vegas poker players, who live by a strict and unsentimental regard for percentages, would regard the risks as totally unacceptable.

"Mister Mac" is fond of describing himself as a "practicing Scotsman" with a "highly developed sense of fiscal responsibility." Yet what he and his shareholders had obtained by purchasing control of Douglas was the right to take part in a game of industrial Russian roulette. Boeing's recent dominance had only been won by confronting extraordinary risks: The legendary gamble of backing the 707 during the mid-fifties was the more remarkable because it was made long after the company lost $13.5 million on the propeller-driven Stratocruiser.* By the time of the jumbo-jet generation the price of the game had risen to new and even more intimidating levels. Boeing's president William Allen and his staff committed $16 million—roughly one-fifth of the company's net worth at the time—to develop the 707 prototype. In order to develop the 747, Boeing had to commit not less than $750 million, which was *90 percent of the corporation's net worth.* In 1968, with the program well advanced, Allen was interviewed by *Fortune.* "I have a good feeling about the 747," he said, knocking on wood. "But a stubbed toe," he added reflectively, "could be disastrous."

James McDonnell and David Lewis hesitated for some time before committing themselves to taking a hand in a game with those kinds of stakes. Naturally, Jackson McGowen, president of the Douglas division, and his team were eager to get started with a new plane, now that they had some money in the bank. They were aware that Lockheed, which had been out of the commercial market since the failure of the Electra, was now pushing to get back and that there was talk of a new jumbo aircraft which might be a bit smaller than a 747 and work shorter routes. It was also clear that the airlines were not very interested in any more "stretched" 200-passenger DC-8s possessing only half the capacity of a 747. The concept of the airbus was already forming. Boeing, deeply committed to the long-range 747, would have trouble getting into that market, and so there was a chance, for the first time in an aircraft generation, to take revenge upon the upstarts from Seattle.

Lewis and "Mister Mac," though, were naturally still unsure of their new division's capacities and even more unsure that it could deal with the backlog of orders for existing planes and at the same time produce a new one. Nor could this be an easy decision for anyone conscious of how many sets of whitened bones there are in aviation's Death Valley. In 1961 General Dynamics posted what was then the largest annual loss in corporate history ($178 million) because the Convair 880/990 series was developed after the 707 and the DC-8 had cornered the early market for long-range jets. And not so long before, the famous Glenn Martin Company, for which both Donald Douglas and James McDonnell had worked as young engineers, was

---

* Passengers still recall the double-decked Stratocruiser, a development of the military C-97, with affection. But it was a troubled aircraft technically, and only fifty-six were built.

wrecked by the failure of the Martin 202–404 series.* Yet all salesmen and most engineers like the idea of a new plane, though it was "Mister Mac" and his chief lieutenant who were going to have to go out and take the financial risks. For a large part of 1967, the newly formed McDonnell Douglas Corporation resembled the hosts of Tuscany, when "those behind cried 'forward!' and those in front cried 'back!' " James McDonnell, by his own account, spent weeks agonizing over the economics of airlines. Could the carriers really afford to buy new aircraft? And if so, how big should the new jumbos be? Not unnaturally, the "practicing Scotsman" wanted to find a rational answer.

But the truth was that, rational or not, there was no choice for McDonnell Douglas. For all the success then attending the company's products, they were for the most part aging ones. The Phantom was not expected to do business for very much longer. Naturally, the McDonnell division hoped to get the contract for the F-15; it intended to succeed the Phantom as an air-supremacy fighter, but to rely upon success equivalent to the Phantom's would be recklessly optimistic. The competition to build a U.S. Navy air-superiority fighter was a very open one, and there was no McDonnell Douglas candidate for the Air Force's latest brainchild, a long-range bomber called Advanced Manned Strategic Aircraft. On the Douglas side, orders for the DC-8 and DC-9 were tapering off.

As *Forbes* magazine bluntly put the point, the DC-10 was a gamble that McDonnell Douglas "could not avoid, and cannot avoid to lose." †

THE FORMAL TAKEOVER of Douglas by McDonnell occurred on April 28, 1967. A chilly shower fell on Beverly Hills as the shareholders gathered to vote the Douglas Aircraft Company out of existence. Donald Douglas Sr., then aged seventy-five, was moved close to tears by the words of a woman shareholder. "Mr. Douglas, if you have a heavy heart today," she said, "remember that Douglas will always be there."

Yet already salesmen for the Lockheed Corporation of Burbank, a few miles away over the canyon roads, were commuting between airline offices with every intention of making it difficult for Douglas to survive even with the backing of McDonnell's cash. The Airbus Race had begun in earnest, and Douglas was not less than twelve months late already.

Design work had begun on a DC-10 in 1965, but other than that designation it had little in common with the airplane of today. It was a four-engined, double-decked, long-range airliner, which would have carried 400 people and

---

* The Martin aircraft, powered by either two or four piston engines, utilized advanced metallurgy techniques in wing construction. There were several accidents caused by metal fatigue, and the reputation of the aircraft was ruined.

† *Forbes,* August 1, 1969. Today the Phantom is still winning orders and McDonnell indeed won the F-15 contract, but thanks mainly to David Lewis much air-supremacy business has gone to General Dynamic's F-16 "lightweight" fighter. Grumman won the air-superiority fighter contract, with its F-14.

might have been "stretched" to carry 600. It would have been a rival to the Boeing 747. But when the 747 itself was announced in April 1966, Douglas —then just beginning the long slide into financial ruin—once again reacted by surrendering. Development of the DC-10, said James Edwards, head of commercial sales, would be slowed down, and there would be no planes in service until 1974–75.

"We realize we might be losing out on the initial orders," said Edwards, repeating a familiar litany. "However, we believe that by waiting three years or so to go ahead with the program we will increase our chances to design a better airplane." Douglas had become one of those companies in which time, like the fruit of the medlar tree, was always going rotten without ever being ripe. Such an attitude was not likely to continue under the new regime of David Lewis and James McDonnell, for all their initial hesitation about committing themselves. But when they first took charge, there was not very much that they could do, because as the Wall Street advisor Stanley Osborne noted, during the takeover negotiations, design work on new Douglas planes had not been provided for financially and had come to a halt.

Still, in 1965 Douglas had been involved, along with Boeing and Lockheed, in the design competition for what was the military ancestor of the wide-bodied airliners—a huge strategic freighter called the C-5A Galaxy. Lockheed won the competition to actually build the C-5A and found the experience nearly as disastrous as winning Helen was for Troy. But, that apart, all three manufacturers were able to explore—at government expense, of course—the new technologies that were necessary to build really big airplanes. For Boeing in 1966 the natural development was the 747. Lockheed and McDonnell Douglas concentrated their attention on the airbus.

According to aviation legend, the idea of the airbus was first born to Frank Kolk of American Airlines. On February 25, 1966, Kolk was flying back to New York from Cincinnati, where he had been inspecting a new General Electric fan-jet engine—one of the new and mighty prime movers which the C-5A contest had brought into existence.* LaGuardia was wet and congested. There were, thought Kolk, too many airplanes. And with traffic growing at 14 percent a year, airfields like LaGuardia were soon going to become impassable. What were needed were fewer and larger aircraft to replace the 130-seat Boeing 727s and 80-seat DC-9s. But the 400-seat Boeing 747 would not be able to land at municipal fields like LaGuardia. Their runways were too short, and many were not strong enough to support it.

The following Monday, February 28, Kolk sent out a one-page specification to Douglas, Boeing, and Lockheed. It asked for a twin-engined aircraft that would use a small field, carry 250 passengers, work economically on routes up to 1,500 miles, and would not require a long-range runway.

Kolk's idea met with most enthusiasm from Lockheed, anxious to repair the debacle of the Electra, and looking for ways of exploiting the fact that it was the firm actually building the C-5A. Boeing, preoccupied with the 747,

* See Chapter 2.

was rather less keen, but thought that future versions of the big plane might be scaled down to meet airbus requirements. The least eager response came from the Douglas team. It was "pick and shovel work" as Kolk recalled later, to get McGowen's men off the subject of stretched DC-8s. That was, of course, reasonable enough, for the salesmen knew there was not then any design effort going on at Douglas that could produce anything else.

ON THE LAST day of December 1966, Daniel Haughton, chairman of Lockheed, was told that his company had lost to Boeing in the competition to develop an American supersonic airliner. On New Year's Day he went into his office at Burbank and started calling up his executives in rapid succession. The message was in each case the same: Get hold of all the designers and engineers you have who have been working on the SST and put them onto the L-1011 project. That was Lockheed's drawing-board title for its version of the airplane that was soon to be known—rather to the distaste of the airlines, who loathe plebeian labels—as the airbus.

Lockheed had plenty of work to do, for it had not built a commercial jet before, apart from the little JetStar executive aircraft. (McDonnell Douglas at one time operated a JetStar: Dan Haughton recalls that "Mister Mac" telephoned him one day and said how pleased he was that Douglas had lost the C-5A contest because "it means we have enough money to buy one of your JetStars.") Much of 1966 had been spent in multisided wrangling among Lockheed, a selection of customer airlines, and—with varying degrees of interest—the Douglas and Boeing teams. Frank Kolk's original two-engined scheme had altered so much that in the view of many other airline men he had lost all paternity in it. ("How American can claim any authorship defeats me," growled Robert Rummel, Kolk's equivalent at TWA.) The main battle had been over the number of engines. A twin-engined airplane would have suited airlines such as American, with relatively few long-range sectors in its route structure. But it had no appeal at all to airlines like TWA, for two engines could not be able to lift 250 to 300 passengers *and* enough fuel to carry them right across America.* It was apparent, by the time that Lockheed began its serious design work, that three engines and a range of 3,000 miles would be required.

On September 11, 1967, Lockheed called a news conference at the Overseas Press Club in New York and announced that the company was ready to take orders for the L-1011. (The name TriStar came later, after a competition among Lockheed employees.) A detailed specification of the aircraft was produced. Two months later, the Douglas division of McDonnell Douglas announced that it, too, would build a three-engined, wide-bodied airbus. Effectively, "Mister Mac's" hand had been forced. It was too late to take on

* FAA regulations say that an airliner flying over mountain ranges must be capable of suffering the failure of one engine and still maintain its altitude. A twin-engined airbus would not have been able to meet that requirement on transcontinental routes.

Boeing in the long-range competition, and if any more time went by, Lockheed would be just as firmly in command of the short- and medium-range market. Naturally, Douglas could not announce a detailed specification of the DC-10, because by November 1967 there had not been time to develop one.

Neither Lockheed nor McDonnell Douglas were at this stage committing themselves to anything more than the production of "paper airplanes." Metal could not be cut until orders were forthcoming. Still, the fact that Douglas was able to go even so far was a considerable tribute to the abilities of David Lewis. His performance in "putting out the fire at Douglas" is one of the celebrated miracles of Wall Street and of aviation history. The company's financial recovery had begun with a large, if temporary, injection of government-guaranteed money—a "V-loan" * of $75 million, which, in view of McDonnell Douglas's subsequent puritanism about Federal guarantees for Lockheed's bank loans, seems a little ironic. But it is only fair to say that the money was used purely as bridging finance until the consummation of the takeover. Once the financial position clarified, Lewis had set about organizing the production problems. And here what was striking was how carefully limited his ambitions were.

The McDonnell takeover had been predicated on the necessity of massive reorganization at Douglas. Stanley Osborne's report, while making allowances for the difficult conditions that prevailed in 1966, had laid the blame for the debacle firmly on the existing Douglas management. Lewis arrived to find demoralized executives and a workforce sitting disconsolately around a plant which they feared would shortly be closed down. With a neat stroke of applied psychology, he gave orders for the plant to be painted. The workers' own deduction that people would be unlikely to repaint a redundant factory carried more weight than any smooth words that Lewis might have found.

Not that Lewis seems to have had any problem finding smooth words. When people asked what had happened to the expected influx of hard-eyed reformers from St. Louis, the new president replied that "rather than do it the Harvard Business School way, and combine operations right away, and drop a lot of people, we decided not to merge together right away, but rather concentrate on helping the Douglas people solve their immediate problems." Apart from the appearance of Lewis and a couple of production executives in California, and the removal of Donald Jr. to Missouri, there were virtually no personnel changes. "This way," said Lewis, "Jack McGowen and I had time to get to know and appreciate each other. As a result, we have a very experienced and talented man, not some outsider, running Douglas now."

* V-Loans were established by the Defense Production Act of 1950 and, basically, are a means of providing U.S. Government guarantees for commercial bank loans to defense contractors. Douglas borrowed $75 million in 1967 on the strength of contracts it had won from the National Aeronautics and Space Administration (NASA). The "V" in V-Loan presumably stands for Victory although the Pentagon says that the exact meaning has been lost in time.

Of course, what these feather-edged sneers about the Harvard Business School diverted attention from was the fact that the justification for the McDonnell takeover was the urgent need to get expert management into Douglas. It had been sold to the antitrust men at the Justice Department on the grounds that the men at Douglas would never be able to cross the river on their own. But David Lewis very soon began to give to aerospace reporters the same analysis that Jackson McGowen and his colleagues had been giving before the crisis. Douglas's problems, he said, were "external," caused by shortage of credit and by the supply problems deriving from the Vietnam War.

On that account, of course, there had never been a need for anything but a "friendly" takeover, such as Signal Oil might have accomplished. But to put it more bluntly than the honey-tongued Lewis ever would have done— the one really urgent need was to remove the Douglases, father and son, from the chieftainship. That done, there was no need to do any elaborate restructuring of the Douglas division, even had the new owners actually been able to lay their hands at short notice on a sizable cadre of experienced commercial-airplane executives.

Whatever the Justice Department might have been led to think, the decisive issue was the Douglas delivery problem. As Lewis explained it, the choice was whether to ignore delivery dates and slow down production until it became really efficient (i.e., actually reform the company) or "pour on resources and try to meet schedules." * After some deliberation, he and "Mister Mac" chose to pour on the resources. They knew it would be costly. But they reasoned that if the airlines had to wait much longer for the DC-8s and DC-9s they had been persuaded to order, then they might never buy any more Douglas airplanes at all. The McDonnell operation was at that time well supplied with cash, and it made sense to use it for clearing the decks. Certainly the application of money to the problems of the Douglas division wrought a remarkable change. During 1967 Douglas delivered 195 aircraft, which was twice as many as the year before. Then in 1968 deliveries went up to 302. And, learning as they went, the Douglas teams became strikingly more efficient. The average time from the joining of a fuselage to the first test flight of a plane came down from 165 days to 90 days, and eleven months after the formal takeover Douglas was running at a profit again. Six months after that, $460 million of debt had been repaid.

Three and a half years after the end of Douglas as a separate firm, *Forbes* pronounced that McDonnell Douglas had become "perhaps the healthiest major company" in aerospace. The key to this exuberant health, said the magazine, was "the decision made by parent company president David S. Lewis and his boss . . . James S. McDonnell, to damn the torpedoes and build the airbus." †

* *Aerospace International,* September–October 1968.
† *Forbes,* July 1, 1970.

THE TORPEDOES WERE numerous. Lockheed had gone out in front with its airbus studies and, of course, had nothing else on its commercial agenda except the L-1011, while Douglas's capacities were straining to deal with the DC-8 and DC-9 backlog. If the DC-10 were to catch up with the TriStar—and it could not afford not to—there was going to have to be a good deal of subcontracting.

Assuming the problems of production capability could be solved, the men from St. Louis still had no experience whatever of negotiating with airlines. Selling fighter planes to a government is a simple task compared with selling passenger aircraft to commercial airlines. "We have absolutely no leverage with these airlines," said David Lewis, with the air of a man discovering one of the great laws of the cosmos. Douglas did have salesmen, of course, but they had tarnished themselves somewhat by the earlier overcagerness in selling DC-8s and DC-9s. "Mister Mac," determined not to let the Douglas people run riot again, tried for a period to make himself the chief salesman for the DC-10. Rather swiftly, he found that he did not have the wheeling-and-dealing talents required.

The estimate of most people in the industry was that the fate of the airbus lay with the managements of the four huge airlines—American, Eastern, TWA, and United. And most of these experts assumed that the "paper plane" which won the decisive share of the Big Four's initial orders would be the only one to be turned into metal.

The U.S. airline industry has grown into a business worth $15 billion a year, supporting some two hundred airline firms. But more than half of the $15 billion is accounted for by the Big Four.

Unhappily, Douglas's relations with two of them were most unfriendly. Once, there had been a "special relationship" between Douglas and American, the first airline to order the miraculous DC-3. But in the fifties and sixties, American had chosen the 707 and the 727 in preference to the DC-8 and the DC-9. At the end of 1967 American had 205 planes in its fleet and not one of them was a DC-anything. (It did, however, still have nineteen Lockheed Electras.) Douglas had responded, somewhat injudiciously, by forbidding its staff to fly American on business trips.

Relations with Eastern were even worse. During the sales competition between the Boeing 727 and the DC-9, Eastern had split its options, ordering both—and suffered grievously, for Eastern's schedules were among those worst affected by the production delays at the Douglas plant. Positions on a delivery schedule are of crucial importance when traffic is increasing, and the rhythm of deliveries matters deeply. Normally, an airline likes to get its new planes in winter, so it can introduce them into the fleet and iron out any problems before the summer boom. Just before the airbus contest began, Eastern sued Douglas for "inexcusable delays" in delivering DC-9 aircraft.

The lawsuit, which resulted in a $24.5 million victory for the airline, made it almost inconceivable that Eastern would order the DC-10.*

And neither TWA nor United were, by this time, particular enthusiasts for Douglas aircraft, judging by their fleets. Altogether, there were 505 Boeing airplanes in service with the Big Four, against 137 Douglas.

Conceivably, this very fact tipped the politics of the situation toward the DC-10. The airlines had become accustomed to a situation in which they could get their way by playing up the competition between rival manufacturers. Yet the Boeing preponderance had become so great that the airlines were near to putting themselves under the rule of a monopoly supplier. Lockheed was mounting a challenge, but there could be no guarantee that Lockheed's return to the commercial market would outlast its enthusiasm for the TriStar. Long term, some people thought, Douglas was the only possible across-the-board rival for Boeing. And so the company must, somehow, be hauled back on to its feet.

Robert Rummel of Trans World Airlines thought he had seen some rugged sales contests during his twenty years with TWA. But the airbus battle, by his account, surpassed them all. "There was hardly a day when there wasn't a McDonnell Douglas man or a Lockheed man in this office discussing specifications. And whichever one *was* in here, the other guy was outside waiting for his next turn." Rummel was one of the first people in the business who began to think that perhaps this was too much of a good thing and that Lockheed and McDonnell Douglas might end by tearing each other to pieces. Most airline executives, however, reveled in the leverage they exercised. *Fortune* quoted a typical, if anonymous, view in 1968: "It was great. The longer the negotiations lasted, the more we got." That is to say, the airlines got promises—paper promises—of higher and higher payloads, more performance, better galleys, custom-built navigation layouts, cheaper prices, and easier and easier financing.

One apparently minor concession which American Airlines extracted from McDonnell Douglas concerned the operation of the rear cargo door in the proposed DC-10. The Douglas engineers suggested hydraulic actuators to close the door. They were thoroughly familiar with this piece of technology, and so were their subcontractors, the Convair division of General Dynamics, who were going to build the fuselage and its doors.

But the American Airlines engineers thought that a hydraulically operated door would be too rich in working parts. Surely it would be superior in principle, they argued, to change to electrically driven actuators, and McDonnell Douglas agreed to do so. It must have seemed a trivial concession to the men who made it. They could not know how far from trivial its consequences would be five years later.

---

* McDonnell Douglas has lodged an appeal against the judgment (given in April 1974) but Eastern has also appealed on the grounds that the award is inadequate. U.S. District Court, SD, Miami, civil case number 70-1129.

THE COMPETITIVE TENSIONS were made all the more severe by the circumstance that neither Lockheed nor McDonnell Douglas had specified engines for their paper airplanes. So the engine firms, Pratt & Whitney, General Electric, and the British firm Rolls-Royce, were also in the field, making promises simultaneously to the airlines and to the airplane builders. Again, for at least two of the competitors, the issue at stake was survival in the market.

Just as Boeing, Douglas, and Lockheed were the only three firms in the non-Communist world who could hope to develop a major commercial airframe,* so these were the only three firms that could be taken seriously as builders of major new power plants. Pratt & Whitney, having won the contract to build the engines for the 747, enjoyed some degree of security. But GE and Rolls were alike in that neither would have a major commercial engine project if excluded from the airbus program.

General Electric had been among the pioneers of the jet engine business in America and the provider of some of the most spectacular military power-units, including the J-79s of the Phantom II. (On that score alone the GE people thought that McDonnell owed them some consideration.) But at the beginning of the 1960s, GE's commercial jet business went down, through no fault of its own, in the collapse of the Convair 880/990 program. Having lost the 747 competition, the GE commercial engine division was threatened with the same fate as Convair's—that of becoming, given luck, a subcontracting handmaiden to others.

General Electric entered the battle, however, with some powerful advantages. One was that GE had built the first big fan-jet, the TF-39, of Lockheed's C-5A. This meant that GE had plenty of fan-jet development experience acquired on test beds paid for by the U.S. Government. Very likely this development background is the reason that General Electric's CF-6 is reckoned to be the smoothest performer among the three big fan-jet engines.

GE's other great advantage was that its commercial engine division is only one part of a $10-billion-a-year industrial empire, whose interests are spread as far apart as light-bulb manufacturing, construction of nuclear power plants for submarines, and marketing Taiwanese transistor radios. Pratt & Whitney †and Rolls are more specialized concerns, whose financial resources are largely tied to the switchback economics of the aerospace industry. General Electric is far too large and well-established to be described by any such upstart label

---

* The career of the Franco-German A-300B Airbus, a twin-engined 300-seater of similar vintage to the DC-10 and TriStar, proves rather than disproves the point. It has a limited application and only thirty have been sold so far—mostly "captive" sales to the French and German national airlines. Airbus Industrie can scarcely hope to recover its investment. (Soviet interest in buying an American widebody suggests that the "non-Communist" qualification may be otiose.)
† Pratt & Whitney's parent company does now have considerable interests outside of aviation—it owns, for example, the Otis Elevator Company—and has, consequently, changed its name from United Aircraft to United Technologies.

as a "conglomerate": best, perhaps, to say that GE by itself is a large part of the structure of American capitalism. Not the least part of GE's power is its close connection with the Morgan financial empire. There have long been Morgan Guaranty members on the GE board, and one thing that meant in the airbus competition was that GE's campaign to stay in the civil jet market would not fail for lack of finance.

Rolls-Royce, for its part, was at a critical stage in its corporate career. The famous, or notorious, automobile (once described by Ettore Bugatti as "the triumph of workmanship over design") ceased to be the main part of the Rolls business well before World War II. During that war the British Government was so anxious not to interrupt Rolls's work on the piston engines of most of the best British aircraft—and some of the best American ones, like the P-51—that they tried to exclude the firm from jet-engine development. There was, perhaps, a certain grim satisfaction at the Rolls headquarters in Derby when the government had to admit, in 1943, that no other British manufacturer had been able to come up to scratch. In any event, Rolls-Royce was the first non-German firm to get a jet engine working properly.

The company went on after the war to build powerful military engines, and to develop the cooled-blade technology that made fan-jets commercially practicable.* But Rolls's technical excellence had been consistently neutralized by its involvement with the British airframe industry. When in 1955 Rolls developed the Dart, the first effective turboprop engine (and, still in production today, one of the most successful), the company signed a remarkable agreement with Vickers, the airframe builders (now part of the British Aircraft Corporation). This precluded for some time sales to any other manufacturer. The deal was encouraged by the British Government in the hope that the Dart's undoubted excellence might create an export monopoly for British turboprop airlines. Lockheed, originally anxious to install the Dart in a commercial transport was baffled and furious to encounter an agreement which would have been incomprehensible—even illegal—in America. Reconstructing Rolls's commercial reputation in America took a considerable time.

In 1958 the company was first to develop the bypass, or fan-jet technology. But almost immediately afterwards Rolls-Royce found itself bound up with the fate of the British Aircraft Corporation's Vickers VC-10, a competitor to the 707 and DC-8, and just as ill-timed as the Convair 880/990. In the curious condition of British aerospace, hovering somewhere between state socialism and private capitalism (but one in which all major projects require some government finance), aeronautical design decisions become a branch of political science. The VC-10 project was given political blessing, though not without misgiving, and Rolls-Royce could not have escaped from it even if some members of the firm might have liked to do so.

Much technological jingoism was expended in Britain upon the VC-10.

* See Chapter 2.

But it had a design characteristic that was most unfavorable to engine development. As all four of the VC-10's Rolls-Royce Conway engines were mounted on the rear of the fuselage, they could not have large diameter fans. Therefore, the Conway was inevitably a "low bypass" engine. Whether freed of the restrictions of the VC-10 configuration Rolls-Royce engineers would have moved more rapidly to big fans is, of course, unknowable. What is certain is that Pratt & Whitney, although not first with the principle, were very quick to realize its advantages. The P & W engine that powered early models of the 707 was a pure-jet. After Rolls had introduced the Conway, Pratt & Whitney brilliantly recovered their lead by clapping a large diameter fan—it was almost as simple as that—onto the front of their pure-jet, turning it into the sweepingly successful JT3D-7, which powers the 707s now in service and, in modified form, most DC-8s.

By the time of the airbus competition in 1966–67, British politicians and administrators had accepted the acid truth that Rolls-Royce's future depended upon getting engines into major American airframes. The realization, however, came late: whereas both GE and Pratt & Whitney had big-fan engines actually running, the British firm had only a "paper engine" to offer for the paper airplane.

All the same, Rolls-Royce did not enter without advantages. First was its excellent technical reputation, expressed practically in the "three-spool" design of its candidate engine, the RB-211.* Another Rolls-Royce advantage was that few American airline and airplane executives, whatever their patriotism, wished to be confined to one, or at most two, engine suppliers. (Pratt & Whitney, the dominant firm, had lost the airlines a lot of money by late deliveries in the mid-sixties.) And again, Rolls could expect to be fairly strongly placed to perform the engine-builder's traditional role—as producer of the most expensive piece of equipment in the airplane—namely, assisting customer airlines with credit. As a source of financial backing the British Government turned out to be somewhat inept operationally, when compared, say, to Morgan Guaranty. But its resources, nonetheless, were considerable.

The RB-211 also promised to be cheaper, chiefly by reason of lower British labor costs. It looked good enough, indeed, for both Lockheed and McDonnell Douglas to be strongly tempted. " 'Mister Mac' really wanted those engines," said Sanford McDonnell (the founder's nephew), "and we tried hard to get them." "Mister Mac's" enthusiasm distressed General Electric so much that they persuaded Representative Robert Taft of Ohio to rise in Congress and complain about McDonnell Douglas's lack of patriotism. Taft alleged that if Rolls-Royce were to get the contract for the DC-10, then it would cost America $3.5 billion in foreign exchange, and perhaps twenty thousand jobs.

There seems little doubt that Rolls-Royce had the initiative and might have been able to get their engines into *both* proposed airbuses had Rolls possessed the manufacturing capacity to promise sufficient production. (One

* See Appendix A.

Rolls-Royce weakness is an inability to expand production rapidly, because of the shortage of engineering subcontractors in Britain capable of doing high-quality aerospace work. Much larger proportions of a Rolls engine have to be made "in-house" than is the case with P & W or GE, who can draw upon the vast subcontracting resources of the American precision engineering industry. In the event, Rolls-Royce signed up with Lockheed, so that the DC-10 business went almost automatically to General Electric.* It now became clear that if only one airbus survived, then one of the big engine builders, as well as one of the airframe firms, would be leaving the commercial market, and very probably for good.

But what was also becoming clear as 1967 turned into 1968 was that whoever could win the race looked like doing very big business indeed. This was particularly so because of the prospect that versions of the big tri-jets might stretch their range to cover transoceanic flights. As George Spater of American Airlines put it, there existed "for the first time since the design of the DC-3 of thirty-three years ago, an all-purpose airplane."

Suddenly, Boeing's slightly offhand response to the airbus idea—that such requirements might be met by "stretching" 727s or "shrinking" 747s—became to look like a grisly error. Their market domination, built up at the cost of so much risk and patience, looked as though it might be wiped out as quickly as Douglas's earlier position had been. In those days of fast-growing passenger traffic, it seemed that the wide-bodied airbuses, capable when fully loaded of remarkably economic rates of cost per seat-mile, might drive every other kind of jet off the short-to-medium routes. (Both the TriStar and the DC-10 were calculated to cost one cent per seat-mile, against 1.38 for the 727.)

Struggling still with the enormous development costs of the 747 program, Boeing had no money available to develop an answer to the TriStar and the DC-10—certainly not by spring of 1968, when the Big Four airlines proposed to place their next large equipment orders. All that Boeing could do was send its then chairman William Allen on a hasty, slightly desperate tour of the airlines offices which was, in its way, the clearest of tributes to the momentum that the airbus idea had generated. On one of his airline visits, Allen encountered a member of the Lockheed sales team. "Have you come to sell airplanes, or to find out what we're doing, or just to muddy the water?" Allen was asked. Allen had to answer yes to all three questions. But in truth the Boeing proposal did not amount to much more than a muddying of the water—or at best, a plea for time.

Allen said that if the airlines would delay their decision for a year, Boeing would be able to offer them an airbus. It would probably be a three-engined version of the 747. Apart, however, from the fact that it would be called the Boeing 757 and would be better than anything seen so far, there was

---

* Pratt & Whitney does make engines for the Series 40 model of the DC-10, because the U.S. airline Northwest wanted a P & W engine and was prepared to order twenty-two aircraft.

very little that Allen could promise about the airplane. Granted that most of the airlines were getting a little uneasy about Boeing's near monopoly position, it was no doubt hopeless to ask them to sacrifice a year's prospective business in order to preserve it. "You may love Boeing as we all do," said one airline man, "but we don't love them that much." *

ALLEN'S INTERVENTION PRODUCED little, if any, effect, and American Airlines announced that on February 16, 1968, it would give its decision first. The Lockheed team were so confident of victory that chairman Dan Haughton traveled from California to lead the delegation and hear the news in person. When they arrived in American's headquarters on Third Avenue, New York, they were told that David Lewis and Jackson McGowen of McDonnell Douglas had just been given an order for twenty-five DC-10s at a total price of $400 million. The decision was attributed to the personal salesmanship of David Lewis. "If it had not been for Dave Lewis, we might have bought the Lockheed airbus," said an American executive. "The two planes were almost identical in price and design, but he made the difference."

Back in Burbank, Dan Haughton swallowed hard and said that TriStar development would continue anyway, because he was "confident that orders would be forthcoming." Tough talk, however, could not dissolve the Lockheed team's dismay at losing their front position. Just after the American Airlines announcement they saw a full-page advertisement in the Los Angeles *Times* which showed what the interior of the new DC-10 would look like. What it looked like was a TriStar interior—indeed, it might just as well have been a copy of the brochures that Lockheed had been distributing to the airlines for its own plane.

The advertisement drummed home the fact that Lockheed could not expect to score points simply because they had been the first to begin detailed design work on the wide-body formula. Anyone else could make the basic calculations as well as they could, and the airlines might have just as much faith in the McDonnell Douglas engineers when it came to turning ideas into metal. Clearly, Haughton and his men would have to produce something fresh if they were to arrest the McDonnell Douglas bandwagon.

First, in David Lewis's words, they "slaughtered the price." American Airlines had agreed to buy its DC-10s for about $15.3 million apiece. That price had been calculated on the proposition that a prospective $1 billion of development costs might be recovered over a total sale of some two hundred airplanes—or, putting it another way, around $5 million per airplane. That was a not unreasonable projection, even if reaching two hundred sales would obviously take a considerable time. After selling 200 airplanes, the program would begin to make a clear profit.

Now Lockheed announced that for a limited time, it would be prepared

* *Fortune,* June 1, 1968.

to take orders for the TriStar at a discount of $1 million each: that is, for $14.4 million per plane. On such a price, the breakeven point for a program would move out to 250 airplanes.

Dramatic events followed the slaughter. On March 29, Dan Haughton announced that Lockheed had made the "largest sale in aviation history": 144 TriStars, for a total price of $2,016 million, at an average price of $14 million each. This was a counter-bandwagon of impressive size: while the American Airlines order only took the DC-10 one-tenth of the way to breakeven point, Lockheed had gone past the halfway point with a single leap.

The larger part of the largest-ever sale was perfectly genuine. This was two of the Big Four, Eastern and TWA, going predictably for Lockheed. However, there was also the curious addition of fifty TriStars to be bought by a British corporation named Air Holdings Ltd.—these to be paid for in dollars, for "resale outside the USA." Air Holdings was not an airline, and it was brought into the deal by Lockheed, Rolls-Royce and the British Government, with two aims in mind. One was to top up the number of TriStars ordered, and so increase the bandwagon effect for Lockheed. Another was to disarm the claims of Representative Robert Taft and others that the TriStar's British engines would be bad for the American balance of payments. It was a device to transfer some part of the burden to the British balance, which in those far-off days appeared to be getting stronger. In effect, what it amounted to was a guarantee that Lockheed would sell fifty TriStars overseas. (In 1971 the Air Holdings order was described in the House of Commons as "a bookkeeping transaction—and a dubious one at that.")

Almost immediately after this coup, Delta Airlines announced for the TriStar and ordered twenty-five. Delta is not one of the Big Four, but its operations are sizable, it is consistently the most profitable airline in the United States, and its reputation for efficient management stands very high.

The DC-10 was now in serious difficulties, holding only twenty-five orders against the TriStar's 168 (or 118 without the rather suspect Air Holdings deal). On April 5, when "Mister Mac" appeared before the first annual general meeting of the McDonnell Douglas shareholders, he was able to report that the DC-8 and DC-9 programs were moving into profit. But he had to say that the DC-10 might not be built at all, in spite of the jubilation over the American Airlines order. "Our boys won that one," he said philosophically, "but in March we found out that we can't win them all."

"Mister Mac" did not at all resemble a man whose greatest corporate gamble might suddenly be heading for failure. On the contrary, he was relaxed, amusing the shareholders with his traditional pretense of being a simple Arkansas farmer. There are plenty of people in the aerospace industry, however, who would argue that James McDonnell is never more formidable than when he is assuming his hayseed manner—and what he and David Lewis were really doing was preparing for a final assault on United Airlines, biggest of the Big Four, with its thirty-two million passengers a year.

Their first move was to shave half a million dollars off the DC-10 price—
"We had no intention of fading away," as David Lewis put it. The DC-10's
overall price could not be cut as low as the TriStar's, because the GE engines
were unavoidably more expensive than Rolls's RB-211. Against this, however,
General Electric was offering to help make financing arrangements through
Morgan Guaranty.

United, whose headquarters are in Chicago, is certainly the largest airline
outside the Soviet Union, and many people would say it has been the best-run
at least since 1931, when it inaugurated modern air travel by ordering
the Boeing 247. By 1968 United was operating a fleet of 337 aircraft. It
had just made a profit of $108 million and had decided to buy eighteen
Boeing 747s for its transcontinental routes. Its decision on the airbus
question would be announced in Chicago on April 25.

There were considerable pressures on United to choose the TriStar. People
in the United engineering department told colleagues that they liked both
airplanes but had some preference for the TriStar. The fact was, though,
that at this stage of the design process there was no real way of making an
engineering judgment between the two paper airplanes. The choice was a
matter of finance, and of aeronautical politics.

Both Eastern and TWA tried to persuade United to choose Lockheed—
in the knowledge that if that were to happen, the DC-10 would die. There
was, of course, obvious self-interest in this, since both of them were high
up on the delivery lists for another plane. But in TWA's case at least, this
went a little further. Robert Rummel's view was that "the industry could
not support the 747 *and* the TriStar *and* the DC-10," because the development
costs were so enormous. "We believed that if the two airbuses were built,
both manufacturers would be weakened by the competition, and events
seem to have proved us right," he said.

TWA felt strongly enough to offer to give up some of its delivery positions
if United would choose the TriStar. There was, of course, much cold logic
to the argument that one of the airbuses should drop out. To get both firms
into profit, not less than five hundred airbuses would have to be sold—some
of them, at least, in the intercontinental market on which the success of the
747 depended. It was legitimate to think that both manufacturers might
lose money, and according to TWA's argument, one financially healthy
airbus builder would be better for the industry than two sick ones. It was
not, in principle, an argument that people at Lockheed or McDonnell Douglas
would have disputed. They would simply have insisted that the other firm
should be the one to give up.

Roughly, the Lockheed argument was that their plane had been first in
the field. The rival had been hastily proposed by an outfit which would not
exist at all but for vast injections of cash from a firm (McDonnell) which
had less claim to be in the commercial airplane business than Lockheed itself
could show. The McDonnell Douglas argument was that Douglas was,
traditionally, a commercial airplane builder: that Lockheed, whatever the

past glories of Constellations and Vegas, had become in essence a defense contractor and should stick to that business. In other words, the McDonnell Douglas–General Electric team was not accepting invitations to retire gracefully. And there was one very solid set of reasons why GE, anyway, should not do so. One of the largest single shareholdings in United Airlines is held by GE's allied bank, Morgan Guaranty.

On April 25, 1968, George Keck, President of United, announced that United had chosen the DC-10, and was placing orders for sixty airplanes. David Lewis—looking, said *Time* magazine, like "a baseball pitcher who had just scored three strike-outs"—was on hand at United's headquarters to announce that the DC-10 program, now with eighty-five orders, would definitely go ahead.

George Keck took the trouble to phone Dan Haughton of Lockheed to try to soften the bad news. United thought the TriStar would be a fine plane, Keck said. "But if the DC-10 isn't built, then Douglas will be out of the business. And I think that would be a bad thing for the airlines, and a bad thing for the industry."

No doubt the decision of Keck and his colleagues was influenced by reasoning on the competitive structure of the aerospace industry, as well as by the concrete facts of the make-up of their shareholders' register. It was a case where corporate logic and ideology could go hand in hand. All the same, United's decision to keep the second airbus alive was one implicit with disaster.

Nobody at that time foresaw the horrors of energy crisis and general inflation that were shortly to end the great airline boom. Had anyone done so, the decision might have temporarily gone against building even one, let alone two, American airbuses. Nobody, indeed, could be expected to make accurate predictions about the effect on airline traffic of Middle East politics, and of the unique circumstance of all the industrialized Western economies becoming depressed at the same time. But it could more reasonably be said that any sensible person ought to be aware that such disasters can occur, and that spells of economic good weather cannot be treated as though they will go on forever. There were, of course, plenty of men in all the companies concerned, not the least "Mister Mac" himself, who were perfectly capable of seeing this. The trouble was that there was nothing, in practical reality, that they could do about the perception. The only logical consequence that could flow from it would have been a decision to stay out of the jumbo jet race, which in the circumstances of the aerospace industry as it then existed—and now exists—would have meant corporate extinction. It is no deterrent to tell a man that he may be running into mortal danger in a few years time if at the same time you tell him that the only way he can avoid it is to cut his throat right now.

Both McDonnell Douglas and Lockheed had airily predicted that there would be, eventually, homes for one thousand airbuses. Now they were committed to going out and proving the point, in order to make profits on top of development budgets that were coming out at $1 billion each. "We're going to have an awful lot of money out for an awful long time," said David Lewis thoughtfully. "It's a bit more exposure than we're used to."

The consequences were dramatic. Within two years of the first airbus orders, and within three years of Douglas's collapse and take over, McDonnell Douglas was being celebrated as the strongest, healthiest, and most profitable company in the big league of aerospace. It had the F-15 fighter contract signed, and was beginning to get new airbus orders faster than Lockheed. By that time Boeing was seen as "dangerously sick"—with 707 and 727 orders down to a trickle, the SST project canceled, and Boeing's financial system still weighed down by the huge development costs of the 747. As the Douglas plant in Long Beach built up its production teams for the DC-10 line, Boeing in Seattle began shedding workers by the thousand in a desperate struggle to remain solvent. (There supposedly appeared a sign outside one of the Boeing plants which said: "Will the last engineer to leave please turn out the lights?")

Looking, for the moment, two years further ahead of our story, we can see Lockheed—approached once as possible rescuers of Douglas—lurching toward dissolution. And yet further ahead, the odds can now be seen to have altered yet again. Boeing, having paid its 747 development costs and having survived the desperate economies of the early seventies without losing its technical capacities ("When you're up to your ass in alligators," said another semiapocryphal notice, "try to remember why you began draining the swamp"), found business conditions transformed. Airlines which could see no prospect of filling their airbus seats were looking again at "stretches" of the twelve-year-old 727. The collapse of the American SST program hardly mattered because with the escalation of fuel prices, the special-performance version of the 747 (smaller and with a longer range) looked like a better prospect for rapid transit at extra-long ranges. And the airbus firms by this time were struggling with a flood of cancellations.

"Businessmen play a mixed game of skill and chance," said John Maynard Keynes in the famous text quoted at the head of this chapter. Keynes also pointed out: The average results of the game,

> are not known by those who take a hand. If human nature felt no temptation to take a chance, no satisfaction [profit apart] in constructing a factory, a railway, a mine or a farm, there might not be much investment merely as a result of cold calculation. . . . Most, probably, of our decisions to do something positive, the full consequence of which will be drawn out over many days to come, can only be taken as a result of animal spirits—of a spontaneous urge to action rather than inaction, and not as the outcome of a weighted average of quantitative benefits multiplied by quantitative probabilities. . . .

These conditions are nowhere more true than in the aerospace industry. Nonetheless, it is not the way that corporate ideology says that executives ought to behave. They are supposed to make careful analyses, seeing far into the future, within which there will be returns on investment sufficient to reward the equity shareholder for his risks and compensate the fixed-interest stockholder for his faith. It is supposed that they who make the correct calculation will be rewarded by the favor of the market, and they who have miscalculated will be punished. On the scale at which private-enterprise business used to be conducted, it was possible to maintain the pretense of sober calculation and the traditional competitive model. The distinction of aerospace is that the pretense is now really difficult to maintain. In terms of influencing the final "reward" chance is the dominant factor (especially as all the competitive groups acknowledge—no doubt with sincerity—that their skill is probably equally matched elsewhere).

Indeed, granted the conditions of the airbus competition, the notion of "profitability" became almost impossibly rarefied. To speak of a five- or ten-year program getting into profit is one thing. But as the marketing conditions for big airliners grew more and more difficult, executives began to talk in terms of fifteen- and twenty-year programs—a time scale, apart from any other historical considerations, well outside the horizons of the sort of people who actually deal in the shares of major American firms. Both McDonnell Douglas and Lockheed were to attempt—with distinctly uncomfortable results—to overcome this problem by the principles of "liberal accounting."

According to this system, money spent on the research and development of a new aircraft is not written off as money that actually goes out of the company, but is retained in the books as an asset. Then, as the airplanes begin to sell, a certain amount of each sale is regarded as going to redeem that calculation, while another proportion is taken as profit. (The largest proportion, of course, is the production cost of the particular airplane.) The whole system is predicated upon the idea that the program as a whole will eventually sell enough airplanes to realize an overall profit, thereby justifying the inclusion of the research costs as an "asset." But in the conditions of aerospace in the first half of the 1970s any such prediction became more comprehensively metaphysical than anything Keynes might have expected. (Incidentally, one reason for the sudden disfavor into which Boeing fell on Wall Street was that the company chose never to engage in "liberal accounting," but stolidly wrote off the whole whopping development costs of the 747 program as they were incurred, and then went slowly back into profit the hard way.)

The chief point of this excursion ahead is to emphasize that Lockheed and Douglas did more than simply enter a contest which, at best, only one of them could win, and which, according to the economic rules obtaining at the time, neither of them could afford to lose. They did so, in each case, with accounting systems which began from the proposition that each company had already attained considerable success in a race which, realistically, it was imminently possible that both would lose. There was, to put it mildly, a considerable premium on being the first firm to actually get an airbus into service, so that it might start earning money and attracting further orders.

Before going on to examine the competitive airplanes and the way they were built, there is a footnote which should be added to this aspect of the story. In 1973 a committee of wise and great men from airlines, aircraft firms, and banks was asked to advise the Federal Government on the future conduct of the aerospace industry. The committee advised that the time had come to prevent unlimited competition in the civil side of the industry: although there should be competition in the design stages, it should no longer be possible to waste the industry's resources by building in competition airplanes which were aeronautically comparable, differing only in the identity of their financial sponsors. Also, the report suggested, it must be recognized that major civil-airliner projects were probably beyond the resources of private industry unsupported by public funds. One of the signatories of the report was that rugged entrepreneur James Smith McDonnell.

6

Fly Before They Roll

It is difficult in one design to meet perfectly all . . . the
requirements for a successful airplane. The designer . . . is
continually faced with compromises, and must exercise his best
judgment in deciding what requirements may be considered
of secondary importance. As with any other branch of applied
science, airplane design is not an exact proposition.
—DANIEL O. DOMMASCH, SYDNEY S. SHERBY, and
THOMAS F. CONNOLLY,
Airplane Aerodynamics

WILLIAM STOUT, DESIGNER of the "Tin Goose," defined an airliner as
"a machine capable of supporting itself financially and aerodynamically at
the same time." Less rigorous engineers are apt to say that such and such
an airplane is "a fine piece of design, but commercially unsuccessful." By
Stout's definition, remarks of this kind are strictly meaningless: The quality
of a transport aircraft design cannot be measured by looking at any one
variable in isolation. Efficiency in airplane design is a complex equation
within which the variables interact upon each other. For instance, the use
of very large amounts of titanium in an airplane may well improve it
structurally but may have a damaging effect upon its financial qualities.

The official ideology in the air transport industry is that within this complex
equation, safety is a parameter; namely, a quantity which is kept constant
regardless of how the others may be juggled. This is, of course, not quite the
case, and at times the safety value must be considered in terms of the cost
to other values if it is increased. There is, for instance, little doubt that
airliners would be safer if all the passenger seats faced backwards, providing
protection for skulls and spines in the event of a crash or sudden deceleration.
The industry's judgment is that backward-facing seats would generate sales
resistance quite out of proportion to any lives or limbs that might be saved,
and so this simple and inexpensive reform has never been made. What the
industry really means, honorably enough, is that the safety value has traditionally been given a much greater degree of inflexibility than any other
variable.

There was, obviously, never any risk that either DC-10 or TriStar would
be unable to support itself aerodynamically. Once, when the question at issue

was whether the airplane would get off the ground at all, maiden flights possessed truly experimental charm. On the first flight of the DC-1 both engines stopped because the fuel supply failed as soon as the pilot pulled the airplane's nose up. But in these better organized days a maiden flight is primarily a public relations occasion.

The important question about the airbuses was which of them would generate the greater financial lift. What counted in this was not what might happen on each maiden flight, but which flight was going to take place first. Shortly after David Lewis's announcement that the DC-10 was definitely going ahead in competition with Lockheed's TriStar, a slogan appeared in the Douglas plant at Long Beach which put the matter in straightforward games-players' terms: Fly before they roll.

The rolling out of a new airliner, when it emerges from its shed to begin taxi trials and engine runs, is celebrated as much if not more than the maiden flight; usually, it occurs a month or six weeks beforehand. The McDonnell Douglas team, having started serious design studies behind Lockheed, aimed to catch up and take a significant lead themselves. This was a formidable task, for the Douglas design effort, although much admired for its thoroughness, had not, in the jet era at least, been notably expeditious.

On July 17, 1970, two years and a quarter after United Airlines placed the crucial order, the prototype DC-10 was rolled out at Long Beach. It was attended by a bagpipe band in kilts, by California Governor Ronald Reagan, and by Vice-President Spiro Agnew, in his capacity as chairman of the now defunct National Aeronautics and Space Council. Mr. Agnew said that the airplane would not only be larger than its predecessors, it would also be cleaner and quieter.

THE LOCKHEED PUBLIC relations department struck back, by promoting a dedication ceremony for the special plant which had been built out at Palmdale in the desert—at a cost of $50 million—for the construction of the vast numbers of TriStars which the world's airlines were shortly expected to order. This generated excellent coverage. But it could not put Lockheed back into the lead. On September 2, the DC-10 flew while the first TriStar was still inside its hangar.

As *Newsweek* said, perfectly accurately, the prototype DC-10 was "embarking, in a sense, on a combat mission." The flight was brought forward two weeks in order to overshadow the Lockheed rollout ceremony, which was due in mid-September, and there were rumors that the DC-10 might carry things still further by overflying Burbank as the TriStar emerged. "If they do," said Lockheed Vice-President Charles de Bedts, leaping into the spirit of the report with an enthusiasm he now regrets, "we'll have our military planes in the air." Mr. De Bedts's remark was made as a joke. Rather less jokingly, he was reported at the same time to be accusing McDonnell Douglas of helping journalists to write articles suggesting that the TriStar program, and Lockheed as a whole, was running into deep financial trouble. He complained that people were trying, prematurely, to write Lockheed's obituary. They would find, he said, that "we make an extremely stringy piece of meat." That was, as things were to turn out, a very accurate piece of prophecy, for Lockheed when pushed to the edge of bankruptcy turned out to have so much political muscle that government and Congress had to bail the company out. For all that, Douglas had engineered a formidable triumph by getting the DC-10 into the air before the TriStar was out of its shed.

Given the state of the company's health at the time of the takeover, it was a remarkable achievement to design and build so huge and complicated a machine in only two and a half years. What makes the feat all the more remarkable is that Douglas was designing two versions of the airplane at the same time. The TriStar was produced only in the basic airbus configuration, with a range of 3,000 miles. But from very early in the program, Douglas offered the DC-10 Series 10, with 3,000-mile range, and in addition the DC-10 Series 30, with 6,000-mile range.

The Series 30, with its much greater fuel capacity, and correspondingly much greater takeoff weight, was made possible because General Electric could produce a version of the CF-6 fan-jet with 5,000 pounds extra thrust.* It was in most respects the same airplane as the Series 10, but, 25 percent heavier and naturally there was extra design effort in fitting the wing with more commodious fuel tanks, and in calculating the extra strength required in load-bearing structures. The effort paid off more than handsomely, in that by the time the DC-10 made its first flight, it was taking a comfortable lead in orders over the TriStar. There were 241 DC-10s on order, *of both models,* worth $3.6 billion altogether, and 173 TriStars, *all strict airbuses,* worth $2.6 billion. The Series 30 DC-10 was the machine George Spater of American Airlines had in mind particularly when he said that what the airbus generation had produced was another "all-purpose" airplane. It meant that a fleet of DC-10s could operate a route network, including flights from New York to Buffalo and also from London to Los Angeles.

Douglas had come back from utter commercial defeat to win what was, by the ordinary rules of market competition, a clear-cut victory. It was one which in the understandable view of Douglas engineers and executives should have entitled them to scoop the pool. To them, and to many other American businessmen, it seemed worse than puzzling when, four years later, Lockheed was rescued by government financial backing from the consequences of having lost the race. Robert Metz, a financial columnist in *The New York Times,* put the point bluntly, as one of those who reckoned that Douglas— having outlasted Boeing's brief interest in the airbus—should not have to deal with a state-backed competitor:

---

* Lockheed did not have a suitable engine for a long-range TriStar for some time, but a 6,000-mile version is now available, powered by an advanced version of the Rolls-Royce RB-211 engine, called the -524.

If Lockheed is not bailed out, McDonnell Douglas will have the market to itself. And the people who hold this view ask, why not?

After all, McDonnell Douglas has proved itself to be more efficient than Lockheed. If Lockheed falls by the wayside, the spoils go to the swift. . . .

McDonnell Douglas had certainly been swifter than Lockheed. And in some kinds of simple artifacts, swiftness of production may be the same thing as efficiency. But this can hardly be the case in aeronautical engineering. If two planes are identical in quality, then the swifter producer is obviously the more efficient. But if the planes differ in the quality of their design—for whatever reason—then the question becomes a good deal more complicated.

TriStar and the DC-10 are identical products in the sense that both have been certificated as conforming to the Federal Airworthiness Regulations (FARs) for large jet transports. But FARs, literally interpreted, produce only a loose definition of what an airplane should be. "I could build you an airplane," says the Boeing engineer Ed Pfafman, "which fitted into every one of the FARs and you still wouldn't fly in it if you knew what you were doing." In other words, the regulations provide nothing more than a general framework. The safety and quality of an airplane depend heavily on the fruitfulness of the dialectic which takes place, during development, between the manufacturer's design team and the officials of the Federal Aviation Administration (FAA).

Questions about the quality of the DC-10 can only be answered by making a detailed comparison between the plane and its rival, the TriStar, and by looking in some detail at the relationship between the FAA and McDonnell Douglas.

THE BASIC MODEL of the DC-10 is a fraction less than 181 feet and 5 inches long—say, 60 yards, or half the length of an American football field. It is 155 feet and 3 inches from wingtip to wingtip, and the top of its tail is 58 feet above the ground. The TriStar is 3 feet shorter, and not quite as tall, but its wingspan is the same. Fully loaded, each airplane has the same weight: 315 tons, which is about the same as two good-sized brick houses.

Probably very few of the passengers who have flown in TriStars and DC-10s could readily distinguish between the two designs. Both airplanes, despite their bulk, have the same external appearance of beefy elegance. Inside, things are even more similar. The passenger cabin in each case is nearly nineteen feet wide, with seats set eight abreast and divided by two broad aisles. The cabin walls seem almost vertical, the ceiling is high, and people walk about without any feeling of restriction. Unlike earlier jet transports and unlike the 747, neither of these airplanes have galleys cluttering up the passenger deck: food preparation is done on a lower deck, and there are elevators to shift the cabin staff and their trolleys up and down. Wide-bodied jets are rarely crammed full to capacity, and this fact, combined with their handsome basic dimensions, provides a form of air travel almost as commodious as the airships of forty years ago. A meticulous passenger with a tape measure would find that the DC-10 has slightly larger windows. "All passengers pay the same price for their seats," David Lewis is supposed to have said. "The guy in the middle should have as good a view as the guy next to the window." The same traveler could also discover that there is a few inches more room in the TriStar's lavatories.

Such minutiae apart, a passenger might be forgiven for assuming the DC-10 and the TriStar to be products of the same minds. And in a sense they are, because both sets of designers bargained ideas with the same airlines at the same time. The airlines dictated the performance they wanted, the number of passengers they wanted, and the relative degrees of comfort that first class and economy class should enjoy. Given those parameters, both Lockheed and McDonnell Douglas started out by fixing the width of a passenger seat. This was made two inches wider than in the 707 generation because, according to airline research, the human race—or anyway, that part of it which regularly flies in jet transports—has become somewhat larger since the 1950s. (Presumably, we shall know things are really bad when the Western butt contracts.) The designers in each case then multiplied the seat width by eight, added the width of a couple of aisles, and drew a circle around the result. That approximately fixed the width of the fuselage, and the desired passenger-carrying capacity fixed the length. "Passenger seating," said McDonnell Douglas, "decided the design of the DC-10," and the statement is just as true of the TriStar. As one of the Lockheed designers explained to us, the other decisive measurement, along with the width of the human seat, was the convenient size of a trolley to carry food and drink along the aisles. Admittedly, this was a designer of cabin interiors speaking, but that was, of course, the significant point. He spoke as a man who had spent a lifetime trying to fit as much comfort as possible into a space dictated by the airframe and power-plant engineers. But the fan-jet revolution, providing huge new power supplies, enabled him and his colleagues for the first time to design an airplane from the inside out.

The fact that two design teams produced seemingly similar results when confronted with similar criteria might suggest that aeronautical design is not, after all, too far from being an exact science. And it may reasonably be argued that in a modern technological system, with so much experience to draw upon, there should only be one best answer to any specific need. Therefore, inevitably, the TriStar and the DC-10 should turn out to verify the excellence of each other.

However, the TriStar and the DC-10, although very similar to look at, are not quite so much alike as it first appears. Their intellectual pedigrees were different. Lockheed came to the task with very considerable design resources and with a tradition of designing complex and novel aircraft. When the airbus race got under way, Lockheed was the builder of the world's largest (though not most successful) aircraft, the C-5A Galaxy, and the fastest, the YF-12A, better known in its strategic-reconnaissance role as the SR-71 "Blackbird." But the company had not participated in the development of the first generation of commercial jets, and so the TriStar design team had less civil aircraft practice to draw upon than Boeing or Douglas. They met the deficiency in two ways: by unashamed copying—chiefly from Boeing technology—and when that was not possible, by going back to first principles.

The Douglas team started in a different position. The DC-10, as its name implied, was regarded as the most recent in a sequence of commercial aircraft. The Douglas executives and designers saw their company as the world's premier builders of commercial aircraft, whatever the recent power of Boeing and despite the unlovely circumstances of financial failure and takeover. The facts of corporate history suggest that the tradition was not quite the seamless web that the company's literature naturally liked to make out, and obviously the company was weakened during the final years of its founder's rule. Nonetheless, the belief in continuity meant a great deal to the people who remained into the McDonnell Douglas era.

Whatever difficulties encumbered their production, the DC-8 and the DC-9, simply as flying machines, were well regarded by the people who used them. And only a design team with a large expertise in commercial transport practice could possibly have put the DC-10 design together as quickly as they did. The question, in the light of subsequent events, must be whether the process may in some respects have been too rapid.

The chief structural difference between the TriStar and the DC-10 is the mounting of the tail engine. Lockheed's engine is buried inside the rear of the fuselage, sucking in its air through a curved scoop which opens up just ahead of the vertical tail-fin. The Douglas engine is mounted well above the fuselage and is supported on the vertical tail itself.

Putting a third jet engine into an airplane is not easy, and when the airbus design parameters settled around the tri-jet concept, neither the airframe designers nor the engine manufacturers could be quite sure how to configure such a layout or how well it would work.

Simply because of their weight, engines have always been difficult for designers to deal with, and the advent of jets, though making many things simpler, certainly worsened the problem of finding mounting places. Propellers can at least be hung on the front or the rear of almost any surface without affecting its load-bearing qualities. But, unless it is slung under the wing, a jet engine demands a large hole (nowadays, one of imposing diameter) through any structure on which it is mounted, and wing spars with large gaps in them are not very practical. Some of the early multiengine jets did have engine installations directly applied to the wing, like a World War II bomber, and the first jet airliner, the Comet, achieved a visually elegant solution with the engines buried in the wing root at the point where it thickened to join the fuselage. Such methods were possible only with the narrow bore pure-jets of the 1950s. They certainly could not be applied to big bypass engines; and they were in any case structurally extravagant.

Paradoxical as it may appear to the earthbound eye, the best place for any heavy part of an airplane is a long way out on the wing. This enables the wing to be a good deal less strong than it would otherwise have to be. In practice, the position is complicated by the fact that the wing bends and twists rhythmically in flight, and depending on how the rhythmic periods work out, there will be some positions on a particular wing which are quite unsuitable.* Nonetheless, it is a sound general proposition that the best place for the engines while the airplane is in flight would be the wingtips, allowing maximum "bending relief" to the wings. There is nothing really paradoxical about this: It is the wing, after all, which holds up a flying aircraft, just as a girder holds up the weight which is carried on a bridge. If all the dead weight —engines, on top of payload and fuel—is packed into the fuselage, the "bending moment" applied to the wing will be at a maximum. Spreading the weight of engines and fuel out *along* the wing means that the structure can be a good deal less weighty. On the ground, of course, the picture is exactly reversed, which is one of the reasons why jet transports do not in practice have wingtip engines.

Anyone passing through an airport can observe that in recent years many aircraft designers have chosen to mount engines as often as possible on the rear of the fuselage. This, however, is a demonstration of the fact that even when its enterprises appear most sternly rational, the human mind is still apt to be swayed by fashion.

In the all-rear-engine jets, the advantages of bending relief were abandoned in pursuit of notional improvements to the internal noise level. The Boeing 727 and the DeHavilland (now Hawker-Siddeley) Trident, each with three engines mounted on the fuselage tail, and the DC-9, Sud-Aviation Caravelle and BAC 111, each with two engines at the tail, exemplify the trend. It was carried furthest of all in the British Vickers VC-10 (and the Russian IL-62), designed with a battery of four rear engines, two on each side of the fuselage.

After a great deal of expensive machinery had been built, it was realized that the virtues of rear mounting were as insubstantial as the virtues of one hair style when compared with another, or one kind of autobody styling. Some skeptics had pointed out from the first, on the question of internal noise, that even the most lavish quantities of soundproofing material were not likely to weigh any more than the extra metal required in the wing spars. And in any case, extension of the bypass principle soon altered the whole question of noise. The noise from the *rear* of a fan-jet is relatively mild. In the take-off and landing phases, when engine power (and therefore noise) is apt to peak, most of the noise in a fan-jet comes out of the *front* of the engine. The

---

* Much airborne misery might be averted if all airlines distributed flight literature explaining that a wing is—must be—"aerolastic," like a tree or a reed that sways with the wind. Passengers watching flexing wings in turbulent air would be comforted by reflecting that the trees which snap in the wind are the ones which don't bend.

airplane itself is well below the speed of sound, but the big fan-blades, adding their spinning speed to their forward speed, are quite likely to go supersonic, and when they do, they project a cone of shock waves forward of the engine intake. (Set around inside the intake rim of a fan-jet, you will see rectangular strips of sound-absorbing material. This is a sad expedient which will be with us until someone can make a fan-blade hold together while being thin enough and wide enough to shift its due quota of air at a suitably low "helical Mach number.")

As we saw in the history of Rolls-Royce jet development, rear mounting carried to the extreme point seen in the VC-10 quarrels seriously with the whole principle of bypass engines, because it restricts engine diameter. But the nastiest consequence of the rear-engine formula was that it made necessary a "T-tail" configuration, leading in turn to the unlovely problem of the "deep stall."

"T-tail" aircraft, such as the 727, the Trident, DC-9, and most business jets, are recognizable because the horizontal element of the tail is not attached to the fuselage, because there is an engine-pod on each side where the tail's horizontal surfaces would normally be attached. Instead, the horizontal tail plane is perched on top of the vertical stabilizer. This requires the vertical fin to be a good deal heavier than would otherwise be the case. But the really worrying effect is aerodynamic and applies when the airplane gets into a nose-up "stall" situation.

In normal flight, the airflow opens and closes smoothly over an aircraft wing, but for any given attitude of the wing, there is a particular forward speed required to keep the airflow smooth. Typically, a "stall" occurs with the wing inclined sharply upward and the airspeed rather low; that is to say, something like the usual landing attitude of a modern swept-wing aircraft. In a stall, the airflow over the wing collapses. Instead of cutting smoothly through the air, the wing leaves a wake of turbulent partial vacuum behind it. (One might compare the effect to the difference between an oar-blade cutting through water edge-on or going broadside through the water.) A stalled wing gives no lift, and if nothing changes the situation, the airplane will fall out of the sky.

By definition, a nose-up stall occurs with the wing higher than the tail. This means that in an airplane with a "normal" tail, the wake of de-energized air from the stalled wing passes well above the horizontal tail. The pilot simply pushes his control column forward, the elevators (situated on the horizontal tail) hinge downward, and the airstream pushing against them makes the rear of the aircraft rise and the nose go down. Airflow over the wing is reestablished, and provided the airplane has a little height in hand, there is no danger.

But the geometry of a T-tail aircraft is such that in a stall the horizontal tail may be caught in the wake of deadened air. This is a "deep stall," in which the wing cannot provide lift, and the elevators cannot alter the plane's altitude and so there can be no recovery.

T-tail aircraft have, nonetheless, carried many millions of passengers safely enough. This is a good illustration of the fact that modern engineering skills and flying routines can triumph over design principles which may be more than a little dubious. After the initial burst of deep-stall crashes, which cost perhaps three hundred lives, there seem to have been very few cases of T-tail planes crashing for this reason (though the exact cause of a crash is not always established). The final solution to the problem is a "stick shaker," an electro-mechanical device which shakes the control column and sounds a hooter when its sensors tell it that the airplane is getting close to stalling point. If the pilot remains unimpressed by these alarms, the machine will eventually force the control column forward with more-than-human power.

By the time the wide-bodied jets were being designed, it was clear that the best way to mount four engines was in pods slung beneath the wings on pylons. But a tri-jet layout was not so simple. There was certainly no case for going back to the rear-triple layout. In a world of ideal engineering, all three engines might be wing-mounted, with two on one side and one on the other. However, the asymmetry problems that would result from an engine failure are probably insoluble and no manufacturer or airline would relish the public relations problems that a lopsided airplane would present.

Therefore, the third engine had in both cases to be rear-mounted. A single rear engine does not compel the adoption of a T-tail, but the engine-makers immediately manifested uncertainty about the way a rear-mounted fan-jet might perform.

In triple-engine T-tail aircraft like the 727, the engines clipped on each side of the fuselage had not produced any great difficulty. But the middle engine, buried in the rear of the fuselage with its long curved inlet pipe snaking up to a scoop in front of the tail, had been more troublesome. Inevitably air that has to go around "corners" will behave in a complex manner, and much attention had to be paid to the inlet design to prevent power loss in the central engine. What was most awkward was that at takeoff the fuselage sloping upwards in front of the central engine's air intake sometimes "starved" it of air, causing power loss at a deeply inconvenient moment.

Cures for these difficulties were found in the 727 generation, but it was hard to say how things might work out with the airbuses. The new engines would have a vastly greater appetite for air, and the whole business of getting air into engines is one where the best available theories give rather poor predictions. Like most things to do with engines, it is in the "suck-it-and-see" area.

Structurally, of course, it is nicer to have the engine inside the fuselage. ("You can really get your arms around it," as Dan Haughton said.) And aerodynamically, it is quite useful to have a fat jet exhaust emerging from the tail cone because it means that the curves at the rear of the fuselage can be much less abrupt, allowing the air to close gracefully after the wide-body's passage. The attitude of the engine firms was that *probably* a 40,000- to

50,000-pound fan-jet could be made to behave consistently at the end of a curved intake, but that it might take quite a while to find out.

The TriStar designers decided to take the risk and go for the curve, and as a result there was much traveling between Burbank, California, and Derby, England, with drawings and with bits and pieces of S-shaped pipe. The outcome appears to be highly successful, in that the structure is elegant and the engine operates smoothly.

The Douglas designers chose to avoid such imponderables, and therefore designed the engine mounting on the vertical tail fin. Given that position, they could be certain that the engine would behave as if mounted on a wing pylon: the problems of design, as they saw it, were thereby simplified.

But "simplicity" in engineering is a slippery concept. Most engineers would say that gas turbines are "simpler" than piston engines. They require very little by way of electrical equipment, and their parts, instead of reciprocating, merely revolve. But that simplicity is attained only by applying metallurgical and aerodynamic techniques far more elaborate than anything required in a piston engine.

What we can be sure of is that every design decision in an airplane affects, however slightly, every other decision, though to be sure, there are many practical occasions when we can ignore the connection, just as we need not bother with the fact that the movement of a distant star must make some tiny variation in the gravitation of our planet. When the Douglas engineers sited their central engine on the vertical fin, this meant not only that they had to design a pierced girder to accommodate the engines but also that there were constraints upon the positioning of the wing engines.

Because the rudder is hinged to the rear of the stabilizer, the engine is cut into the space available for the rudder. This meant that the wing engines had to be fairly close in, because one of the chief tasks of the rudder is to check the effects of "asymmetric thrust" caused by engine failure. If one wing engine of a tri-jet fails on take off, the aircraft will yaw violently. The further out the engines are placed, the greater will be the leverage of the remaining engine, and the more violent the yaw. The TriStar, with the whole of the area at the rear of the vertical fin available, was able to mount a more powerful rudder, and could therefore have more widely spaced wing engines. Thus it acquired advantages in bending relief, and more protection against the much canvassed dangers of engine "shrapnel." In these high thrust engines, the stresses on fan-blades and turbine-blades are gigantic, and they are big enough to do considerable damage if the blades fly off and hit anything. The engine casing is supposed to contain errant blades, but in practice it cannot be expected to do so every time, and certainly the "shrapnel" danger was taken seriously enough in the TriStar and DC-10 designs to lead the control cables through channels well away from the fuselage skin.

Without doubt, the Lockheed designers adopted a layout that tended to generate more complexity in the design process. The end product was, however, advantageous, even if one advantage was added protection against shrapnel damage which might not, and should not, occur. In other ways, the Lockheed team appear to have chosen solutions which required much refinement to fulfill their promise, but which would, if successful, exploit the state of the art more fully than the solutions adopted in the DC-10.

There is, for instance, a problem in transonic flight which is called "jet upset." It has several times been lethal and remains dangerous. It first became apparent in civil aircraft with the 707 generation of airliners, with cruising speeds approaching Mach 0.9. The manifestation is dramatic. From level flight, the aircraft throws itself into a steep dive, and the controls become virtually immovable. Even if the pilot can drag the control column back, the elevators may have no effect, so that the aircraft continues its precipitous descent.

Nobody doubted that jet upset was due to some paradoxical airflow phenomenon, but it was quite another thing to explain it. Airflow in the subsonic speeds is well understood. It can be analyzed by methods based on the fluid mechanics of nineteenth-century physics, and even before Orville Wright took off at Kill Devil Hill, Ludwig Prandtl and Frederick Lanchester had virtually completed a theory of subsonic flight which still serves today. Airflow at supersonic and hypersonic speeds also behaves like a respectable fluid to which workable theories can be attached. But at "transonic" speeds, close to the speed of sound, the picture is chaotic.

On a schematic level, it is not hard to see why. A sound can only travel through air at Mach 1: the "speed of sound" in any given set of conditions is simply the rate at which an energy impulse, caused by a shout, say, or a shot, ripples through a free mass of air molecules. The colder the molecules are, the slower they react. Therefore, in the cold of the stratosphere, Mach 1 may be less than 600 miles per hour, while on the floor of the Red Sea basin it may be close to 800 miles per hour.

The flying surfaces of an airplane going at subsonic speed send a sound "message" ahead of them, which travels at Mach 1, and this gives the air-molecules a little warning, so that they can part and flow around the airplane in an orderly manner. Air molecules will also behave in a perfectly orderly —if slightly tackier—way, provided they are taken totally by surprise, that is, if the airplane is traveling faster than Mach 1. (The molecules are simply hurled aside before they can start any movements of their own. Of course, they make up for this by forming a large shock wave, the cause of the sonic "boom," but that's primarily a problem on the ground: it doesn't bother the airplane unduly.)

Disorder is likely to occur when the airplane gives some, but not sufficient, warning of its passage: enough, as it were, to throw the crowd into a panic, but not enough to allow the members of the crowd to work out what to do. Quite likely, the airflow at the front of the wing will suffer the most disruption and lose a large part of its lifting capacity. Suddenly, the wing's center of lift shifts backwards, which has exactly the same effect as shifting the whole wing back. The aircraft becomes nose heavy and plunges violently downward.

Jet upset, and similar effects, caused several known crashes in the early jet-transport days, and may well have been the cause of some mysterious disappearances. High-altitude, high-speed flight produced aerodynamic effects which sometimes disoriented pilots brought up in the thick, cozy air layers close to the earth's crust. A propeller-driven transport cruising at 11,000 feet is probably flying 100 percent or more away from its stalling speed; only some strange and improbable flight-path convulsion can make it come to grief. This is not the case with a jet transport cruising in the thin, cold air above 30,000 feet. Thin air makes less drag, which produces economy. But it also gives less lift, so that the jet, in its smooth, ethereal flight may be only 10 percent away from the stalling speed at which it will fall out of the sky.

At these ethereal heights, the speed of sound is much slower in miles per hour. Early jet pilots, influenced by the traditional desire to keep well away from stalling speed, were apt to edge too close to the sound barrier and discover something worse than a stall. (The "coffin corner" effect, which made the XB-47 bomber notorious, resulted from the fact that it was quite easy to get combinations of speed and atmospheric conditions in which slight deceleration would produce upset.) More than once 707 crews found their airplanes going into "unstoppable" dives; but the fact that the phenomenon was reported and analyzed shows that it was survivable. Provided the aircraft did not build up too much speed in its dive, the airflow would restore itself in thicker, low-level air where Mach 1 is in excess of 700 miles per hour.

There are several layers of defense against jet upset in present-day jets. The most important of which is that piloting instincts have been reconstituted so that jet transports are flown, not by the air-speed indicator, but by the Mach meter, which calculates the ever varying speed of sound and the airplane's relation to it. Pilots speak of cruising at "point eight-five Mach" rather than at so many miles per hour or knots.* Control systems and aerodynamic forms have also changed. To the casual eye a DC-10 or a TriStar does not look anymore "streamlined" than a 707. In fact, the wing section is much more refined, and much less vulnerable to the compression effects which occur around Mach 1. The 747 is remarkably well armored against jet upset, because aerodynamic juggling has produced a wing whose lift center shifts *forward* with the approach to Mach 1. Therefore, a 747 will rear up and bleed off its excess speed, instead of going into an upset dive. Upset remains theoretically possible with a TriStar or DC-10. But the Lockheed designers erected a powerful defense against it by building an "all-flying" tail. In the usual system, elevators are hinged to the rear of the horizontal stabilizer. In an "all-flying" tail there is no division into stabilizer and elevators: the whole surface is movable and changes its angle to control the airplane's angle of flight. The traditional stabilizer-and-elevator system is a well-tried

* The memory of the physicist and philosopher Ernst Mach is thus preserved. In 1887 he perceived that airflow at the speed of sound would disobey established mechanics.

picce of engineering. But in a critical situation, it cannot generate the huge control effects possible with an "all-flying" tail which guarantees that the TriStar can always be forced out of a high-speed dive (unless, of course, the speed is so high that the structure reaches breakup point, when all would presumably be lost anyway).

The concept and the final result have a certain elegance and the TriStar's tail appears to work successfully. But fairly obviously, the mechanism of an "all-flying" tail required a great deal of working out. The Douglas team, on the other hand, stuck to the traditional elevator design, relying on the DC-10's general flying qualities to insure it against trouble.

THERE ARE OTHER respects, some minor, in which the Douglas and Lockheed designs diverge. At the rear of the passenger cabin, where the pressurized section of the fuselage comes to an end, each plane has a pressure bulkhead. In the DC-10 this is a flat disk, whereas in the TriStar it is a segment of a sphere, bulging towards the tail. This is the reason why the TriStar's lavatories, ranged on the inside of the curve, are slightly roomier. The purpose, however, is not to give extra elbow room to the passenger deploying his electric razor. The point is that a curved bulkhead, though it takes rather more design time, can be made slightly lighter.

There is also subtle difference in the way the fuselages are constructed: in the TriStar the side walls are thinner. This particular refinement was pioneered by the Convair design team for the unhappy 880/990. Although that airplane was a commercial disaster, the sophistication of its fuselage structure was generally admired. One might have expected the method to be used in the DC-10, since Convair (a division of General Dynamics) was subcontractor for the fuselage manufacture and part of its design. The idea is simple but requires much detailed working out. Each of the hooplike fuselage ribs is thinned down between cabin ceiling and cabin floor, and the loads thus shed by lightening the ribs are transferred to lengthwise fuselage members which are made correspondingly stronger at this level. The method requires not only the usual riveting techniques, but also chemical bonding of metal surfaces. Douglas, however, did not choose the new technique their subcontractor had evolved. The DC-10 stuck to the traditional constant-thickness rib.

Just as the TriStar designers chose the more advanced—or, as Douglas might say, more elaborate—solution at the rear and sides of the fuselage, so did they at the flight-deck end. A basic need in the flight deck is to give the crew windows which will provide a wide and undistorted view of the world, and like many other things concerned with aircraft, it is rather more involved a process than it looks. The established method—used in the DC-10 and traceable back to the DC-1—is to fit a number of flat glass plates into the upper nose, minimizing as far as possible the inevitable discontinuities between flat surfaces and the compound curves of the streamlined skin.

Naturally, such a marriage has its problems and the flat windows, unless they are to do unacceptable violence to the airflow, must cut up the pilot's view with a series of metal struts. Molding the windscreen to follow the ideal airframe curve might seem a better approach: unhappily, however, it produces insuperable optical problems. Nevertheless, a compromise can be made by using sections of cylinders, or cones, for the basic shape of the windscreen— rather as a conic projection can be made to give a better account of the globe than the rectangular Mercator projection. The Boeing 747's windscreen is a section of a downward-sloping cone: the TriStar's is a section of a cylinder, also with its axis sloping downward. The practical result, after a good deal of fancy solid geometry at the design stage, is that in both these aircraft the pilots have a wider angle of uninterrupted view than in the DC-10 with its traditional windows. (One disadvantage of the TriStar windscreen is that it is considerably more expensive to construct and replace.)

In summary, those places where divergence can be discerned, it appears that the DC-10 designers usually went for the known technology, for the thing which they had done before. It was a natural enough thing to do in a firm which had plenty of experience on which to draw, and it seems an almost inevitable corollary of the determination to "fly before they roll."

But that leaves the questions: Is the DC-10 as good a plane as it could have been, given the state of the art at the time it was designed? And, is it as good as its contemporaries?

The question is impossible to answer in absolute terms because the DC-10 benefited, as do all American aircraft, from the sheer competence of the U.S. aeronautical tradition. More airplanes have been built in America, over a longer period, to higher standards, by better-educated people, and used and serviced by more highly trained pilots and maintenance men than in any other part of the world. The individual skills and organizational disciplines in the American airline industry are so high that almost any design can be made to "work" and almost any airplane made to perform with impressive reliability.

One highly experienced British test pilot, summing up the qualities of all three wide-bodied jets, purely as flying machines, said that they came out roughly equivalent, though in different ways. The 747, he said, had the best flying controls and the TriStar had the best cockpit. (By this he didn't mean physical comfort as such: the fate of an airplane depends on the crew's ability to turn vast quantities of information into swift, correct decisions, and the design of cockpit systems is critical to this.)

The great strength of the DC-10, in this pilot's view, was the superb performance of its engines. The appeal of the DC-10 to pilots is that it is a strong, well-shaped basic airframe, supplied amply with power from three superbly tractable engines. As George Cayley discerned a century and a half ago, flight is a matter of applying power to drive a plane surface through the air. This is a truth which is obvious to every working pilot as he lifts his plane off the runway or juggles with throttle settings on the approach path. The classic problem of a jet engine asked suddenly to deliver more power is "surging." A sudden opening of the throttle increases the speed of air going through the engine, and the airflow around the turbine or compressor blades may disintegrate for a moment, causing a sharp drop in thrust after a sudden increase. Long and skillful development engineering has given the CF-6 engine especially good internal aerodynamics; therefore, the DC-10 has an unusually smooth and flexible power supply.

But, on this superb base, the Douglas team built an aircraft that was less advanced than it might have been, less advanced in small but significant detail than either the TriStar or the 747. McDonnell Douglas would undoubtedly represent this as a hard-headed refusal to become involved in needless elaborations. Indeed, since the heady days of the DC-3, technological caution has always been part of the Douglas philosophy. In Alexander Pope's words, quoted in the company's literature: "Be not the first by whom the new are tried,/Nor yet the last to lay the old aside." Or as the steelmaster Andrew Carnegie put it a little more cynically: "Pioneering don't pay."

One other way in which this philosophy manifested itself in the DC-10 was the design team's decision to install just three separate hydraulic systems to operate the control surfaces. The DC-9 and the very similar British Aircraft Corporation (BAC) 111 were among the first aircraft to employ three hydraulic systems (as opposed to one or two), each one capable in an emergency of handling all of the plane's hydraulic needs. Aircraft like the DC-9 and 111 might just be flyable without any hydraulic power, if the pilot could exert enough brute strength to move the control surfaces by means of the manual reversion system. But in the jumbo-jet generation even this vestigial capacity vanishes. Without hydraulic power a 747, a TriStar, or a DC-10 would be as certainly doomed as a rudderless sailing ship on a lee shore.

Both Lockheed and Boeing decided it would therefore be prudent to equip their wide-bodied airplanes with four hydraulic systems, each one more or less capable on its own of providing enough control for the pilot to be able to land. In Boeing's case, of course, the wisdom of that decision was amply demonstrated when Pan Am's Flight 845 smashed into a pier at San Francisco on July 30, 1971.*

McDonnell Douglas claims that in the DC-10 the three hydraulic systems are so lavishly provided with crossover points that their combined redundancy is as good as the redundancy provided by the four hydraulic systems in both the TriStar and the 747.† That may be so, although the engineers from Lockheed and Boeing would argue otherwise.

What is certain is that a DC-10 can only afford to lose two of its three

---

* See Chapter 2.
† Basically, a redundant system is one which an airplane can afford to lose, because its functions will be automatically taken over by other systems. In a DC-10, each of the three hydraulic systems normally power different controls but, in an emergency, any one system can take over the essential functions which require hydraulic power.

hydraulic systems, no matter how magnificently conceived they may have been. Both the TriStar and the 747 can afford to lose three. In that context it is difficult to believe that the DC-10 would not be better served if the designers had elected to add one more system.*

It is true that many of the points of difference between the DC-10 and the TriStar can fairly be seen as a matter of sophisticated argument in which different engineers would simply take differing views. But in one respect the original design of the DC-10 was seriously inadequate. And not only did the design show, in this respect, marks of haste and commercial anxiety. So, too, did the company's response to the failure of that design in action. However, by that time the competitive screw had taken up yet another turn.

---

* By the same argument, of course, five systems would be better than four, six better than five, and so on. But the weight of each extra system would have to be traded off because the maximum operating weight of every airplane is firmly fixed by the power of the engines. Our point is that Lockheed and Boeing both believed that the arrival of the jumbo-jet era was the appropriate moment to move up from three systems to four and the logic seems insuperable.

7

Daniel Haughton's Possum Hunt

> The payment of debts is necessary for social order. The
> non-payment of debts is quite equally necessary for social
> order. . . . Unfortunately, the second of them violates a great
> many seemingly legitimate interests. . . .
>
> —SIMONE WEIL,
> *On Bankruptcy*

ON FEBRUARY 1, 1971, Dan Haughton, then chairman of Lockheed and chief prophet and inspiration of the TriStar, took the afternoon plane from Los Angeles to London feeling quite justifiably pleased with himself. That morning he had finally settled a dispute with the U.S. Government that had been dragging on for two years and which had plunged Lockheed into dire financial difficulties. The settlement had cost Haughton $480 million but, with the company's very survival at stake, it had not seemed an exorbitant price to pay and now, at last, the way seemed clear for Lockheed to get on with building the TriStar.

Haughton was going to London to supply some much-needed reassurance to Rolls-Royce, his partner in the TriStar project. The British company, and more especially the British Government, which had a fair amount of tax-payers' money tied up in the RB-211 engine, had become increasingly nervous during Lockheed's protracted dispute with Washington over four military contracts. Lockheed's bankers had frozen the company's credit until the dispute was settled, and at one time it had seemed very likely that the TriStar program would grind to a halt through lack of ready cash. When Haughton boarded the London plane, he was convinced that his bankers would now provide whatever money was needed. "For fourteen hours I felt pretty good," he said.

The next day Haughton, and a handful of other Lockheed executives, were due to have lunch with Lord Cole, chairman of Rolls-Royce, in a private room as the Grosvenor House Hotel on Park Lane. Cole had become Rolls's chief executive a year before at the insistence of the British Government after the company had run into serious liquidity problems because of the development costs of the RB-211. The Government had agreed to bail Rolls out with loans of $143 million, but first it installed Lord Cole and another director, Ian Morrow—an accountant with something of a reputation as a company doctor—to review the RB-211 contract and assess the cost of curing the technical problems that had shown up during testing of the prototype engines.

In the middle of the morning Haughton, who was staying at the Hilton Hotel, got a telephone call from Lord Cole: Would he mind arriving at the Grosvenor House thirty minutes early, and alone, for a private chat? Haughton did as he was asked and, over a glass of sherry, delivered his good news that the Lockheed crisis was over. Now, he asked, how was the Rolls-Royce review going? How long would it take to solve the technical problems and begin delivering production engines? Lord Cole replied that the problems seemed insoluble. In two days time, he said, Rolls-Royce—the most prestigious symbol of British industry—was going to formally declare itself bankrupt. When the other Lockheed executives arrived for lunch—cold salmon—they found Haughton looking "as if he had got a bullet between the eyes."

The decision had been taken, in great secrecy, by Rolls's board a week before on January 26. Ian Morrow's review had shown that the RB-211's development costs had risen from the original estimate of $156 million to an estimated $600 million. Even worse, the development schedule had never recovered from an early setback when Rolls's plan to manufacture the massive blades of the main fan out of a new material called Hyfil had proved unworkable. Hyfil is a composite of carbon fibers and epoxy resin, and if it could be used for the fan blades, instead of titanium, it would save three hundred pounds of weight in each RB-211 engine—enough to give a TriStar the ability to carry five extra passengers. But, as Rolls's designers had discovered at vast expense, Hyfil will not stand up to much punishment. When the first prototypes of the RB-211 were flight-tested—mounted on the fuselage of a Vickers VC-10—hailstones, and even raindrops, pitted the tips of the Hyfil blades. Strengthening the tips of the blades with titanium had only weakened the roots and, eventually, Rolls had been forced to revert to all-titanium blades.*

The setback was costly but had not seemed fatal. As a precaution, Rolls had simultaneously been developing the technology necessary to manufacture titanium blades. But now, in January 1971, Morrow saw the failure of Hyfil as a root cause of the delays that had put the development program far behind schedule. The test engines had failed for a long time to deliver the performance that Rolls had promised Lockheed in the RB-211 contract. The power problem had been solved, but now the engines were drastically overweight and neither Morrow, nor the Government's auditors, had much confidence that the designers could meet their promises.

* The designers now claim that they can solve the problems of Hyfil but the knowledge is, for the moment, academic.

On January 22, Lord Cole had told the British Prime Minister Edward Heath that the company could not go on with the RB-211 contract without financial help from the Government. If that help was not immediately forthcoming, the company would have no other choice but to call in a receiver. Heath was unmoved. A year earlier, and probably against his better judgment, he had by agreeing in principle to the Rolls loan broken the Conservative party's creed—of which Heath was very much the architect—that the "lame ducks" of industry should be allowed to go to the wall. But now, the Prime Minister said, the Government would nationalize the parts of the company that were important to national defense and sell off the profitable bits, like the motor car manufacturing division; not another penny of taxpayers' money would be invested in the RB-211 project. Faced with this intransigence, Lord Cole and his fellow directors had no alternative, under British company law, but to tell Rolls's preferential creditors—the long-term debenture holders—that it was broke.

The news, when it was announced in the House of Commons on February 4, had a stunning impact. It was, as the U.S. business magazine *Fortune* said, akin to an announcement that Westminster Abbey had become a brothel. For Haughton, Rolls's bankruptcy was a cataclysmic blow for, overnight, all his carefully laid plans to refinance Lockheed were in ruins. His bankers would most certainly not lend him money to build a plane which, without the RB-211 engines, had become the world's biggest glider. And without TriStars rolling off the production line to provide desperately needed income, Lockheed was doomed.

As Haughton walked down Park Lane from the Grosvenor House to the Hilton (whose more functional style of hospitality he much prefers), he must have pondered the irony of the situation he found himself in. By character and inclination he is an apostle of the doctrine of free enterprise and of its primary commandment that only the fittest should survive. He believes, at least in theory, that anybody who gets into the aerospace business must be prepared to take on all the risks unaided. This is a conviction Haughton illustrates with a folksy anecdote: "Boy runs out to his Paw who's ploughin' the field and says: 'Paw, there's a mountain lion got in the kitchen where Maw is.' Paw stops the horse, shifts his chaw, and replies: 'Well, the lion got hisself into the kitchen: he'll just have to get hisself out.' " Both Lockheed and Rolls had got themselves into the kitchen and, under Haughton's rules, neither company was entitled to look for outside help when the going got too hot. But given Haughton's buccaneering personality and his utter dedication to Lockheed's survival, there was never the slightest possibility that he would reject *any* legitimate recourse.

The survival of TriStar depended upon Haughton's ability to organize a rescue operation on a scale perhaps unprecedented in the business histories of either the United States or Britain. He had to cajole two governments, twenty-four banks, ten customers, and twenty-five major suppliers into helping. Of course, Lockheed's customers and suppliers who had invested a total of $550 million in TriStar—the airlines through advance payments and the subcontractors through building up inventories—were as anxious as Haughton to save the plane; many of them also faced bankruptcy if the rescue operation failed. And Lockheed's bankers, who had loaned the company $350 million, stood to lose most or even all of it.

But the two governments were in an ambivalent position. The administrations of both President Nixon and Prime Minister Heath were certainly committed to the ideology of free enterprise and therefore sympathetic to the argument that both Lockheed and Rolls-Royce should be allowed to fail and their assets redistributed through the British and American economies. There was some merit in the opposing argument that both companies were, because of their defense work, "special cases" and that their liquidations would harm national interests, but the evidence to support this was ambiguous and far from overwhelming. The British Government thought that work on military jet engines could continue through the setting up of a new and nationalized company, to be called Rolls-Royce (1971), that would simply take over the facilities of the bankrupted Rolls. And, in the United States, it was fairly certain that after liquidation some form of restructured Lockheed could be set up to continue work on military contracts. So, the real point at issue was whether the two governments should bail out the TriStar program— and the only justification for that was that it would save jobs.

Politicians, of course, have a powerful incentive to be seen to be saving jobs. And while redundant Lockheed workers might *eventually* find employment within the U.S. aerospace industry, the prospects for those working on the RB-211 project were grim. In Derby, a bleak industrial town in the English midlands where Rolls had its main plant and employed about 20 percent of the working population, trade unionists, scientists, business executives, civic leaders, politicians of multiple persuasion, and shopkeepers united in the fight to save Rolls-Royce. The appeal to the British Government for help was neatly reflected in a bitter parody of the Lord's Prayer that circulated around the factory:

> Give us this day our receiver
> And forbid our redundancies
> As we forgive them
> That did nationalize us.
> But deliver us from creditors,
> For ours was the Merlin,
> The Spey and the Conway,
> Dart and the Avon,
> RR-men.*

But, however strong the appeals, the British Government could not—or would not—countenance the rescue of the RB-211 project without water-

---

* The Merlin, Spey, Conway, Dart, and Avon were all classic Rolls-Royce aircraft engines. The Merlin powered the RAF's Spitfire fighters which fought in the Battle of Britain. The Dart was probably the most efficient and one of the most successful turboprops ever built.

tight guarantees that the TriStar was going to be built—and built in sufficient quantities to make production of the engine worthwhile. Guarantees of that quality could only be supplied by the U.S. Government. For its part, the U.S. Government would need guarantees that Rolls could solve the technical problems and deliver the engines—guarantees that only the British Government could deliver. Neither government would act unless Lockheed's banks, its customers, and its suppliers shouldered the burden of putting up the immediate cash that was needed to resurrect the TriStar program.

Haughton's task, which he began on February 3, 1971, was to reconcile all of these conflicting interests. John B. Connally, then Secretary of the U.S. Treasury—who was to become one of Haughton's most powerful allies—described the problems vividly: "Dan, it's no good chasing one possum at a time up the tree," he said. "What you've got to do is get all those possums up the tree at the same time."

DANIEL JEREMIAH HAUGHTON was born in 1911 and raised on a red-dirt farm in the backwoods of Alabama. He spent his early years looking after the calves and goats which, he thought, was infinitely preferable to going to school. To help make ends meet, his father, besides being a farmer, also ran a small grocery store and worked as a timekeeper at a coal mine; and young Dan helped out by selling newspapers and cutting timber for mine supports. He began school when he was eight years old with considerable reluctance, but he soon developed an aptitude for figures and was able to exploit his talent by lending money to his school friends at what he now considers to be outrageous rates of interest.

At the age of seventeen his mathematical ability took him to the University of Alabama, where he studied accounting and business administration. He supported himself and paid for his studies by working vacations as, variously, a bus driver, a dynamiter, and a coal loader in a mine. When he graduated in 1933, the depression was at its depth, and although Dan had no clear idea of where his future lay, he determined that the prospects, and the climate, were better on the West Coast and he moved to California.

For six years he tried various jobs, without much distinction, until in 1939, he was hired as a systems analyst for $3,300 a year by the Lockheed Aircraft Corporation of Burbank, near Los Angeles. The job had a drawback in that not too many people knew what a systems analyst was and Dan was even once refused credit at a furniture store because the owner simply wouldn't believe that such a job existed. That minor irritation apart, his recruitment into the aviation business was from the start an unqualified success. He could not, of course, have joined at a more propitious time, for with the onset a few months later of World War II in Europe, Lockheed like other plane makers began to expand at an unprecedented rate. With America's entry into the war, the demand for planes became insatiable while the number of young men with management potential who were still civilians became increasingly scarce. Haughton was soon spotted by the brothers Robert and Courtlandt Gross, who had bought a bankrupted Lockheed for $40,000 in 1932. He was given a series of more important jobs in production and proved himself to be an ideal company man, according to Courtlandt Gross, always willing "to subordinate his own time and rest to the overall good of the company."

When, in 1949, he was switched from production to the administrative side of the business, Haughton also proved to be a tireless salesman. He was made president of two Lockheed subsidiaries that manufactured equipment used to service planes and which had been losing money, mainly through lack of orders. President Haughton packed his briefcase with samples and barnstormed the country, utilizing his stubbornness and his folksy charm in about equal proportions. In not much more than a year, both of the companies were making profits.

His reward was to be given the job of reopening a massive production plant in Marietta, Georgia, which had been used during World War II by the Bell Aircraft Company to produce Boeing-designed B-29 bombers under a pool arrangement. In 1950 America had just entered the Korean War and the U.S. Air Force once again needed every plane it could get. At Burbank, Lockheed concentrated on producing P-80 "Shooting Star" jet fighters, while Haughton's first task in Georgia was to de-mothball and modernize 120 B-29s, which had been stored for five years in a Texas desert. Within a year he had built up a workforce of ten thousand and begun building Boeing-designed B-47 "Stratojets," the first plane designed specifically to carry nuclear bombs. By all accounts, he ran an exemplary operation and by 1952 he had gained the necessary corporate clout to win for the Georgia company production of the new C-130 Hercules military transporter which was to become, without doubt, Lockheed's biggest money maker. Today C-130s are still rolling off the Georgia production line, having earned Lockheed—at a conservative estimate—$300 million in profit.

Haughton's five years in Georgia provided him with deep satisfaction, not only professionally but also in his personal life for he was able to spend far more time than before with his wife, Jean, who suffers from multiple sclerosis. They were prosperous enough to buy a farm of 425 wooded acres—very different from the red-dirt farm Dan was raised on—and they ran a small herd of Black Angus cattle. In their spare time they fished the farm pond for bass and bream.

But, inevitably, Haughton's future lay at the Burbank plant and in 1956 he was recalled to California, where the Gross brothers continued the careful grooming of their heir apparent. In 1961, when Robert Gross died, Haughton became president. Six years later Courtlandt Gross retired at the age of sixty-two, handing over the reins of Lockheed to his protégé. The boy from the Alabama backwoods was now head of the biggest defense contractor in the Western world.

IN RETROSPECT WE now know that Haughton became chairman of Lockheed at a critical point in the history of both the company and the U.S. defense contracting industry. During the early 1960s, the relationship between the U.S. Government and the industry had undergone radical change, primarily because of a hardening of public and political attitudes towards military spending. The Government—under continual prodding from Congress—was forced to become a far more selective customer and, in general, defense contracts became harder to win and a great deal tougher to fulfill. By 1967, when Haughton took over Lockheed, the wind of change was blowing so strongly that the major contractors were scrambling to *reduce* their reliance on military business. It was no coincidence that McDonnell and Lockheed, who both relied on the Government for upwards of 90 percent of their business, chose almost the same moment—1966—to enter the commercial aviation market—McDonnell by taking over Douglas and Lockheed by developing the TriStar.*

It had all begun innocuously enough with the appointment by President Kennedy of Robert McNamara as Defense Secretary. McNamara's ambition was not so much to cut defense spending—indeed, he increased it—but to insure that the Pentagon got better value for its dollars. His methodology was to clamp down on "cost overruns" which, traditionally, defense contractors had been able to extract from the U.S. Treasury without much difficulty. All too often, it seemed, manufacturers would win a military contract with a low bid and then escalate the price of the plane, gun, missile, or whatever during production by claiming that costs had overrun, on the usually accurate assumption that it was too late for the Pentagon to go elsewhere. McNamara had joined the White House from the Ford Motor Company and believed that if automobile manufacturers could produce motor cars more or less within a predefined budget—which, of course, they have to do to stay in business—then the same should be expected of defense contractors. He therefore introduced the fixed-price contract which limited the Government to paying only a small percentage of any "cost overruns": the manufacturer had to pay the rest out of his own pocket.

However, in 1963 the Pentagon got a new recruit who believed that the fixed-price concept was too modest and that the defense procurement system was in need of far more drastic reform. Robert Charles had spent eighteen years with the McDonnell Aircraft Company, seven of them as an executive vice-president, and his inside knowledge of the business made him a formidable Assistant Secretary to the U.S. Air Force. He believed that rival

---

* Lockheed had been knocked out of the market by the failure of the Electra—see Chapter 3. Boeing's defense and commercial business had been well balanced since the introduction of the 707 in 1959.

manufacturers competed for defense business only up to the stage where the research and development contract was awarded. Whoever won the research and development (R and D) contract was, generally speaking, home and dry and could subsequently quote pretty well any reasonable price for actually building the hardware. The obvious suspicion was that manufacturers were putting in low bids for R and D contracts in the almost certain knowledge that if they won they could make up their profits during production. On the average, research and development accounts for only 20 percent of a defense contract's total cost and the nub of Charles's argument was that competition among manufacturers was ending too early.

Under Charles's guidance the weapon the Pentagon devised to make military procurement more efficient was the Total Package Procurement (TPP). This radical concept dictated that manufacturers competing for a contract had to submit one bid specifying in advance the price of R and D, testing, evaluation, *and* production. The winner got a contract that dictated the total price, the schedule for delivery, the specifications the hardware would have to meet, and the maximum profit the manufacturer would be allowed to make on the deal. There were, at least in theory, no loopholes: each contract was divided into segments and each segment had a "target" and a "ceiling" price. During the first segment, costs *had* to come within the ceiling price or the rest of the contract could be canceled. The actual costs incurred during segment one fixed the ceiling price for segment two, and so on. In other words, there was no longer any percentage in putting in a low bid for, say, research and development because there would not be any "excess" profits from later stages of the contract, to subsidize the R and D work.

The first TPP contract put out to tender, in 1965, was for the design, development, and production of what was—and, so far, remains—the world's biggest aircraft, the C-5A military transporter. The C-5A was the first wide-bodied aircraft conceived, and the new technology it inspired led directly to the development of the commercial jumbos. Lockheed, Boeing, and the then independent Douglas company were the major contenders for the contract. Lockheed won. The consequences were disastrous for the U.S. taxpayer and almost fatal for the corporation.*

Other TPP contracts were subsequently accepted by Lockheed, with awful results, but the C-5A contract is most germane to our story, because it was to make by far the greatest contribution to the decline of both Lockheed's fortunes and its reputation—which, in turn, led almost to the elimination of the TriStar from the airbus race. The saga of the C-5A is also important because it throws a little light on the complex relationship between

---

* The defense industry eventually persuaded Congress and the Pentagon that the TPP concept was "unworkable" and the policy was dropped after McNamara's departure to the World Bank in 1968. As far as military aircraft are concerned, the Pentagon now operates a "fly before we buy" policy under which one or two prototype planes are built and extensively tested before the main contract is awarded.

the U.S. military establishment, the industry that serves it and Congress, which sanctions the wherewithal.

THE C-5A CONTRACT that Lockheed, Boeing, and Douglas all competed for, called for the invention of an aerial Goliath. The U.S. Air Force wanted a plane that could carry prodigious loads over considerable distances and operate in the most unfavorable weather conditions. Each C-5A would need sufficient capacity to carry enough military hardware—from tanks to helicopters—to equip an army division. On some missions it would be required to airlift 50 tons (U.S.) of payload 2,875 miles, land, and subsequently take off from a dirt runway of less than 4,000 feet in any weather and return to base without refueling.*

It must have been obvious to the rival manufacturers that meeting such a specification was going to demand considerable innovation in both design and engineering and, eventually, production techniques. And while the replacement of slide rules by computers as the main design-engineering tool has vastly improved the quality of predictions that designers can make about how well their "paper planes" will perform when actually built, long and sometimes bitter experience should have taught them—and aerospace executives—that there is many a slip between drawing board and production line. Dan Haughton, for one, has pointed out that plane-makers are in the business of taking risks: "Engineering is not an exact science and we have to invent as we go along." †

Yet, by bidding for the C-5A contract under the conditions laid down by the TPP concept, Lockheed, Boeing, and Douglas all made firm promises to build a plane which was at that stage little more than a gleam in the Pentagon's eye. What's more, they were promising to do it within strict time and cost limitations.

Boeing's private estimates showed that the likely cost of developing and then building, 115 C-5As could be around $2,800 million. But the bid the company actually submitted was for $2,300 million—*half a billion dollars lower*.‡ Boeing knew but did not reveal that it had no more than a 30 percent chance of fulfilling the contract at that price. In other words, in an attempt to win the contract, Boeing was prepared to take an almighty gamble, with astronomical stakes, when on its own calculations the odds were three-to-one against.

Douglas submitted a bid of $2,000 million. We don't know what Douglas thought the odds were of pulling it off at that price but we do know that the Pentagon, in an ungenerous mood, had privately estimated that the

---

* For comparison, the airbus version of the TriStar carrying its maximum payload of 43½ tons needs 7,600 feet of concrete runway to get airborne, 5,300 feet to land, and can fly only 2,683 miles before refueling.
† *Fortune,* August 1969.
‡ Senate Armed Services Committee, May 1969.

C-5A program would cost *at least* $2,200 million.* And it must be said that Douglas was making its C-5A promises at a time when its production lines were under considerable strain and DC-8s and DC-9s were being built late at and a loss.†

Lockheed's bid was for $1,900 million—$100 million below the Douglas bid, $300 million below the Pentagon estimate, and $400 million below Boeing's chancy offer. To this day the company insists that the bid was genuine and included a hoped-for profit of $177 million. Haughton has said: "I approved our bid with confidence. I sincerely felt we could design and build the airplane at the price we quoted." ‡ It is now a matter of record that C-5A costs overran by almost $1 billion—$247 million of that paid by Lockheed and the balance by the taxpayers. For that money the Air Force received only 81 of the 115 planes it originally ordered, and none of the 81 C-5As will be able to achieve the "flying life" which the Air Force had hoped for (30,000 hours) unless modifications costing $1 billion are made to the wings which have developed fatigue problems. The question that remains is whether Lockheed simply got its sums hopelessly wrong, or whether Haughton and his fellow executives were guilty of woeful mis-management.

The defense advanced by Lockheed, and by the company's few consistent supporters in Congress and the Pentagon, usually went something like this. Development of the C-5A posed formidable technical problems which were bound to take some time to solve. Unfortunately, the TPP contract demanded that production of the first fifty-three planes (Production Run A) must begin while five prototype models were still being tested and evaluated—in other words, before all the bugs had been ironed out. This involved Lockheed in a great deal of expensive modification work during Run A and costs, naturally, overran. Eventually the technical problems were solved. And, under the terms of the TPP contract, cost estimates for producing the remaining fifty-seven planes (Production Run B) were revised with new "target" and "ceiling" prices. All would have gone tolerably well but for the Air Force's arbitrary decision—under enormous political pressure—to cancel thirty-four planes.**

It is certainly true that the technical problems were formidable. The C-5A was conceived as a plane that could fly into combat zones where its survival might well depend on its ability to be loaded and unloaded quickly and simultaneously. The fuselage therefore had to open at both ends and the plane had to be able to "kneel down" on its undercarriage to near-ground level to allow helicopters and tanks to drive straight on and off. In addition, in order to give the C-5A its phenomenal short-landing capability, Lockheed

* Senate Armed Services Committee, June 1969.
† See Chapter 3.
‡ Senate and House Armed Services Committee, June 1969.
** Dan Haughton testified along these lines to the House Committee on Banking and Currency, July 1971.

had to develop wings that could flex by as much as twenty-five feet at the tips. To cope with cross winds during landing, the designers had to come up with an undercarriage that would allow the fuselage to swivel head on into the wind while the landing gear followed the runway. It is also true that the C-5A met all of its specifications, eventually. But how far should Lockheed be congratulated for achieving what it had *promised* to do— and then only in partial measure? The C-5A's specifications were clearly laid down in the contract and, presumably, nobody tried to persuade Lockheed that meeting them would be easy. But while the problems were predictable, the cost—in terms of time and money—of finding solutions could not possibly have been and that fact provides the fulcrum of the case against Lockheed. To paraphrase Dan Haughton, making any new plane involves inventing as you go along, and the C-5A was, from its conception, far from being an ordinary aircraft. Yet, by accepting a contract which, in the company's own words contained "a stringent accumulation of guarantees, requirements and specifications" Lockheed gambled on being able to predict the unknown. It seems more than a little disingenuous that when the gamble failed the company should cry foul because the terms of its contract were "rigidly enforced." *

The Pentagon itself does not emerge unblemished from a rational examination of the C-5A affair. Lockheed's bid for the contract was originally rejected by the Air Force's Source Selection Board on the grounds that the estimates were probably too low. And, anyway, the Board preferred Boeing's proposed design for the plane. But, in the upper reaches of the Pentagon, the Lockheed bid looked irresistible, and the Board was overruled. The Pentagon has never satisfactorily explained why it believed that Lockheed could build the C-5A for $300 million less than its own experts had estimated.

When—between 1966 and 1968—the extent of the development problems began to emerge, the Pentagon again displayed dubious judgment by hushing things up. One of its civilian employees, Ernest Fitzgerald, a cost-control expert working for the Air Force department, eventually became so concerned—and frustrated—that in November 1968, he disclosed the situation to the iconoclastic senator from Wisconsin, William Proxmire. Proxmire set up a Congressional economic subcommittee to hear, in public, Fitzgerald's charges that Lockheed was attempting to conceal the ever-rising costs of the C-5A project and that very senior Air Force officers were assisting in the cover-up by telling outright lies. The resulting publicity and subsequent investigation into Fitzgerald's charges by the Government Accounting Office and the Securities and Exchange Commission prompted the Pentagon to cancel thirty-four C-5As—thereby "saving" $1 billion.† The Pentagon took its revenge on Mr. Fitzgerald by phasing out his job after deciding that his

* Written testimony of Haughton to House Committee on Banking and Currency, July 1971.
† "Saving" is hardly the correct word. The eighty-one C-5As which the Air Force got still cost about $1 billion more than the price Lockheed originally quoted for producing 115, and, as we have said, wing modifications will cost another $1 billion.

appointment as a permanent civil servant had been awarded through a computer error. It turned out that the computer had never erred before. Fitzgerald has since been reinstated on the orders of the Civil Service Commission and awarded three years back pay. He is currently suing the Air Force for punitive damages.

By the beginning of 1970 Lockheed was in very deep water indeed. In addition to the losses the company was suffering on the C-5A contract, it had also run into trouble on three other military programs.* In March Dan Haughton wrote to the Defense Department asking for $600 million in "interim financing" to keep the contracts going. He implied that if the money was not forthcoming, production of the C-5A would be halted and the U.S. Air Force would be left with nine planes: the rest were—in Lockheed's words—in "large bits and pieces." †

To many of Lockheed's critics, including Congressman William Moorhead of Pennsylvania, Haughton's threat smacked of blackmail. Representative Moorhead compared the company, not to a lion that had got itself caught in the kitchen, but to "an 80-ton dinosaur that comes to your door and says, 'If you don't feed me, I will die.' And what are you going to do with 80 tons of dead stinking dinosaur in your yard?" The Defense Department refused to pay up and the Senate's Armed Services Committee began investigating the disputed contracts. Lockheed's public image was not helped by evidence from Government accountants who revealed that although the C-5A was being built by the Lockheed–Georgia company, $181 million of the progress payments had been credited to the Lockheed–California company. The insinuation—strenuously denied by Lockheed—was that the money had been siphoned off from the C-5A to help pay for the development of TriStar. Lockheed claimed that the money was used to fund C-5A development work that was carried out at the California plant, but the explanation fell on many deaf ears.

TO ADD TO the company's troubles TriStar too was looking increasingly unhealthy by 1970. Lockheed's once commanding lead in the airbus race—in terms of both development progress and sales—had been obliterated. Under the inspired leadership of David Lewis, the Douglas division of the new McDonnell Douglas Company had geared up for production of the DC-10 in what must have been record time. Work on Ship One began at

---

* Two were TPP contracts—one from the U.S. Army for development and production of a rigid-rotor helicopter called the Cheyenne, and one from the Air Force to build the propulsion system for a short-range air-to-air missile, designed by Boeing, and called SRAM. The third contract was between Lockheed's shipbuilding subsidiary and the U.S. Navy for production of twenty-two ships at a fixed price. The Cheyenne ran into serious technical problems during development and a prototype machine crashed, killing the pilot. Shortly afterwards, in May 1969, the army canceled the contract on the grounds that Lockheed was "incapable" of correcting the helicopter's deficiencies.
† Lockheed press statement, March 1970.

Long Beach in July 1969, just fifteen months after "Mister Mac" had given the DC-10 his go-ahead.

And Lewis had also personally led an aggressive sales campaign which, in the same month, had captured a crucial order for the DC-10 from a consortium of European airlines.

Once it had been established that there were going to be two U.S. airbuses, what mattered most—in the long term—was not how many planes each manufacturer could sell in the initial stages of the sales competition, but how many airlines each could sign up. The logic was simply that once an airline was committed to either the DC-10 or the TriStar—and to the buildup of the necessary repair and maintenance facilities—the winning manufacturer was virtually assured of getting the airline's future airbus orders. There was also a bonus to be won in that an airline which chose, say, the DC-10, would then have a very positive incentive to persuade uncommitted airlines to follow suit in the hope that expensive maintenance facilities and stocks of spare parts could be shared. (When Northwest Air Lines ordered the DC-10, for example, it encouraged Japan Air Lines (JAL) to do likewise because both carriers fly between Japan and the United States and could share their respective facilities. JAL eventually ordered the DC-10.)

Lewis therefore devoted a great deal of energy in the summer of 1969, making four trans-Atlantic journeys, to selling the DC-10 to the European consortium KSSU, which represented KLM-Royal Dutch Airlines, Swissair, Scandinavian Airline System (SAS), and the French carrier Union de Transports Aeriens (UTA). He later told *Forbes* magazine: "The number of planes [36] they wanted to buy between them wasn't worth all the fuss. What made it important was that all the other big overseas airlines were watching."

Lockheed's rival campaign to sell the TriStar to KSSU was led by the company's president Carl Kotchian, who had proved himself to be no mean salesman by successfully handling many of the important negotiating sessions that led to the sale of TriStars to TWA and Eastern Airlines. But in Europe Kotchian found himself completely outgunned: McDonnell Douglas had an offer of *three* versions of the DC-10—the basic model very similar to the TriStar in performance, and two longer-range versions which could both fly 1,500 miles further. The basic DC-10, called the Series 10, has General Electric CF6 engines, which can deliver 40,100 pounds of thrust each, giving the plane a range of around 4,000 miles. The Series 30 has General Electric CF6-50 engines, delivering 51,000 pounds of thrust each and the Series 40 has Pratt & Whiney JT9D turbofans—similar to those which power Boeing's 747—which are rated at 50,000 pounds thrust each. The Series 30 and the Series 40 both have a range of between 5,500 and 6,000 miles. Lockheed was promising a long-range model but that would have to wait until Rolls-Royce could produce an engine with around 50,000 pounds of thrust—8,000 more than the RB-211 was due to deliver.

Back in California, McDonnell Douglas was clearly winning the production race too. On July 23, 1970, the first DC-10 rolled out of its hangar, powered by its own engines, and on August 29, Ship One made its maiden flight two weeks ahead of schedule. (The first TriStar got off the ground eleven weeks later on November 17.)

The only solace for Lockheed came from the Senate which, at the end of August, grudgingly sanctioned payment of $200 million to continue work on the C-5A. But that, of course, was $400 million short of what Lockheed had asked for and the company had to borrow money from its twenty-four banks to keep things going while it continued its argument with the Defense Department. The bankers chose that moment to display their lack of confidence in Lockheed by demanding collateral (mainly land and plant) for all the company's loans, which amounted by then to $350 million.

As Douglas had discovered in 1966, a corporation that is considered unworthy of credit by its own banks usually is doomed to a takeover or bankruptcy. To avoid those twin perils Dan Haughton had to repair Lockheed's credit rating, and he could do that only by settling, once and for all, the disputes with the Defense Department. Haughton sued for peace. Washington dictated the terms:

> (1) Lockheed would have to accept a "fixed loss"—in effect a fine—of $200 million on the C-5A contract. In addition, $47 million which Lockheed had spent keeping the program going—primarily in paying interest on short-term loans—would be "disallowed" and would have to come out of the company's pocket;
> (2) On the other three disputed defense contracts, Lockheed would have to accept losses totaling $237 million;
> (3) In return, the Air Force would provide the necessary funds to complete the C-5A program—but the thirty-four canceled planes would remain canceled.*

Leaving aside the question of blame for the C-5A disaster, the peace terms were hardly generous from Lockheed's point of view. Even after getting tax rebates, Lockheed would be left with a net loss for 1969 and 1970 of $119 million and the company's equity would be reduced, overnight, by one-third. But, with his urgent need to borrow $250 million to keep the TriStar program going, Haughton had little choice other than to settle. He agreed to the terms and then arranged for new credit agreements to be drawn up. Those agreements were just days away from being signed when Rolls-Royce delivered its mortal blow.

LOOKED AT FROM Lockheed's viewpoint, the bankruptcy of Rolls-Royce was an act of commercial piracy devised by the British Government to allow the disavowal of a contract which had been fiercely competed for and joyously entered into. Rolls had received considerable help and encouragement

---

* Letter from Deputy Secretary of Defense David Packard to Senator John Stennis, Chairman, Senate Armed Services Committee, December 30, 1970.

from the British Government, which provided 70 percent of the estimated development costs. After the deal with Lockheed had been signed, David Huddie, then managing director of Rolls's Aero Engine Division was rewarded with a knighthood. To all intents, Rolls-Royce would continue in business, thinly disguised as Rolls-Royce (1971). It would continue operating the same plants under much the same management, and it would continue work on all of the contracts that were deemed to be "in the national interest." The only major project not considered to be of interest to the nation was the RB-211 contract.

To Rolls-Royce's designers and engineers, the bankruptcy was an act of sheer insanity, precipitated not by the company's financial state—bad, in all conscience, though that was—but by erroneous reports on the technical merits of the RB-211.

It is a fact of life that aero-engines are unpredictable beasts that can only be developed empirically. No experienced engine designer expects other than blowups, burnouts, and similar disasters during the development of a new engine. The process, inevitably, is one of trial and error and, because the RB-211 encompassed so many unique features—and was twice as powerful as any engine Rolls had previously produced—the trials of its development were, from the outset, bound to be exceptionally severe.*

That kind of truth does not of course appeal to certain people, such as accountants who spend most of their time in a business environment where events can generally be predicted with some accuracy. Yet it was accountants, from the highly respected London firm of Cooper Brothers, who in November 1970 were hired to examine Rolls-Royce and review the RB-211 contract. The review and the appointments to the Board of Lord Cole and Ian Morrow were a precondition to the grant of the $143 million loan which Rolls had requested from the Government and its banks.

Cooper's report, completed in January 1971, showed that the RB-211 contract, once celebrated as a great triumph for Rolls, had been weighted overwhelmingly in favor of Lockheed. The price for each engine had been firmly fixed; with the subsequent escalation of the development costs, Rolls now stood to lose $264,000 on every engine it delivered. Even more alarming was the revelation that first production models of the RB-211, then being tested at Derby, were performing far short of the specifications laid down in the contract. The accountants forecast, at best, considerable delays in delivery to Lockheed which could make Rolls liable for $120 million in contractual penalty payments. (It is intriguing to speculate what the accountants would have made of the situation if "Mister Mac" McDonnell had achieved his ambition to acquire the RB-211 for the DC-10. His nephew Sanford McDonnell, now president of the company, told us that Rolls chose to go with Lockheed because "we were driving a harder bargain.")

To the British Government it looked as though the RB-211 project was

* See Chapter 2 and Appendix A.

a pit of unfathomable depth into which it was being invited to pour incalculable amounts of taxpayers' money. Frederick Corfield, the British Minister of Aviation, told Dan Haughton that there was no rational alternative to total abandonment of the project, unless Lockheed was willing to pay *all* of the future development bills and about 50 percent more for each engine produced.

However, events were to quickly prove that the accountants—unused to the empirical nature of aircraft engine development—had overreacted to the adversities they had witnessed on the engine test bed. On February 4, the day that the formal bankruptcy was declared, the main test engine at Derby ran to all of the parameters laid down in the contract.

In mid-February, Lord Carrington, then Minister of Defense, sent what he liked to call "my three ferrets"—aircraft engine experts—to reinvestigate the status of the RB-211. The ferrets, more familiar with the technical perspective, reported back in much more optimistic terms than the accountants had done.

By early March Lord Carrington was sufficiently reassured to offer Lockheed a deal. The estimate for completing development was now down to $288 million (from approximately $450 million) and the Government said it was willing to put up half of the money on condition that Lockheed provided the balance. In addition, Lockheed would have to pay about $340,000 more for each RB-211.

Haughton rejected the proposal, saying: "I have as much risk as I can say grace over now." * But at least the Government now obviously shared, in some measure, Haughton's desire to save the program. Prime Minister Heath agreed to supply $4½ million a week to keep development going while the two sides continued negotiating.

In between his meetings with the British, Haughton crisscrossed America meeting with suppliers, bankers, government men, and the heads of the three airlines, TWA, Eastern, and Delta Airlines, who stood to lose the $200 million they had put up as progress payments if Lockheed went bankrupt. (On March 18, Delta ordered five DC-10s, just in case the rescue operation failed, but it didn't cancel its order for twenty-four TriStars.)

Haughton found time to order some drastic pruning at Burbank: nine thousand workers were laid off; a pay increase due to 31,000 salaried staff was suspended; senior executives took a pay cut averaging 12 percent, and Haughton reduced his own salary of $153,000 by 25 percent.

In the third week of March, the British Government sent a delegation, headed by the Secretary of the Cabinet, Sir William Nield, to Burbank. After three days of negotiations, both sides removed to Washington where the British Ambassador Lord Cromer, and U.S. Treasury Secretary John Connally joined in the talks. On March 24 the British sent reinforcements, Lord Carrington and Attorney General Sir Peter Rawlinson, while in London Prime

* "The Salvage of the Lockheed 1011," *Fortune,* June 1971.

Minister Heath and his ministers monitored progress daily. President Nixon was kept equally well informed by Connally.

By the beginning of May there were few areas of disagreement left. At Derby the first production engines were running beautifully and were almost ready for delivery to Burbank. TWA, Eastern, and Delta were ready to re-affirm their confidence in both the engine and the TriStar by agreeing to pay out around $100 million in extra advance payments. Even Lockheed's lending banks had provided an overdue fillip by lending the company an additional $50 million—albeit against a pledge of what little negotiable collateral the company had left.

The only possum that remained untried was the U.S. Government. On May 10 Lord Carrington told the British Parliament that the Government would provide $240 million to complete development of the RB-211. In return, he said, Lockheed had agreed to buy 555 engines at $1,020,000 each—an increase over the original contract price of $180,000 per engine. (In their turn, all of the TriStar's customers had agreed to pay an extra $643,000 per plane to cover the increased cost of the engines and the cost of the six-month delay in production which the Rolls-Royce crisis caused.) Lord Carrington emphasized, however, that the deal was conditional on Lockheed's ability to raise yet another $250 million to guarantee production of the TriStar. Lockheed's lending banks had promised to put up the cash, but that deal was also conditional. It depended on a guarantee from the U.S. Government, no less, to repay the money if Lockheed went broke. On May 13, President Nixon sent the necessary draft legislation to Capitol Hill, urging that it be made into law as soon as possible.

TriStar's fate—and the fate of Lockheed—now depended on the will of Congress.

THE U.S. AIRCRAFT industry has an unwritten convention which dictates that when one of its number runs into trouble, rival manufacturers will re-frain from exploiting it, at least, publicly. No matter what has caused the crisis, the troubled manufacturer will remain—in the official view of its competitors—a fine company, building fine planes. It is permissible, of course, for a rival to point out that its planes are just a little bit finer but that, supposedly, is the limit. (The philosophy behind the convention was explained when Dan Haughton refused to discuss the DC-10 crash with the authors: "Hell, next year we might be the ones with the problem.")

Consequently, while just about everybody else with an opinion joined in the debate over President Nixon's draft bill, McDonnell Douglas, which had the most to gain from the demise of TriStar, maintained a discreet cor-porate silence. The company's stated policy was that it had "no comment" to make on Lockheed's dilemma, and "Mister Mac" issued a warning to his people: "There will be no dancing on dead men's graves." *

* *Forbes*, June 1971.

Robert Gross

Courtland Gross

Alabama farm boy Daniel Haughton (previous page) began his career with Lockheed Aircraft in 1939, the same year James McDonnell started in business. Haughton's executive potential was soon spotted by the Gross brothers—who in 1932 had resurrected Lockheed from bankruptcy—and he was carefully groomed to become chief executive of America's biggest defense contractor. Lockheed's prosperity and technical reputation were mainly based on military products like the Hercules transport, but it also built airliners until the commercial failure of the Electra Two in 1959 drove the company out of the civil market for a decade. Haughton's attempts to break back into that market with a $1-billion investment in the L-1011 TriStar signaled the start of the great airbus race—and brought Lockheed to the verge of a second bankruptcy.

Electra One

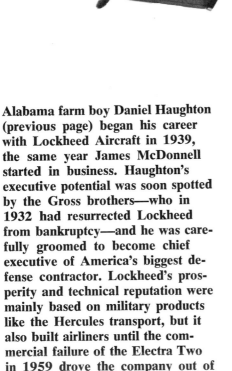

Electra Two

The L-1011 TriStar—
Lockheed's entry in the Great Airbus Race

C-130 Hercules military transport

Constellation

William Boeing   **William Allen**

The Boeing 747—
the first, and biggest, jumbo

In 1935 Boeing's 247 airliner was overwhelmed by the DC-3 and for the next
twenty-five years Boeing had little success
in the commercial market. Instead the
company concentrated on advancing the
state of the art and eventually produced
jets like the B-47 and, later, the KC-135
tanker, the "father" of the 707. With its
family of jets Boeing has turned the tables
on its old rival and become the dominant
commercial plane maker. Surprisingly,
neither the founder, William Boeing, nor
Bill Allen, who succeeded him, were
engineers.

Boeing 247

KC-135 tankers

B-47 nuclear bomber

B-707

B-727

B-737

Boeing, Douglas, and Lockheed all competed for the contract to build the huge C-5A transport for the U.S. Air Force. Lockheed won but found, because of technical problems, the experience almost as disastrous as winning Helen was for Troy. Nevertheless, the C-5A competition gave all three companies their first experience of jumbo-jet technology, and for Boeing it led directly to development of the 747. At one time Douglas planned to produce a civilian version of its unsuccessful C-5A entry, which would have been a double-decked DC-10 with private compartments, a dining room, and nursery. The practicalities of airline economics produced the more modest DC-10 and its almost identical rival, the TriStar. From the outside the only readily visible difference between the two aircraft is the position of the tail engine.

Douglas's first concept of the DC-10, based on its entry for the C-5A competition

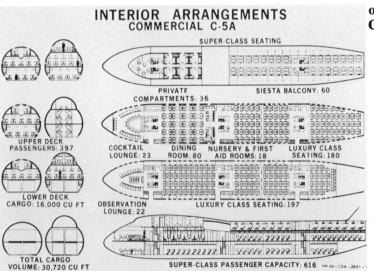

INTERIOR ARRANGEMENTS
COMMERCIAL C-5A

SUPER-CLASS SEATING

PRIVATE COMPARTMENTS: 36   SIESTA BALCONY: 60

UPPER DECK PASSENGERS: 397

COCKTAIL LOUNGE: 23   DINING ROOM: 80   NURSERY & FIRST AID ROOMS: 18   LUXURY CLASS SEATING: 180

LOWER DECK CARGO: 16,000 CU FT

OBSERVATION LOUNGE: 22   LUXURY CLASS SEATING: 197

TOTAL CARGO VOLUME: 30,720 CU FT

SUPER-CLASS PASSENGER CAPACITY: 616   PR 65-C5A-2621-1

The C-5A, called by Lockheed the Galaxy

Back to back, a DC-10 and a TriStar

The TriStar was the first of the wide-bodied jets to crash and kill passengers. On December 29, an Eastern Airlines TriStar flew itself into the swamps of the Florida Everglades, killing 101 of the 176 people on board. In 1971, a Boeing 747 crashed into a pier at San Francisco International Airport during takeoff, but the pilot managed to get airborne and, an hour later, land with all of his passengers alive. The 747's impeccable record was finally spoiled in November 1974 when one crashed at Nairobi, Kenya, killing fifty-nine people.

But while the corporation was constrained by convention, its employees—especially some of its executives—were not. Stephen P. Dillon, who was in charge of special procurement for the DC-10 program, and Donald Malvern, a vice-president at McDonnell Douglas's St. Louis headquarters, both acting as "private citizens," took on the task of writing to congressmen, and argued passionately against the loan bill. Dillon wrote: "The question now is whether the United States and its economic system will go down the drain of history in order to preserve for only a short time a mismanaged company (Lockheed) and a mismanaged program (TriStar)." * Malvern argued that if the Government backed Lockheed it would have a vested interest in persuading the airlines it regulated to buy TriStars rather than DC-10s.† (Malvern now says that he acted out of self-interest; he owned McDonnell Douglas stock and appreciated that the demise of Lockheed could only improve its value. "Mister Mac" sent him a note suggesting that he "become less visible and shut up.")

Fred Carlin, a buyer in the procurement department, distributed copies of a fifteen-page letter, opposing the loan guarantee, to Douglas suppliers—some of whom were also Lockheed suppliers—urging them to sign it and send it off to Washington.‡

And even official company spokesmen engaged in a little surreptitious backstabbing, according to *Business Week* magazine. Reporting that the official response to questions was "no comment" the magazine went on to say:

> Practice has been somewhat different from policy, however. The same spokesmen . . . speak freely "off the record" about the Lockheed affair. "There's no way Lockheed can make money on its airplane in the next four years," one spokesman told *Business Week* recently. "The L-1011 is never going to make a nickel." **

It is, of course, hardly surprising in the circumstances that the normal conventions of the industry should have been so comprehensively abandoned. McDonnell Douglas stood to gain 100 immediate orders for DC-10s worth approximately $1,500 million if the TriStar was canceled. And McDonnell Douglas would have been left with a monopoly of the tri-jet airbus market which, on the most conservative of estimates, will be worth $7,500 million by 1980.

What was questionable about parts of the campaign against the TriStar was that it was furtive and that it traded in some dubious half-truths. Without doubt, the most serious attacks against Lockheed were contained in a dozen carefully prepared papers which began circulating in Washington in the spring of 1971. The papers, which became known as the "fact sheets,"

* *Business Week,* July 31, 1971.
† Letter dated May 11, 1971, from Donald Malvern to Representative Henry S. Reuss, Democrat, Wisconsin.
‡ *Business Week,* July 31, 1971.
** *Business Week,* July 31, 1971.

gave no indication of authorship but they were expertly argued and uncompromisingly critical, and they were quickly seized upon by Lockheed's congressional enemies. They were written without the knowledge or blessing of McDonnell Douglas, but their content strongly suggests that a considerable amount of the information came from inside the Douglas plant at Long Beach.

In attempting to destroy Lockheed's already shaky credibility, one of the papers, entitled *L-1011 Technical Status,* suggested that the TriStar might be unsafe and described the "all-flying" tail system (actually a safety point—see Chapter 6) as a hazard:

> The L-1011, undoubtedly influenced by the predominantly military aircraft background of Lockheed, has departed from proven safe commercial transport design in several areas.
> Lockheed has chosen a military fighter-type one piece horizontal tail, when the safest commercial transports in the world use conventional redundant horizontal stabilizer and elevators. Such a one-piece design gives the pilot a control that can easily overstress the airplane and a system difficult to make fail-safe. Certification for this is still unsettled within the Federal Aviation Administration.

The point about certification was spurious because TriStar had not been presented for certification at that stage. (Neither, for that matter, had the DC-10.) The paper went on to cast doubt on what it called "other questionable commercial design practices":

> The design of commercial airplane structure for long life is a highly specialized technology radically different from military aircraft. Lockheed has not demonstrated full knowledge of that technology. For example, the L-1011 wing-fuselage intersection design is conducive to introducing residual stresses during manufacture. The stringer-frame-skin joint creates stress concentrations. Wing carry-through structure tends to induce high loads in fuselage. All these contribute to reduced fatigue life and, not surprisingly, several failures have already occurred during ground fatigue testing.
> L-1011 makes extensive use of bonded structures in primary structure even though in today's operating aircraft most corrosion problems have occurred in the bonded structure.

Earlier, the "fact sheet" had cited Rolls-Royce's development setbacks and raised another alarming prospect:

> Should Rolls-Royce not be successful (in preventing an increase in the weight of each engine) additional total airplane overweight problems could develop. Wing flutter—a dynamics problem—is increasingly aggravated by increases in engine weight and could require major wing redesign. Only high speed flutter testing will determine this.

These claims were, of course, aimed at laymen—senators and representatives—who stood little chance of being able to understand, let alone evaluate, the merit or otherwise of the technical argument. But the message was clear enough: On top of everything else, the TriStar might have to be redesigned, might not get certificated, and could even be unsafe. By way of emphasis the "fact sheet" went on:

> The certainty of airline service structural problems cannot be stated now. That remains to be seen but it should be remembered that Lockheed has had major structural problems on their last commercial transport, the Electra, and their most recent military transport, the C-5A.
> At present less than 300 hours of the original planned 1,695 hours of flight is still to come and the problems to be uncovered are still unknown. . . . A modern transport airplane is a complex product and the best of engineering staffs are imperfect. The program and technical management has been changed extensively during the critical creative period of the L-1011 and the last new passenger-carrying airplane that Lockheed certified was the Electra, thirteen years ago. There is a sizable unknown technical risk remaining in the L-1011—perhaps one as critical as the dynamics problem which resulted in airline losses of three Lockheed Electras through wing failure. . . .

In setting out the economic arguments against TriStar, the "fact sheets" were designed to appeal to all men. For Californians there was the paper which claimed that total employment in the state would *increase* by between 5 percent and 10 percent if TriStar were canceled because more of the bits and pieces which go to make up each DC-10 are produced, by subcontractors, within California's borders.

For those interested in the national impact there was the promise that cancellation would lead to more U.S. jobs because "the L-1011 airplane has a high foreign labor content"—besides British engines the TriStar has British-made engine nacelles and Japanese-made doors. (This point had not been lost on the DC-10's engine-makers, General Electric. When the Rolls bankruptcy was announced GE and Pratt & Whitney both moved quickly to try and recapture the order that had been lost to the British in 1968. Lockheed and the main TriStar customers in effect reran the engine competition but decided that, despite all of the problems, the RB-211 was still a better bet than its U.S. competitors. P & W accepted the defeat philosophically but GE reacted fiercely with the company's president, Fred Borch, firing off angry letters to the White House. GE proposed what it called "an American approach" to the crisis, namely, that Lockheed should get the loan guarantee but only on condition that American engines were used to power the TriStar.)

The "fact sheet" authors admitted that some Lockheed workers would suffer—"employment discontinuities would be inevitable"—but they predicted that most of those laid off would eventually get a job with McDonnell Douglas as it expanded to cope with the new rush of DC-10 orders.

On the other hand, they warned, if Lockheed stayed in the commercial plane business, the consequences would be disastrous for everybody:

> Lockheed has . . . created a major competitive impact with the L-1011 but the effect has been largely destructive. . . . In cooperation with Rolls-

Royce the business tactic adopted, either by ignorance or by deliberate management design, was to undercut prices and terms—essentially a loss-leader,
competitor-breaking policy; an attempt to dominate a market by deliberately
selling at a loss. . . .

This underpricing competitive strategy, compounded with apparent major
cost miscalculations, has bankrupted Rolls-Royce forcing nationalization;
brought Lockheed to mandatory reorganization; lured U.S. airlines into
advancing funds which may well be lost and, most important of all from a
national viewpoint, largely removed the future profits from the entire commercial aircraft manufacturing industry. . . .

The claim that Lockheed had invaded sovereign territory, as it were, and
damaged the industry was, perhaps, the most dubious and certainly the most
self-righteous argument advanced in the "fact sheets"; it ignored the inconvenient truth that the TriStar was conceived at a time when it appeared that
Douglas was largely bankrupt of money and ideas and when the airbus
market seemed to be wide open.

While neither the industry nor society could afford the resources which
went into creating two such similar and vastly expensive products as the
DC-10 and TriStar, it is also probably true that by 1970 most of the damage
had been done. By any rational measure, the time to cancel one or another
of the projects was when both were still on the drawing board—before scores
of thousands of jobs and billions of dollars had been committed.

Ultimately, the major issue was whether the United States Government
should intervene to save jobs which were threatened because a commercial
adventure had come unstuck. It was not as unique a proposition as some
of its critics liked to pretend. After all, the rescue of Douglas was partially
financed by a V-loan which is nothing more or less than a bank loan guaranteed by the Government.

And it was the threat to jobs which finally decided the matter. On August
2, 1971, the Senate met to vote on President Nixon's proposed bill. Three
days before, the House of Representatives had approved the legislation but
only by three votes and the Senate was even more equally divided.

Senator Alan Cranston of California, Lockheed's congressional champion,
had been lobbying his colleagues for days, and he calculated that the vote
would go against the loan by 48 votes to 49. As the roll was being called,
he therefore set out to convert one of the "no" votes belonging to a Democratic senator from Montana, Lee Metcalf.

At first Senator Metcalf refused to believe that his vote would be decisive.
But as each senator's name was called off, he saw that Cranston's list of
predicted aye and no votes had been completely accurate. With just half a
dozen votes to go Metcalf told Cranston: "I'm not going to be the one to
put those thousands of people out of work." And he stood up and voted aye.*

The news was announced over loudspeakers at Lockheed's plants. That
night people celebrated by dancing, quite literally, in the streets of downtown

---

* *The New York Times,* August 5, 1971.

Burbank. The company made a movie, called, *Dan's Days at Lockheed* to pay tribute to his achievement.

And people celebrated in Derby, too, fully aware of the irony that they owed their jobs and the survival of Rolls-Royce as a front-ranking engine manufacturer to the sheer tenacity of an American.

Dan Haughton's view of his possum hunt is that he did whatever was necessary to save the corporation to which he dedicated thirty-six years of his life, receiving in return, of course, money, power, and success. His philosophy was, perhaps, best summed up by a framed memento which hung on the walls of his Burbank office. It contained the photograph of a cat clinging precariously to its perch on the branch of a tree. The caption read: "Hang in there, baby."

IT WAS THE revelation that Lockheed had been bribing foreign government officials to buy military and commercial aircraft, including TriStars, that finally dislodged Haughton from his perch. On Friday, February 13, 1976, Haughton and the company's president Carl Kotchian resigned saying: "We have decided that the time has come when the fortunes of this great corporation and its people will be best served by a change in its top management." The two men admitted that between the beginning of 1969 and February 1974, they had sanctioned bribes or, as they put it, "questionable payments" of $22.4 million. In doing so, they had been following a Lockheed tradition that goes back to at least 1962 when, in Kotchian's words, a "high Dutch official" was paid $1 million as part of Lockheed's successful campaign to sell F-104 Starfighter jets to European air forces. (The official was alleged to be Prince Bernhard of the Netherlands, who was also accused of receiving a further $100,000 as part of Lockheed's unsuccessful campaign to sell Tri-Stars to the KSSU Consortium, which includes the Dutch airline KLM. Prince Bernhard strenuously denied receiving any money. At the time of writing the Dutch Government was still investigating the allegations.)

One particular bribe of $2 million that Lockheed paid to Japanese Government officials is an important part of the story of the DC-10 crash, and we shall return to it later. However, it should be said now that Lockheed was not unique among U.S. aircraft manufacturers in employing corruption as a sales aid: McDonnell Douglas and Boeing have both admitted making "questionable payments" to foreign government officials to sell commercial jets.

Haughton's personal martyrdom was, of course, very much in character. As Courtlandt Gross—now living in retirement in Pennsylvania—said, Dan has always considered his own interests to be subordinate to those of the company. It was not too painful a sacrifice either because he was due to retire, in any event, in September 1976 and now gets a pension of $65,000 a year. He will also act as a consultant to Lockheed, at an undisclosed fee, for the next ten years.

8

The Ordeal of Bryce McCormick

Nearly every accident contains evidence which, if correctly
identified and assessed, will allow the circumstances and the
cause to be ascertained so that corrective action can be
undertaken to prevent further accidents.
—INTERNATIONAL CIVIL AVIATION ORGANIZATION
ANNUAL DIGEST OF AIRCRAFT ACCIDENTS, 1974.

AT ABOUT 7:20 in the evening of June 12, 1972, American Airlines Flight
Number 96 took off from Detroit Metropolitan Airport in Wayne County,
Michigan, bound for Buffalo and eventually New York. Despite the season
the plane, a DC-10, was lightly loaded with only fifty-six passengers, a little
freight, and a corpse, traveling in a coffin in the aft baggage compartment,
located under the floor of the rear passenger cabin. First Officer Peter Paige
Whitney, aged thirty-four flew the takeoff. The Captain, fifty-two-year-old
Bryce McCormick, handled the radio communications, first with the airport
control tower and then, as the plane climbed out over Lake Erie, with the
air traffic control center at Cleveland. When the DC-10 had reached six
thousand feet, Whitney switched on his autopilot, but the plane was still
flying through clouds and out of habit he kept his hands on the control
column. At ten thousand feet, as a matter of routine, Whitney lessened the
plane's rate of climb to bring the air speed up gradually from 250 knots to
340 knots. Above, through the thinning clouds, he could just see the first
evidence of the sun.

Back in the passenger cabin, the "fasten seat belt" and "no smoking" signs
had been turned off. In the first-class compartment Al Kaminsky and his
friend Hyman Scheff left their wives to play gin rummy in the forward
lounge. With so few passengers on board, the eight flight attendants were in
no hurry to begin their chores. Cydya Smith, the chief flight attendant (whom
American Airlines likes to entitle the "First Lady"), went to the service
center to begin making coffee, but the others remained in their seats, chatting
or reading.

At about 11,500 feet, and five minutes or so after takeoff, the DC-10 finally broke through the last of the clouds. Captain McCormick could see a Boeing 747 far above them and said: "There goes a big one up there." Whitney leaned forward, close to the windshield, to get a better view. At that moment, two miles above Windsor, Ontario, the plane revealed a basic design flaw that had been built into every DC-10.

McCormick heard a loud bang. To Whitney it sounded like a book being brought down on a table very solidly. To the flight engineer, Clayton Burke, it sounded like the thud of a human body hitting the ground.

The rudder pedals under McCormick's feet sped, with great velocity, in opposite directions—the left pedal down to the floor and the right upwards, jamming McCormick's right knee against his chest. Simultaneously, a cloud of charcoal-grey dust particles, collected from every crack and crevice in the cockpit, sprayed into McCormick's face, temporarily blinding him. The earphones were torn from the back of his head.

Whitney, leaning forward to get a better view of the high-flying 747, was spared the worst effects of the dust storm. But the three throttle levers—one for each of the DC-10's engines—had moved back, unaided, to an almost idle position and, robbed of its power, the huge aircraft decelerated: Whitney was hurled backwards and his head hit the back of his seat.

The door leading from the flight deck burst open, releasing the dust storm and with it the crew's hats, which flew through the first-class compartment in head-height procession. All of the passenger cabin was filled with a damp, white rolling fog which seemed to pour from the ceiling and from the galley deck below. It enveloped everything.

Two metal hatches in the cabin floor, normally concealed by carpet, were thrown into the air and one of them struck Mrs. Kaminsky in the face. The wound began to bleed profusely. Parts of the cabin ceiling fell down and one section hung suspended from an electric cable.

In the rear passenger compartment, empty on this flight except for two flight attendants, the cabin floor partially collapsed into the cargo hold below. A circular cocktail bar at the back of the plane fell into the cavity. So, briefly unconscious, did one of the flight attendants, Beatrice Copeland.

The other woman, Sandra McConnell, found herself fighting for balance on a section of the floor that was progressively giving way. She felt as though she were slipping down into the baggage compartment and towards a great tear where the rear cargo door had been in the side of the fuselage. The coffin and the corpse it contained had already fallen through that hole and were tumbling two miles to earth. Sandra could see the clouds rushing by beneath her feet.

Most of the other flight attendants were lifted out of their seats. Cydya Smith, the "First Lady," hung onto a grab rail in the service center, experiencing the sensation of weightlessness. The wounded Mrs. Kaminsky and her friend Mrs. Scheff were hysterical, both convinced that they were going to die. Mrs. Scheff screamed repeatedly that she would never see her children again.

Unaware of these particular dramas—and of the hole in the fuselage—the flight crew fought to regain control of their machine, which had yawed savagely to the right. The autopilot had disengaged itself. As McCormick's vision returned he yelled, "Let me have it," and took over the controls from Whitney. The DC-10's nose fell steadily towards the horizon.

THE HULL OF a jetliner is, like that of a submarine, a pressure vessel. In a wide-bodied aircraft flying at high altitude, the fuselage walls must contain a pressure of some 20,000 tons, and the fact that this is normally done without incident is no small tribute to the resourcefulness and cunning of aeronautical engineers.

Early aircraft, of course, were not pressurized, and most small, light aircraft are not even today. The most immediately noticeable difference between a pressurized and an unpressurized aircraft is that flying in the former does not make one's ears hurt when it rises or descends a few thousand feet; however, the changeover to pressurization was not made simply to relieve discomfort in the ear chambers of passengers.

Above eleven to twelve thousand feet, the height at which airplanes like the DC-3 and DC-4 cruised comfortably, human life becomes progressively harder to sustain, for the thin air provides too slight a concentration of oxygen. One problem about staying at ten thousand feet is that mountain ranges like the Rockies rise a good deal higher. But the significant point is that, even though humans don't like it, airplanes perform better at higher altitudes. The thin air produces less resistance, while the thirst of the engines for oxygen is readily accommodated by increasing the rate of air intake. It emerged as early as the thirties that, whereas it was relatively easy to force-feed an engine with oxygen (the super-charger system) the same thing was not quite as easy with the human body. Oxygen masks were all right as an expedient for military pilots, but not for airtransport passengers.

The essentials of pressurization are based simply enough on the constant-leak principle. Engine-driven pumps drive air into the cabin at a rate faster than that at which it leaks away through the minor apertures and imperfections of the hull. Cabin booster pumps were already powerful and efficient in the forties. In April 1944, USAF Major Charles Hanse, taking the first B-29 bomber into air combat, watched bursts of fire from a Japanese "Oscar" fighter stitching holes in his cabin wall and simply turned up the cabin boost rate to maintain pressure. In the hijacking era, the idea has spread that a bullethole through a pressurized hull will lead to disaster; in fact, small holes are easily tolerated.

Building up pressure is no great problem, but making sure that the hull can withstand its effects is not so easy. Our sea-level atmospheric pressure of fourteen pounds per square inch does not sound much, but it becomes more impressive when it is expressed, correctly, as nearly one ton per square foot. We do not notice the pressure because there are no "differentials" in our ordinary existence. Pressure on one surface of any object is compensated by equal pressure on the other surfaces. An airliner flying high has, of course, *some* air pressure outside its hull, but in order to retain near sea-level pressure within, it will have to endure a pressure differential between seven-and-a-half and nine pounds per square inch.

Generally, the strength that must be built into the hull to withstand flying loads more or less meets the pressure demands. But there are special difficulties, such as "hoop stress." With high pressure inside and low pressure outside, the long tube of the airplane hull distends slightly. On the ground, with pressure equalized, it relaxes. The succession of tension and relaxation complicates the fatigue patterns in the metal of the fuselage, the problem, mentioned in Chapter Three, that destroyed the early Comets.

Small punctures in a pressurized hull, such as bulletholes, only accelerate the rate of the constant leak. But a large hole, rapidly appearing (the exact lethal size depends upon fuselage diameter and pressure differential), causes an outward acceleration of air so rapid as to resemble a bomb explosion. Aircraft doors, which may open in flight, therefore naturally represent a potential hazard, and consequently a classic problem in pressure hull design.

There is a simple solution: make each door a "plug" door. A plug door opens inwards and is larger than its frame. When it is closed, the pressure differential seats it more and more firmly into its frame. Passenger doors in most commercial planes are plug doors, and there is no way in which they can fly open at high altitude. However, plug doors do have disadvantages which become very severe if a large door is required. Plugs are necessarily weighty objects because a safe plug has to be totally rigid. (The plug principle is often illustrated by a bathtub plug, held in place by water pressure in the tub. In the same illustration, however, it can be seen that if the plug is too soft or flexible, the water will drive it clear through the hole.) And in modern aircraft, the flight loads are transmitted by the fuselage shell. There is no underlying chassis or skeleton. If a big hole is cut in this skin, then strong reinforcement must be built around its perimeter to carry the flight loads around and beyond the hole.

A hull full of large plug doors will soon become prohibitively heavy. Yet the economic value of a large aircraft will be seriously impaired if it has only small doors. In particular, this applies to cargo carrying, where the plug door has the further disadvantage that, by opening inward, it seriously lessens the amount of cargo space available. The problem is at least as old as the pressurized transport itself. For instance, in the KC 135, jet fuel tanker, which was the model for the Boeing 707, it was necessary to provide doors which were too big to be made into plugs.

A layman may fairly ask why a door should matter so much. Clearly, it is bad in principle for pieces to fall off an airplane in flight because there is always the risk, for instance, that they will strike and damage the control surfaces. But flaps, undercarriage doors, and other items fall off aircraft quite often without cataclysmic results. Indeed, any well-built airframe should, in principle, hold together for a while even with several large holes blown in it.

The trouble is that loss of a door—depending how the aircraft is designed—may cause damage far away from the original explosive rupture, and potentially much more serious. Most aircraft hulls are divided up into different chambers: flight deck, passenger cabin, cargo hold, instrument bays, and so on. Obviously, the main hull walls must be able to withstand all the differential pressures likely to be met with in flight. But how many of the internal partitions need to be equally strong?

If there is a partition between flight deck and cabin that is roughly as strong as the average automobile roof, sudden depressurization—say, through a passenger door flying open in the cabin—will certainly burst it. The obvious solution, one might think, would be stronger partitions. But any increase in strength inevitably costs weight which soon begins to add up to hundreds of pounds. So the usual approach is to make sure that doors will *not* fly off the pressure hull. Over the past twenty years, much ingenuity has been devoted to this end.

ONE SOLUTION TO the problem was provided by the development of a "tension-latch" door, which Boeing originally designed for the KC-135 tanker, and which, in refined versions, has been used in subsequent Boeing aircraft up to and including the 747. Unlike a plug door, which carries no flying loads, a tension-latch door is so designed that when in place it twists and bends with the rest of the airframe. It can thus be reasonably light, even though it takes out a very large segment of the hull diameter. The big Boeing door is hinged outward at the top and swung down against a rubber seal. It is then pulled shut by a hydraulic power unit.

On the bottom edge of the door area are a series of "C-latches." A C-latch may be thought of as a hard steel tube, which for about one inch of its length, has had half its barrel cut away, so that the section which would originally have looked like a thick-walled "O" comes to resemble a letter "C." (Imagine a rather stark piece of typography in which "C" is an exact semicircle. For diagrams see pages 296–303.)

When the door is right down against its seal, each C-latch fits up against a round steel bar (its latch spool) attached to the doorsill in the hull proper. Then an electric motor revolves all the C-latches together through a half-circle, so that the "belly" of the C in each case slides around to the other side of the bar, or spool. So long as the C-latches are in place, the door cannot open. Indeed, internal pressure will merely increase the difficulty of moving the latches.

Problems remain, however. Should the electric motor work poorly, the C-latches might not revolve through a full semicircle, which would make it easier for them to slip loose later. Even without that, the strains and vibrations of flight might cause the latches to "creep" loose. In order to prevent such things happening, Boeing devised a locking system. Just above the row of C-latches, they mounted a rod on the inside of the door that can slide fore and aft, parallel to the doorsill, latches, and spools. The rod carries a locking pin for each C-latch. While the C-latches remain "open," the road cannot move, for there are metal lugs on each latch against which the locking pins will jam. But when the latches turn through their half-circle movement to grip the spools, the lugs turn with them. And once the movement is complete, the locking pins can slide across behind the lugs. This horizontal movement of the locking tube is achieved, through several linkages, by the vertical swing of a locking handle on the outside of the door. Large amounts of leverages are involved here. In order to obtain a locking-pin travel of about an inch, the door lever moves through a long sweep. The forces involved thus make the movement of the linkage irreversible except to an equivalent and opposite force applied with equal deliberation.

This all seems neat enough. The person closing the door—normally an airport baggage handler—first operates a series of power controls, closing and latching the door. With those cycles complete, he pulls down the manual locking lever. If it refuses to go all the way home, he knows there is something wrong (for instance, if C-latches are only partially home, it would result in the locking pins being stopped by the lugs.) The plane cannot be cleared for takeoff until the handle is all the way down into its slot.

But the system is not yet infallible. Suppose something were to break in the linkages under the skin of the door. The handle might go all the way home, yet the locking pins would not move. Therefore, a plane *could* take off with the locking handle fully stowed, but with its C-latches only partly closed. Unglamorous as the work sounds—and indeed is—the whole business of maintaining human life in the air comes down to thinking and rethinking about curious and fiddlesome problems of this order. The Boeing engineers devised two further levels of warning. First, in the middle of the main door a small vent door was made, opening inwards—a miniplug, as it were. It has two functions. After every landing the vent doors are opened to allow the escape of any pressurized air that might be trapped inside the fuselage. However, and more important, before every takeoff the vent door serves as a sentry. With the locking handle in the "open" position, this little door stands open, making a fair-sized opening in the main-door surface. It is closed by a series of linkages driven *not off the locking handle itself,* but remotely, *through the rod carrying the locking pins.* These linkages are attached to the "far end" of the locking-pin rod. If the rod itself—due to some failure earlier in the sequence—does not move, then the small door cannot shut. Quite possibly, the locking handle may go right home. But the baggage handler will still have, right in front of his eyes, a warning that the door is not safe.

That warning is available only to people outside the airplane. Baggage handlers and other tarmac personnel at airports are, of course, given some training and expected to demonstrate considerable levels of responsibility. But at the same time it is a good principle of aircraft design that final judgments about the essential safety of the machine should only be made by the flight crew, who are the most highly qualified people around. Conceivably, a drill could be established in which a crew member checked each door from the outside before takeoff. However, good design also takes account of human nature, and the possible reluctance of a flight engineer to leave an air-conditioned flight deck for a wet, cold airport ramp, and poke about without protective clothing, examining the underbelly of the aircraft. Therefore, small electric switches are built into the door in such a way that the movement of the locking-pin tube throws their levers over, breaking the circuits feeding small warning lights on the engineer's flight-deck panel. So long as the door remains unlocked, and therefore potentially dangerous, the light will glow, warning the crew not to take off. Even if the warning light system failed, and the crew took off, the airplane would not be in great jeopardy. With a vent door open, there would be a constant and significant leak, and the crew would be warned long before the pressure differential had built up to a dangerous level. The C-latch system seems to have proved its worth over the years. Up to the point in 1974 when the Boeing 747 fleet had completed two billion passenger hours, there had been seventeen recorded failures of the door-latching mechanism. In each case, the "fail-safe" elements in the system alerted the crew in time.

After the C-latch, another elaborate system was devised for cargo holds—the "semiplug" door, first fitted to the Boeing 727 and later adapted for the TriStar. Lockheed's engineers make no secret of the fact that the TriStar door was copied from the Boeing 727, for although at corporate level, competition is ferocious between the great aerospace companies, there is still much practical interchange between the design staffs. Indeed, the unwritten rule is that each firm has an obligation to tell its rivals about any new safety problem which it encounters. This generous impulse normally crosses the boundary between outright competitors. Curiously, and sadly, as later parts of our story show, it did not in the case of the DC-10 cross the boundary between subcontractor and contractor.

A semiplug door opens outward, like a latched door. Again, it is hinged at the top, and an electric motor swings it down to close. Along its sides are hard metal projections, rather like widely spaced, square cut gear teeth. Projecting along the door frame, there are matching steel pads. As the door shuts, the "teeth" on the door pass between the "teeth" on the frame. Then the closing mechanism slides the door a few inches downwards, until the spurs on the door fit over the pads on the frame. The effect is then similar to a plug door: pressure inside the fuselage will press the door harder and harder against the pads.

This makes an excellent solution for a door which is not so big that it must for structural reasons use a tension latch. Its only real disadvantage is that a good deal of design work must go into the complex gear which hinges and slides the door, and even with the most careful design there is usually some weight penalty. It is not surprising that Lockheed required a large team—some 150 engineers at peak—on the door systems.*

Behind these ingenious mechanisms and their development stood the force of Federal Airworthiness Regulation (FAR) 25, which says that all internal partitions in a pressure hull must be strong enough to resist decompressions which will result from a hull opening "not shown to be extremely remote." In other words, they were meant to convince everyone—but especially the officials who certify aircraft for the Federal Aviation Administration—that the chances of an accidental door opening were so small as to be negligible.

The main "partition" in a wide-bodied jetliner hull is the floor of the passenger cabin which is a long platform slung across the full diameter of the fuselage tube. It consists of a series of light-alloy beams running from side to side, with plates laid across and riveted to them (not much different in principle from the wooden floor of a house). The trouble is that to make a flat surface like this resistant to bending pressure is intrinsically harder than it is to make a tube resist a distending pressure. The hooplike frames of a cylindrical fuselage resist internal pressure well because the thrust comes outward against each part of the circle equally. The geometry of the situation means that all the metal in the hoop is being *stretched;* that is, it is in tension, which is the mode of stress which metal resists most effectively. Anyone can *bend* a paperclip, but to snap the wire by pulling it would require extraordinary power. No designer thinking about the 3,000-square-foot expanse of a jumbo-jet floor would delight in the prospect of having to make it resistant to the kind of pressure differentials a fuselage wall must withstand, that is, almost twelve pounds per square inch. Three or four pounds per square inch is sufficient to support a planeload of professional U.S. football players. To go beyond the requirements of passenger weight and design a full-pressure, wide-bodied floor would incur a weight penalty of 3,500 pounds or more, cutting ten or fifteen passengers off the payload. None of the design teams working on wide-bodied jets in the 1960s went for that cautious option.

One might ask again why the design of the floor should matter so much, since it is not per se crucial to the aircraft's survival. Just how important it is depends on other design decisions, and the chief of these is the route taken by the control cables.

The principle of linkage between cockpit and flying surfaces in a modern airliner is, in fact, the same as it was for the Wright Brothers. Long, taut wire cables, directed by meticulously sited pulleys, direct the pilot's will along the fuselage length and out along the wings to the vital control surfaces: the rudder, elevator, and ailerons. In the old—and not-so-old—days, the wires were rigged directly to the flying surfaces. The pilot's own muscular energy hinged the elevator up to raise the nose, hinged it down to dive, and pushed the rudder left or right to correct direction. But now in large aircraft the lines

---

* Special versions of the TriStar with extra freight potential have a C-latch door similar to the 747.

do not move the control surfaces directly. Instead, they operate the valves of hydraulic cylinders within which fluid under high pressure stands ready to move rudder, aileron, or elevator by thrusting at a piston.

Until the hydraulic revolution began during World War II, big airplanes were exhausting and clumsy to fly. The hydraulic principle had been known, of course, for years and used in automobile brakes: liquids like oil and water will transmit impulses with superb precision. The great advance was to make reliable hydraulic systems really light, with efficient engine-driven pumps to maintain pressure, and reliable actuating cylinders that could be tucked neatly into wings and tail surfaces.

In the wide-bodied generation of 747, DC-10, and TriStar, hydraulic controls reached a level of real mechanical elegance, and it was largely the reason, along with the sheer power and docility of their engines, why pilots took to these airplanes so readily. Power systems were evolved for them which not only moved surfaces precisely against high slipstream pressures, but which did so while giving a delicate "feedback" into the pilot's hands. The control function in all three of the U.S. wide-bodied aircraft represents a huge advance over the effective but relatively crude first- and second-generation jets. By general piloting consent, the 747 is the best controlled of all. Huge and clumsy as it looks on the ground, a 747 can be flown with an ease and delicacy comparable to the smallest and most beautifully balanced of light aircraft.

But this astonishing feat of technical virtuosity depends upon slim taut wires making clean, uninterrupted passage through the airframe, and upon long, thin pressure-filled hydraulic tubes finding their own complex way, to be ready to deliver power at each site where it will be needed. They are, one might say, comparable to the nerves, arteries, and veins of the animal body. (As with the bloodstream, each hydraulic system has an out tube—artery—and return tube—vein.) Like nerves and arteries, these cables and tubes are vulnerable. They cannot be left loose and they must be attached to—and have the protection of—some strong and uncomplicated surface.

The cabin floor is a natural enough choice in some respects, especially if the floor runs right along the length of the plane and through into the flight deck. Looking at the control cables first, they can pass under the pilots' seats, straight back under the floor, and thence into the tail assembly. This was the route chosen by the designers of the TriStar and the DC-10.

The other and perhaps more traditional route is along the ceiling of the hull: attached, in other words, to the fuselage wall proper. Cables so sited, said the advocates of under-floor cables, would be more vulnerable to debris from outside the hull—flying turbine-blades from a broken engine, missiles of various kinds—than ones tucked away several feet from the outer skin. There was, perhaps, something to the argument, provided that the security of the floor could be guaranteed. Boeing, as it happened, stayed with the ceiling-mounted system. It was convenience, in any case, for a 747 has its flight deck perched up above the main deck to give more space in the passenger cabin. Therefore, control cables run easily out at the back of the flight deck and along the inner surface of the hull's ceiling to the tail. But, as a result, even if the floor of a 747 were to collapse, the control cables would not suffer.

Of course, cables alone will not fly the airplane: hydraulic power must be available. And the hydraulic tubes run on the underside of the floor. If the floor were to collapse, would the aircraft become unflyable? It seems unlikely that it would. There are, as described earlier,* four hydraulic systems in a 747, any one of which can in an emergency handle the vital flying tasks. The out-and-return tubes for two systems run along under the port side edge of the 747 floor, and the tubes for the other two systems along the starboard side. Each floor beam has a support pillar some eighteen inches from the point where it joins the hull wall, and all hydraulic tubes are tucked into the "triangle" between floor beam, hull wall, and support pillar (stanchion).

Imagine a door blowing off on one side of a 747 hold. The floor above the missing door will collapse, certainly bending and probably rupturing the two sets of hydraulic tubes on that side. (Hydraulic tubes, unlike control cables, can stand some considerable distortion, however, and they have loops built in to help them survive stretching.) A door ripping off, or a bomb tearing a ten- or twelve-foot hole will collapse thirty or so square feet of the cabin floor in a 747 or any other wide-bodied jet. But a hull opening of that size could not collapse the whole width of the 747 floor—especially as the far side would be protected by its row of stanchions. Almost certainly, it would leave the two hydraulic systems on the other side of the hull in working order. Of course, catastrophe could occur on a scale which would destroy the whole floor—and all four systems. Again, a missile impact could tear away enough of the hull roof to destroy all of the widely spaced control cables.

Essentially, the claim of the 747 design team is that any disaster large enough to destroy all four hydraulic systems, or all the flying controls, would be of such magnitude that the airframe itself would be wrecked. No aircraft can be protected against *all* hazards. Certainly a salvo of heavy antiaircraft rockets, properly placed, would destroy a 747, and so would a very big bomb. But it is probably fair to say that a 747 in flight could hardly be brought down by any accident which left the main airframe intact. Can the same be said of the other two wide-bodied jets? As the TriStar and the DC-10 run their control cables under the cabin floor, the answer depends upon the integrity of that floor. And leaving aside for the moment the threat of bombs or missiles, the floor depends upon the integrity of the doors. The TriStar, perhaps, does not have quite the many-layered defenses of the 747. Its control system would be unlikely to survive the loss of a large cargo door. On the other hand, the TriStar's semiplug door, once seated in place, is highly unlikely to move so long as there is pressure in the hull. The extra complexity of the closing-and-sliding door gear gave the Lockheed designers a system with a very impressive degree of invulnerability.

* See Chapter 2.

The Douglas–General Dynamics team approached the problem differently when they came to design their cargo door. They chose neither the semiplug, nor the C-latch, but designed a quite different system based on "over-center" latches. It appeared, in theory, to be strikingly simple and elegant.

EVERYONE IS FAMILIAR, often unawares, with the over-center principle. It appears in one form in an old-fashioned electric switch. Once the lever of the switch has gone past the critical midpoint of its arc, it cannot "creep" back, but can only be forced to return over its "center" by an effort equivalent to that which moved it in the first place.

The DC-10 cargo door was designed, like the TriStar and 747 cargo doors, to open outwards, and is hinged at the top. It is lowered onto its rubber seal by an electric motor. So far, it resembles the 747 door. But the latches, instead of being half-circles like the Boeing C-latches, are shaped somewhat like the leg and talons of a bird. Instead of revolving *after* the door is shut, they reach out and snap *over* the latch spools—doing the work of binding the door into its rubber seal. Each DC-10 cargo door has an electric actuator that revolves a torque tube to which the latches are attached. (The size of the door dictates the number of latches: a DC-10 rear cargo door has four.) As the torque tube revolves, the talon of each latch is driven down over its spool. The top part of each latch then moves through an arc of about 90 degrees until it comes to rest against a metal stop, fixed to the door. In theory the advantage of over-center latches is that they are irreversible except by the application of power from the actuator. Therefore an over-center latch cannot "creep" whereas a C-latch can. But in engineering there are few absolute gains.

To be safe the top part of an over-center latch *must* pass beyond the center point of the 90 degree arc. If it does then all of the forces created by the build up of pressure inside the aircraft after take-off will be transmitted from the latches to the door structure, which can happily absorb them. But if, for any reason, the latches fail to pass beyond the center point of their arc, the loads created by pressurization will be transmitted not to the structure but to the electric actuator. The acutator is held to the door by two titanium bolts, each one-quarter inch in diameter. If the bolts give way, as eventually they must, the latches will release their hold on the spools.

As it happens, the closing of an aircraft provides a situation in which relatively minor failures can stop an over-center latch before it has gone the whole way. A C-latch *cannot* move until the door is fully shut because the line about which the latch revolves must be exactly the same as midline of the latch spool. (That is why a C-latch requires a separate mechanism to force the door down onto the seal.) With over-center latches the talons may grip the spools but the top part of the latches may still fail to swing through their arc—if, for instance, the door is a little out of true or if the power in the actuator should drop slightly.

Yet, treacherously, because they are irreversible, the talons will still hold onto the spools until the pressure inside the plane builds up enough to be really dangerous.

In an over-center door, the back-up locking system is even more crucially necessary than in a C-latch door. Its prime function is no longer to lock in the strict sense. Over-center latches, once home, are home for good. The back-up system is a *detection* system, designed to tell the operator whether the door is indeed properly shut. Confronted with this need, the DC-10 team came up with a locking system which not merely failed to be superior to the well-tried Boeing design, but which was inferior and frighteningly primitive.

Some of the same principles applied. A sliding bar carried a set of locking pins, and there were lugs attached to each latch in such a way that the pin ought not to slide into the "lock" position until the latches were over-center. Again, the locking-pin bar was pushed through its brief horizontal movement by a set of linkages driven off a handle projecting outside the door. (For diagrams see pages 296–303.) The handle had to be swung through more than ninety degrees in order to "stow" it in a neat slot in the door surface. The same handle movement would close a small vent door more or less in front of the operator's face.

At first glance, the system looks foolproof. The baggage handler would first press an electric switch to swing the door shut, then another switch to drive the latch actuator. He could hear it hum as it drove the latches. Then he would swing down the handle to drive the locking pins. At the end of its journey the locking-pin bar would trip a mechanical switch: on the flight deck, the engineer would see a warning light go off.

In reality, however, the system was almost absurdly unreliable. In the "Windsor incident," as Captain McCormick's flight was soon known, and in the Paris crash of 1974, the mechanical switches failed to work properly. Engineer Burke in McCormick's crew testified that the warning light went out bfeore takeoff from Detroit. And tests made later by the National Transportation Safety Board showed that misalignment in the door mechanism could make the mechanical switches useless as guides to the position of the locking pins.

Much worse, the linkages between the external handle and the locking-pin bar were far too weak and flexible. Such a detection system would only make sense if it could withstand a good deal of force. A strong man must feel himself encountering an irresistible force if the locking pins hit the lugs of unclosed latches.

In fact, the linkages on the DC-10 door as it first went into service would fail under a pressure of from 80 to 120 pounds. In other words, a baggage handler of normal strength could push the handle fully down, thinking that he had thus insured the closing of the door, when all he had done was bend the internal rods and bars out of shape.

Nor was this all. As designed, the last line of defense, the vent door, was considerably worse than useless. As described earlier, the 747 vent door is closed by a linkage drive off the "far end" of the locking-pin bar. Therefore, unless the locking pins go fully into position, the bar cannot move and the vent door will not close. In the DC-10 design, the vent door was driven straight off the locking handle. (For diagrams see pages 296–303.) It therefore provided no check at all on the working of the locking mechanism. Pushing down the handle could shut the vent door, even if there were no connection at all with the locking pins below.

It was, by any sensible standard of safety engineering, a gimcrack piece of design. Yet, because of the decisions taken about floor strength and control-cable routes, the safety of every man, woman, and child who went aboard the DC-10 was dependent upon the efficacy of the linkages from the moment the plane went into service. It is perhaps remarkable that no DC-10 cargo door blew off until Captain McCormick's plane passed over Windsor, roughly one year later.

BRYCE McCORMICK IS, both by habit and inclination, a conservative man. In his personal life he is most comfortable with the familiar, like his modest house in Palos Verdes, southern California, which he designed and built twenty years ago "with these two hands." Professionally, he is cautious, a stickler for crew discipline, and he flies, unwaveringly, by the book.

It was this conservatism which led McCormick initially to regard the DC-10 with some caution when he was chosen to fly one by American Airlines early in 1972.

The DC-10 was obviously radically different from the other planes— DC-6 and 7, Convair 990 and Boeing 707 and 727—that McCormick had flown for the airline during his twenty-eight-year career. But it was not the huge difference in size and engine power which concerned him so much as the total absence in the DC-10 of any mechanical means of operating the control surfaces. In smaller jets, like the 707 and DC-8, there is a reversion system standing by to operate the flaps, rudder, and elevators by hand if the hydraulic systems should fail. But the DC-10, and the 747 and the TriStar, rely exclusively on hydraulics. What would happen, McCormick asked, if all the systems were knocked out?

He found the answers through experimentation on a DC-10 flight deck simulator at American Airlines training school in Fort Worth, Texas. Most airline pilots, and especially those in the United States, now do the bulk of their training on simulators that can reproduce flight with amazing accuracy and realism. Their overwhelming advantage—besides cost—is that they allow a pilot to be relentlessly confronted with almost every kind of hazard he is likely to meet in real flight: multiple engine failures, fires, electrical failures, and so on.

On the Forth Worth simulator McCormick was able to test his alarming hypothesis of total hydraulic failure and, with the help of an instructor, he gradually learned how to exploit the DC-10's exceptional ability to "fly on its engines." Without assistance from the rudder or the ailerons, the plane can be turned to port by increasing the thrust of the starboard-wing engine; similarly, accelerating the port-wing engine (or decelerating the starboard engine) will turn the plane to starboard. If the thrust of both wing engines is increased, or the thrust of the tail engine decreased, the nose of the DC-10 will rise, while the reverse procedure will cause the nose to drop. (Most jets do have this capability to some degree, but the DC-10 is exceptionally responsive because of the positioning of the three engines.)

McCormick became so adept at the technique of using differential engine power to steer by that, he claims, he was eventually able to "fly" the Fort Worth simulator from climb-out to the approach phase without touching any of the controls, save for the throttles.

On June 12, 1972, two miles above Windsor, Ontario, the lives of the sixty-seven people on board Flight 96 depended on McCormick's ability to perform much the same trick, this time for real.

Ironically—in view of McCormick's original fears—the DC-10's hydraulic lines survived the explosive decompression and the partial collapse of the cabin floor. At strategic points the lines are coiled, which in this case allowed them to be stretched enough so that the fluid could still reach the actuators of the tail surfaces. But most of the commands from the cockpit could not, because the wire cables which transport the signals to the tail were severed or jammed.

The rudder was worse than useless. The pedals in the cockpit were immovable and the rudder itself trailed to starboard, causing the plane to yaw continuously to the right. The elevators—the flaps on the back edge of the horizontal tail which control pitch—were sluggish and it took considerable force from the pilots to move them up and down. The stabilizers— the front parts of the horizontal tail which "trim" the aircraft into the correct attitude—did not appear to work at all. When McCormick pulled the lever, shaped like the handle of a suitcase, which is supposed to operate the stabilizers manually, it came away in his hand.* And there was no power available from the number two (tail) engine, because the control cable had snapped leaving the pilots' throttle lever slack and impotent.

All of the wing surfaces were working properly but the ailerons—which allow a plane to bank by breaking up the airflow over the wing, causing it to drop—need very delicate handling. The trailing rudder was forcing the DC-10 to yaw to the right and overly vigorous use of the starboard ailerons would have sent the plane into a spin.

McCormick's first act, after he had recovered from the dust storm, was to arrest the DC-10's dive. With the engines running at near idle speed, the plane was falling steadily and—of much more deadly significance—the nose was dropping toward the horizon. McCormick knew that if the crippled plane got into a nose-down attitude there could be no recovery from the

---

* The investigation by the National Transportation Safety Board showed that the control which puts the stabilizers into automatic operation was working but the indicator was broken, leading the crew to believe otherwise.

dive. Not a moment too soon, he pushed the wing-engine throttles (whose controls had not been affected) fully forward. Without hesitation, or surge, the two engines delivered a burst of enormous power and the DC-10's nose pitched up.

With the immediate danger passed, there was a temptation to make a controlled emergency descent. Indeed, this is the procedure laid down in most flight manuals for the victims of midair collisions, which is what the crew of Flight 96 thought they were. But, in the past, too many pilots have taken that option before fully diagnosing the extent of the damage, with disastrous results. At any altitude less than 14,000 feet there is ample oxygen for humans to breathe and, anyway, McCormick is a firm believer in the cautionary maxim which says that pilots who hesitate will probably survive. He rejected an emergency descent with a sharp "negative" and instead began to experiment, very gingerly, with the controls.

He discovered that he could counteract the starboard bias of the rudder by setting the ailerons on the port wing at an angle of forty-five degrees. And by cutting out the idle tail engine altogether (by switching off the fuel boost pump) he lightened the loads on the elevators, making them marginally more responsive.

But the DC-10 could not be banked in either direction by more than a very gentle fifteen degrees without risking a spin and to steer the aircraft McCormick had to use the differential engine power technique that he had so diligently rehearsed on the simulator at Fort Worth.

Very slowly, he turned the aircraft around and headed back towards Detroit. As he began a gentle descent he pondered the problem of how to land.

WHEN BEATRICE COPELAND recovered consciousness she was lying on top of the wreckage of those sections of the cabin floor which had partially collapsed into the baggage compartment. Ceiling panels had fallen down on top of her and one of her feet was trapped by the twisted metal, preventing escape. She began calling for help.

Sandra McConnell, the other flight attendant stationed at the back of the DC-10, had managed to prevent herself from falling all the way into the cavity and had climbed into one of the rear lavatories to take refuge and await rescue. One of the women's colleagues telephoned the flight deck to ask for help, but flight engineer Clayton Burke, mindful of American Airlines regulations which demand that crew be properly dressed in public, was reluctant to appear in the passenger cabin without his hat. While he searched the cockpit in vain, unaware that all three crew hats had departed during the decompression, Beatrice wriggled her foot out of the trapped shoe and she and Sandra saved themselves. One of them called the flight deck and told Burke not to bother.

Mrs. Kaminsky was given a Kleenex to stem the bleeding from her face wound and Mr. Kaminsky tried to calm her hysteria. He told her there was nothing to worry about. Neither of them believed it: One of the flight attendants had let slip that there was "a great hole" in the side of the plane.

However, Captain McCormick was able to find time to provide his passengers with some reassurance. Over the public address system, he said that there was a mechanical problem, and they were returning to Detroit where he was sure American Airlines would provide another plane with the minimum of delay. He apologized for the inconvenience. It all sounded so routine that even Mrs. Kiminsky took comfort.

Cydya Smith took over the PA system to prepare the passengers for an emergency landing. She couldn't find the script, written for just such an eventuality, so she ad-libbed and described the "brace position"—head down, hands clasped under knees—which is reckoned to reduce the physical shock of a heavy landing.

The other women, armed with plastic bags, collected the passengers' spectacles, pens and pencils, combs and jewelry, which can become dangerous missiles at the moment of impact. They also collected the passengers' shoes which can cause foot and ankle injuries. One man, not fully understanding the procedure, offered Coleen Maley his socks. She felt that if she had asked for his wallet he would have handed it over willingly.

About twenty minutes after leaving Detroit, Flight 96 was back on the airport controller's radar screen. The DC-10 was approaching the runway at 160 knots (180 mph)—30 knots faster than the normal landing speed. McCormick reduced power to cut down his speed but the plane's rate of descent—known as the sink rate—went up from three hundred feet a minute to eight hundred feet a minute. When the undercarriage was lowered and the wing flaps (which normally increase the amount of lift provided by the wings) were set at an angle of thirty-five degrees, the sink rate nearly doubled to fifteen hundred feet a minute. McCormick could do nothing but push the throttles forward. The sink rate reduced to eight hundred feet. The DC-10 touched down at 186 miles an hour.

Immediately, under the influence of the wayward rudder, it veered to the right, departed from the runway, and headed off across the grass toward the airport buildings. As the 150-ton aircraft bounced over the rough ground and across taxiways, McCormick began to have doubts—for the first time since the drama had begun—of his ability to save the plane. He had put both wing engines into reverse thrust, but they showed no sign of being able to stop the DC-10 in time.

It was the first officer, Whitney, who provided the solution. Without waiting for orders, he pushed the throttle lever of the port-wing engine into the full reverse thrust position and beyond. The engine responded by delivering 10 percent more power than the permitted maximum.

At the same time Whitney took the starboard wing engine out of reverse. With the influence of the rudder totally overwhelmed by the power of the port engine, the DC-10 abandoned its collision course, swung to the left, and headed back for the runway. It came to rest half on the concrete, half on the grass, a mile and a half from the runway threshold.

McCormick, understandably concerned about the risk of a fuel explosion, ordered an emergency evacuation. Using six doors and emergency chutes, it took less than thirty seconds for everyone to reach safety, although some of the passengers were slightly injured in the process. The unfortunate Mrs. Kaminsky hit the ground at the bottom of an escape chute so hard that the skin was ripped from her right foot, exposing the bone.

That was not quite the end of the day's excitement. The FBI had been alerted and suspected that Flight 96 may have been the victim of sabotage. Agents—fifty of them, according to Mrs. Kaminsky—were on hand to detain all of the passengers and crew for interrogation, and nobody was allowed to even go to the toilet without an FBI man or woman for company. But explosives experts who examined the plane could find no trace of a bomb and everybody was eventually released.

The crew were put into a hotel near Detroit Airport. McCormick, Whitney, and Burke are not inveterate drinkers but that night they had no trouble killing a quart bottle of Scotch.

THE FORMAL INVESTIGATION into the cause of the near disaster began the next morning. American Airlines set up an inquiry panel, headed by its Director of Safety Mack Eastburn, to hear evidence from the crew of Flight 96 and from the ground staff at Detroit.

William Eggert, who worked as a cargo loader for the airline, described the trouble he had encountered in closing the aft cargo door of the DC-10 just before takeoff.

At first, everything had worked normally. Eggert had closed the door electrically and then held the button while he counted to ten and kept his ear pressed against the door, listening to the latches being driven into place. He was quite sure that the four latches had fully closed, although there was, of course, no means for him to make a visual check.

However, he was unable to pull down the external locking handle which was supposed to drive the locking pins across the latches and simultaneously close the vent door. He checked for obstructions but, finding nothing, pulled on the handle again—this time adding weight with his knee—and succeeded in moving the handle into the fully locked position. But although the vent door closed, Eggert noticed that it was slightly crooked and not positioned firmly against its seal. Not satisfied, he unlocked the handle and repeated the process all over again, with the same result.

Eggert consulted with an American Airlines supervisor and a mechanic, who were both standing near the plane, and was told not to worry: crooked vent doors were, apparently, a regular problem on DC-10 cargo doors. The mechanic, Al Lucas, told the Windsor inquiry: ". . . Due to the nature of the seal around the [vent] door, sometimes there's a little crack or a little opening in it. If you get up there and mess around with it, it will seal permanently but eventually we rely on the pressure of the aircraft to push [the vent doors] right out, completely shut." In other words, there was nothing sufficiently unusual about the state of the cargo door on Flight 96 to alarm American Airlines' ground staff.

The door itself was found shortly after the inquiry in a field about twenty miles from the airport. Examination showed that the four latches had failed to go over-center by one third of an inch.

In theory, of course, that discrepancy should have been sufficient to prevent the locking pins from moving across the top of the latches and into the safe position. And consequently, Eggert should not have been able to move the locking handle which would have left the vent door open. But although the locking pins had indeed jammed against the lugs on the top of the latches, Eggert *had* been able to move the handle. Further examination of the door revealed that when the pins jammed, the vertical rod connecting the locking-pin bar to the handle had simply broken under pressure—and there had then been nothing to prevent the vent door from closing.

At Long Beach McDonnell Douglas conducted a series of tests which showed that the force required to overcome the system was 120 pounds— no great feat of strength for a full-grown man. Other tests carried out independently by General Dynamics, the subcontractors who had built the cargo door, showed that as little as eighty pounds of force was sufficient.

Those tests took a little while to perform. But, even before they were completed, it was clear—within three days of the Windsor incident—that the DC-10 was unsafe. One of the criteria upon which the plane's certificate of airworthiness depended was that the locking mechanism of the rear cargo door was fail safe; in fact, it was positively dangerous. It could, with little difficulty, be improperly closed and yet, through the "safe" position of the external handle and the vent door, give a totally false indication of its status. In addition, the Windsor incident had proved that the cockpit warning system could fail, misleading the crew into taking off with a hazard as lethal as a largish bomb.

The aviation industry does not normally receive such manifest warnings of basic design flaws in an aircraft without cost to human life.* The Lockheed Electra, the British Comet, and the Boeing 707—to mention just three —all claimed many lives before their respective "bugs" were identified and ironed out.

Windsor deserved to be celebrated as an exceptional case when every life was saved through a combination of crew skill and the sheer luck that the plane was so lightly loaded. If the 120 seats in the rear passenger compart-

---

* The cost, in monetary terms, was modest: McDonnell Douglas paid $288,500 to American Airlines for damage to the plane and $11,000 to Mrs. Kaminsky in compensation for her injuries.

ment had been occupied, seven tons of extra weight would have been pressing down on that section of the cabin floor.* In that event, damage to the control cables would undoubtedly have been more severe, and it is highly questionable if any amount of skill could have saved the plane.

Windsor was also exceptional in the amount and clarity of the evidence it provided about its cause—and about the remedies that would prevent it from happening again.

Yet, twenty months later, it was possible for a DC-10 to take off, from Orly Airport Paris, with its rear cargo door improperly closed, with the crew unaware of the hazard, and with the floor and the control cables as vulnerable as before.

Bryce McCormick is a supreme loyalist to both American Airlines and to the DC-10 which he still regards as "the finest plane there is." But even he finds it incomprehensible that the lessons of Windsor should have gone unlearned. After his near escape he was invited by McDonnell Douglas to make recommendations in the light of his experience. He suggested that every DC-10 pilot should be told in detail of the consequences of an explosive decompression and of the flying techniques that he and his crew had adopted to save Flight 96. McDonnell Douglas did no such thing. The company attributed the Windsor incident almost solely to human failure on the part of the baggage-handler Eggert and not to any failure on the part of its designers and engineers. (Similarly, after the Paris crash, it initially blamed the disaster on Mohammed Mamhoudi, the baggage handler at Orly Airport.)

The view at Long Beach, and presumably at the St. Louis headquarters, seems to have been that Windsor was a freak and would not be repeated. There was, therefore, no point in alarming customers and—more importantly —potential customers with vivid details of Bryce McCormick's ordeal. McDonnell Douglas resolved that all it had to do was come up with a fix that would dissuade the Eggerts of this world from using their knees to overpower the so-called safety system.

It approached the problem with great confidence.

McCormick had offered one other, very succinct, piece of advice: "Get that damn door fixed." McDonnell Douglas replied: "Mac, that's a promise."†

---

* On the conservative assumption that each passenger would weigh 120 pounds.
† This account of the Windsor incident is based on the transcript of the American Airlines inquiry, the NTSB report, published in January 1973, and on interviews with Captain Bryce McCormick and Mr. and Mrs. Kaminsky.

9

Fat, Dumb and Happy— The Failure of the FAA

". . . When you have a well-constructed state with a well-framed legal code, to put incompetent officials in charge of administering the code is a waste of good laws, and the whole business degenerates into farce. And not only that: the state will find that its laws are doing it damage and injury on a gigantic scale."

—PLATO,
Laws (Book VI)

THE FEDERAL AVIATION ADMINISTRATION, which regulates the aviation industry in the United States, has eleven regional offices scattered throughout the country, each one responsible for watching over the manufacturers, airlines, and airports operating in its area. The Western Regional Office of the FAA is responsible for California which contains the greatest concentration of airplane-building ability in the world. Besides Douglas, Lockheed, Convair, and North American Rockwell, there is a vast supporting cast of smaller firms, providing everything from airborne lavatories to navigation avionics.

In 1972 the head of the Western Region was a career public servant named Arvin O. Basnight. His office on Aviation Boulevard, not far from Los Angeles International Airport, had on one of its walls a large picture of B-17 bombers in combat—a reminder that Basnight's early flying experience was gained in harsh circumstances during the great daylight air battles at the climax of World War II. Basnight projected an affable, relaxed manner, but it did not quite conceal the watchful gaze and studied speech of the experienced bureaucrat.

The FAA is typical of U.S. public services in that its professional administrators know that the titular head of the agency will usually be a political appointee, nominated by the White House. Sometimes these imported chieftains excel the professionals; sometimes they are incompetent at running the agency they are chosen for; and sometimes, they are a capricious mixture of the two. For professional public servants like Basnight the principle of survival, naturally, is to get through the bad periods with as little personal damage as possible. It is probably fair to describe Basnight as a man who applied the tactics he learned in wartime to the conditions of public service life. B-17 crews were trained to keep close formation, eschew individual action, and make sure that they had covering fire at all times.

There cannot be much doubt that at the time when Mr. Basnight was trying to cope with the aftermath of the Windsor incident of June 12, the FAA was going through a bad period—one from which it has not fully recovered even now. Like many other federal agencies at that time it was suffering from the fact that the White House was inhabited by a group of men who meant to dismantle the complex counterbalances of American government and put in their place the personal rule of President Richard Nixon.

The revelations of criminal and covert behavior overshadow, in memory, the fact that the Watergate era brought at the same time an overt attack on the power and traditions of the federal civil service. Some of the energy of this attack derived from Republican sentiments which were legitimate and traditional, however much some people might disagree with them. Republicans dislike "big government" on ideological grounds. The great federal agencies are often cumbersome and expensive in operation, and are largely staffed by Democrats who sometimes fail to serve Republican administrations —and ends—with undivided enthusiasm. In the case of Nixon and his White House staff, there was no doubt something more—the authoritarian's instinctive dislike of a system which diffuses administrative power among many different centers. In any event, the Nixon White House, beginning in 1969, set out to make federal agencies "more responsive" to the President's will. The result, in terms of our particular story, was the crippling of the FAA and of the other federal agency chiefly concerned with aviation safety, the National Transportation Safety Board.

The FAA owes its decline, in large measure, to President Nixon's first choice of an administrator. In 1969 the President nominated—and Congress approved—John Hixon Shaffer, then fifty, an ex-Air Force officer with rock-solid Republican loyalties, but with very little experience of commercial aircraft to qualify him to head the most powerful and influential aviation-regulating body in the world.

The crippling of the NTSB was accomplished by the denigration and, eventually, the resignation from federal service of Charles (Chuck) Miller, head of the NTSB's Bureau of Aviation Safety who was, and is, a genuinely distinguished air-safety engineer. Had Miller, and the values he stood for, been dominant at the time of the 1972 Windsor incident, then the story of the DC-10 might well have been different. But in Mr. Nixon's Washington, men like Chuck Miller did not generally win their battles.

IN LAW, THE FAA has great powers over the American aviation industry. It regulates, and must approve, every stage of the design and manufacture of every aircraft which is built in the United States. It can inspect manufacturing

plants at will and certificates as airworthy each aircraft that is built. If the
FAA decides, for any reason, to withhold certification, then the particular
aircraft may not leave the ground even to remove itself from American air
space. The FAA also regulates and licenses the pilots that fly the airplanes
and most of the airports they use and operates the air traffic control systems
that rule over the airways. The agency does not formally investigate airplane
accidents because that is the job of the NTSB. But if an NTSB investigation
reveals the need for change in the design of an aircraft, or the way in which
it is operated, the FAA can order the change—no matter how expensive or
awkward—by issuing an Airworthiness Directive, which has all the force of
federal law.

But there is and always has been a potential weakness in the constitution
of the FAA. By act of Congress its responsibility is not only to regulate the
safety standards of American aviation but also to promote its commercial
success. It probably did not occur to very many people when the FAA was
established in 1958 that there was likely to be any serious conflict between
the two aims. After all, it must have seemed obvious that mass air travel
could not establish itself commercially unless it gained a reputation for safety.
And, equally obviously, no aircraft manufacturer's long-term interests can be
served by building airplanes with anything less than the greatest possible
safety standards.

Less obvious, perhaps, is the problem that the strains of commercial competition may be so acute that the men and organizations subjected to them
can cease to be accurate judges even of their own interests, let alone the interests of other people. In theory, of course, it is for just such occasions that
the FAA, in its role as the industry's policeman, exists. But the agency can
only work effectively if the man appointed to head it can cope with the
schizoid nature (part policeman, part promoter) of the Administrator's role.
He must be able to identify with the industry's viewpoint, while standing
slightly apart. The trouble with John Shaffer was that he identified totally
with the industry. Indeed, he was an embodiment of it.

Shaffer served with the U.S. Air Force from 1946 to 1953 and spent most
of that time working with aircraft manufacturers developing new bombers
such as the Boeing B-47. When he resigned from the Air Force, with the rank
of lieutenant colonel, he deliberately took a job outside the aviation industry
with the Ford Motor Company to avoid, he says, accusations that he had
been "feathering my own nest." But after three years with Ford, he deemed
it proper, in his own words, "to go home" and he became an executive—and,
eventually, a vice-president and small stockholder—of TRW, Inc., one of the
largest engineering subcontractors in the aerospace industry.

Among many other things, TRW manufactures engine parts and in particular it makes turbine blades for Pratt & Whitney's JT-9D engines which
power the Boeing 747 and which were the first high bypass fan-jets to go
into airline service. As we explained earlier fan-jet turbine blades are fiend-
ishly difficult to design and make because the turbine runs at much hotter

temperatures than in a pure-jet. It was not surprising then that in 1969–70, when the 747 went into service, the JT-9D should have encountered some troubles. But they were much worse than anybody had expected.

Turbine life was alarmingly short and airlines were using up large sums of money shipping spare engines to stranded 747s, primarily because TRW-made blades were failing under stress. The danger was not so much the loss of engine power, although Boeing is far from recommending it (or even mentioning it widely for fear of emulation), a 747 can and has been flown on just one of its four engines. And even at the rate at which the turbines were failing, it was almost inconceivable that more than two engines would be lost during a single flight. But so long as the problems remained, there was the very real danger of shrapnel damage, with burnt-out turbine blades and other parts of the engine exploding through the casing and ripping holes in the fuel tanks, the flying surfaces, or even the fuselage.

No 747s were lost in these incidents but the investigators at the NTSB became more and more concerned about the potential threat. Matters came to a head in August 1970 when one engine of an Air France 747 exploded after the plane had taken off from Montreal, bound for Paris. Shrapnel damaged the fuel tanks heavily, but the plane could not land immediately because it was overweight. The pilot therefore chose to fly the 350 miles from Montreal to Kennedy Airport, New York, with streamers of jet fuel pouring out behind the airplane. The Aviation Safety Bureau of the NTSB concluded that turbine failure was now a major danger to the 747 fleet, and the NTSB recommended that the FAA should impose mandatory sanctions. It suggested, among other things, that airlines flying 747s should more closely monitor the condition of their JT-9D engines and, as soon as possible, install "vented" turbine blades. As head of the Safety Bureau, Chuck Miller had no power to take action himself but in the days before Shaffer's reign at the Federal Aviation Administration, recommendations from the NTSB had been adopted either totally or in part almost as a matter of routine—and, more to the point, without delay. For example, in 1968—the year before Shaffer's appointment—88 percent of the NTSB's current recommendations had been dealt with by the end of the year. But by 1970 the picture had changed considerably with the NTSB's success rate down to 64 percent and falling rapidly.*

One of the recommendations turned down by the FAA that year was that an Airworthiness Directive should be issued against the JT-9D. Although he had discerned the change in climate Miller was, nevertheless, by his own account, amazed at the decision. Ironically, his outrage forced him into actions which made it generally easier for his position to be undermined.

On leaving TRW to join the FAA in February 1969, Shaffer had sold his stock in the company and resigned as a vice-president. He was able to main-

---

* In 1968 the NTSB made 33 recommendations of which 29 were adopted in whole or in part by the FAA. In 1970, 110 recommendations were made and 70 adopted by the end of the year. In 1971, 99 recommendations were made and 56 adopted. In 1972, only 43 of 113 recommendations (38%) were accepted by the FAA within the year.

tain, therefore, that his assessment of the turbine-blade problem—a situation that manifestly affected TRW—was entirely independent.

At least some people on Capitol Hill were not convinced, however, that Shaffer was the most suitable man to decide whether very tough action was required to deal with the shortcomings of a product made by his old firm.

An investigator from a House of Representatives subcommittee was instructed to find out, discreetly, if Shaffer was laboring under any conflict of interest. But the investigator could not discover anything improper and the FAA remained adamant for inaction. On October 1, 1970, after yet another 747 incident—this time the plane belonged to American Airlines—the NTSB said publicly that the situation was "potentially catastrophic" and urged the FAA to take action before it was too late.

Two days later, the FAA called a news conference and, in effect, accused the NTSB, and therefore Miller, of showing poor judgment and of causing unnecessary alarm. Certainly Miller's behavior was eccentric, judged by the official code. Both the FAA and the NTSB were then parts of the same bureaucracy, the Department of Transportation, and for the head of one section to attack the head of another in public was to run headlong into danger. Miller was at once in Cassandra's difficult position, in which his reputation for judgment could be restored only by the occurrence of the catastrophe of which he had forewarned and which he dreaded.

The FAA took no real action until December 1970—by which time the fuss had died down, at least publicly—when it did issue an AD against the 747 but even then only one requiring regular inspection of the JT-9D's turbine blades. That half-hearted safeguard failed to prevent sixteen further incidents of turbine failure during the following few months, but eventually the engine makers (goaded, no doubt, to some degree by the NTSB's public action as well as by pressure from the airlines) managed to improve the breed of their turbine blades and the JT-9D attained to its present status as a supremely reliable power plant.

Events had proved Shaffer to be "right": no 747 had crashed which, of course, seemingly justified the implicit faith the FAA had shown in the will and the ability of the manufacturers to fix the turbine-blade problem voluntarily.

However, as was bound to happen, Shaffer's faith in the "voluntary system" was eventually ill rewarded. The aircraft involved was much less important than the 747, but this time people had to be killed before a solution was obtained.

Beechcraft builds light aircraft for the sport and executive travel market. In June 1971 a Beechcraft E18S crashed just after takeoff at Cleveland and the NTSB investigators discovered that fatigue failure had occurred in a crucial portion of the wing support. They recommended that the FAA demand more careful and earlier inspection of all similar Beechcraft wings in service, and modifications to deal with the fatigue problem.

All airplane manufacturers, of course, modify their aircraft in the light of service experience. In practice, this means a flow of bulletins going out to all known users of the airplane in question, with directions on how to accomplish the recommended change. Where necessary, kits of parts will be sent out to the operator, though major reworking may require that the plane go back to the factory. The majority of these service changes are concerned with operating convenience or economy, rather than safety. They are therefore optional, and although the manufacturer may provide kits for little or no charge, the onus of cost remains formally with the operator.

An Airworthiness Directive from the FAA is quite different. There is nothing optional about it, because if it is not complied with, it becomes illegal to fly the airplane. As the name implies, an AD is only issued when airworthiness —and therefore safety— is directly concerned.

In practice, ADs do not normally "ground" airplanes, or even interrupt operations very drastically. They merely say that certain changes must be accomplished at particular moments in the airplane's regular maintenance cycle. But they are public documents, which the manufacturer's own service bulletins are not, and they are circulated automatically to the news media, to the aviation attachés of all foreign governments, and to the FAA's overseas offices. Naturally, they place the onus of cost firmly on the manufacturer, because the contract of sale always says that what the manufacturer agrees to supply is an *airworthy* machine. Airworthiness is defined by the existence of FAA certificates, and the issuing of an AD makes those certificates invalid if the directive is ignored.

Instead of toughening up an earlier AD against Beechcraft, the FAA allowed the company to send out a service bulletin *recommending* changes in inspection procedures and saying that kits to deal with fatigue failure would be available to any operators who wanted them. The response to this was leisurely, and only a few airplanes had been thoroughly inspected and modified when another Beechcraft with the same fatigue failure problem crashed in April 1973, on landing at Davenport, Iowa, killing all six people on board. Immediately afterwards, the FAA issued an AD making the necessary changes and inspections mandatory.

There was no connection between John Shaffer and Beechcraft while he was Administrator at the FAA. But after he left the Administration, Shaffer took up several posts in the aerospace industry, and one of them was a directorship at Beechcraft. Shaffer's links with TRW and with Beechcraft were never concealed, and neither he nor anyone else in the Nixon Administration thought they were inappropriate. They were, however, indications that Shaffer was much better fitted for the role of promoter than that of policeman to the aviation industry.

IN MARCH 1971, the chief administrator of the NTSB, Ernest Weiss, was retired and replaced, on the recommendation of the White House, by Richard L. Spears. Mr. Spears's qualifications for the $36,000-a-year job of running an agency that is exclusively concerned with public safety were minimal: He had worked for an aviation company, Aerojet General Corporation, but had no technical qualifications, and more recently he had been a Republican political aide. That might not have mattered if Spears had contented himself with running the administrative side of the NTSB but, according to Chuck Miller, he grossly interfered with the highly specialized work of the Bureau of Aviation Safety.

It now seems clear that between 1969 and 1971 the NTSB and, more specifically Chuck Miller, had incurred the disfavor of White House aides— perhaps through the public row with the FAA during the 747 turbine-blade affair. In 1971, Frederic V. Malek—a West Point graduate and a former Green Beret—was assigned by H. R. Haldeman to "politicize" the executive branch of government. The operation was called the Responsiveness Program and its aim, of course, was to make all the twigs of the executive branch, including federal agencies, more "responsive" to White House wishes. In that cause Malek developed what became known as the Malek Manual (Federal Political Personnel Manual) which, among other things, detailed methods for getting rid of "dissident" public servants. By Malek's definition, Miller and Ernest Weiss would most certainly be considered dissidents. Moreover, Weiss was a notably active Democrat.

Weiss, a career civil servant, was made executive director—its senior administrator—of the NTSB on its inception in 1967. According to the NTSB's chairman John Reed, Weiss did his job "very effectively" and he was "a very fine gentleman, a very efficient gentleman." The two men worked closely together, meeting ten or fifteen times a day, and discussed every major administrative decision. Weiss was, therefore, a little surprised to arrive at work on Monday, January 5, 1971, to find Richard Spears sitting at a desk outside the Chairman's office: unannounced, Reed had acquired a "consultant."

Within weeks of the new arrival, Weiss began to hear rumors from his own staff that Spears was there to replace him—and was already sending and receiving mail in the name of the Executive Director. Weiss confronted Chairman Reed: "He said your work has been good, it's been outstanding. I enjoy working with you. I now have to change the job and I have to put someone else in it. . . . He also said, incidentally, the Department [of Transportation] would take care of me for ninety to 120 days."

What Reed planned to do was change the classification of Weiss's job from G (career civil servant) to C (political appointee). The recommendation, he told a subsequent Senate inquiry, was made by the White House. "I felt this was reasonable. . . . It didn't seem to me extraordinarily rare that with a change of [White House] administration, that certain positions with administrative function would change." The change he had in mind for Weiss was that he should become the NTSB's Program Review Officer, losing in the process two grades in his civil service ranking.

Weiss was not going to take that lying down and he asked to see William Heffelfinger, assistant secretary at the Department of Transportation. (The

NTSB was then part of the DoT for "housekeeping purposes.") Heffelfinger and Weiss had a series of meetings: at one of them the assistant secretary appeared to be very angry. Weiss later told the Senate: "He told me he had just gotten a call from the White House which wanted to know why he had not gotten me out and Spears in."

With very little other choice available, Weiss finally accepted his demotion on March 16, 1971, and ceased to be the executive director the same day. Spears immediately took over his job. (The title was changed to general manager but the job description remained precisely the same.) A month or six weeks later, Spears and Weiss discussed what had happened. According to Weiss: "He [Spears] said that during the time he was a consultant he had come to the conclusion that I was doing an outstanding or excellent job. He went back to the White House to tell them so. . . . He said the White House told them they didn't care what kind of job I was doing, they wanted me out and wanted him in." Spears's apparent regret at Weiss's demotion did not prevent him from taking the job anyway.

Even before he officially began his new career with the NTSB, Mr. Spears displayed a novel interpretation of what his duties should include. On February 8, 1971, using NTSB notepaper and signing himself "executive director," he wrote to the President of Continental Airlines, Robert Six, asking for a job for a White House aide, Harry Fleming. Spears pointed out that Fleming had "a kind of rapport with Pete Flanigan that can be very productive." *
Mr. Flanigan was at that time advising President Nixon on the distribution to U.S. airlines of international air routes: the obvious implication of Spears's letter was that Fleming could help Continental obtain new routes. (When the incident eventually came to light, in 1973, the then assistant U.S. Attorney General Henry Petersen said that Spears's conduct "appears to clearly constitute a violation of the standards of ethical conduct for government officers and employees. . . ." Spears could have been fired; instead, he was given a written reprimand by the NTSB's Chairman, John Reed, a former Republican Governor of Maine.)

* The full text of the letter read:
> Dear Bob: I have just learned my good friend Harry Fleming of the White House staff is interested in making a connection with "one of the air carriers." Harry's situation is briefly that he will be leaving the White House as of the 26th of this month for no reason other than his business interests demand his personal attention. However, this won't take more than half his time. Thus, he could be established on a sort of a part-time or consultant basis. Harry has expressed his interest to me quite casually, and I can tell you he will be a valuable man for someone in Washington since he has such a strong entree at the Department of Justice and State and enjoys a kind of rapport with Pete Flanagan [sic] that can be very productive. I have not mentioned his availability to any of the other carriers yet since I wanted you to be the first aware of this because of my high regard for you personally and for Continental Airlines. In any event I would appreciate hearing from you on this as I do want to help Harry make a good connection one way or the other. Warmest personal regards. Sincerely, Richard L. Spears, Executive Director.

Mr. Fleming did not make "a connection" with Continental but he did become a consultant for Spears's former employers, Aerojet.

More seriously, soon after he joined the NTSB Spears began giving orders to the Bureau's accident investigators: they were told, for example, that reports on accidents had to be produced more quickly. One month was set as the deadline for reports on small-plane crashes and investigations into major disasters had to be completed and written up within six months. Spears's stated aim in doing this—to "cut down on the backlog"—seemed on the face of it laudable enough. But the investigation of accidents is not an exact science that can be governed by arbitrary deadlines. The only rational reason for investigating an air crash is to discover the cause in the hope that repetition can be prevented and that requires, more often than not, painstaking detective work: sifting through the wreckage for clues, rebuilding parts of the plane, making sense of garbled radio messages, and so on. Demanding that the cause of an air crash be determined within a fixed time limit is analogous to insisting that the police capture each murderer within six months of the crime.

And to make the proposition even more absurd Spears ordered investigators to cut back on the amount of paid overtime they worked. Miller was constantly refused authority to replace staff who resigned or retired from his relatively small workforce of technical experts. On the other hand, the number of administration staff almost doubled. The inevitable result was that the backlog got smaller but the quality of NTSB reports—at least in Miller's eyes —declined.

There were other examples of apparent interference. The most serious, perhaps, was Spears's campaign against a small Accident Prevention Branch which Miller had set up within his bureau. The job of the branch was to do exactly what its name suggested, and it produced five important special studies on different aspects of air safety. Those studies concentrated on potential hazards and were, inevitably, critical of the FAA. Shortly after being appointed, Spears initiated a so-called management survey of the Accident Prevention Branch and found it sadly lacking. According to Miller, the morale of the staff was then systematically destroyed: they were, among other things, refused the opportunity to attend training courses and overlooked for promotion. (The branch was abolished on March 1, 1974. A month later, Miller testified to a Senate Committee, which was investigating the activities of the NTSB: "Specifically, the Accident Prevention Branch was maligned and rendered ineffective . . . because they were one of the principal 'overview' activities and were in a good position to be critical of the FAA.")

By the middle of 1972 the hostility between Miller and Spears had grown so considerable and open that John Reed was obliged to call the two men together and ask them "to let bygones be bygones." A kind of accord was reached but it soon broke down, and in April the following year, Spears filed a Bill of Particulars, in fact, a list of allegations against Miller, accusing him of incompetence with the NTSB board. Miller drew up a list of countercharges. In August the board (all five members are political appointees) sat in judgment and decided that while Spears had overstepped his authority,

Miller was largely at fault and had displayed a number of "deficiencies." He was given ninety days to "improve his performance"; otherwise, he would be fired.

The ignominy of that verdict provoked Miller to go to the Civil Service Commission and later the Senate Commerce Committee began an inquiry. However, before the matter could be properly resolved, Miller's doctors discovered that he was suffering from heart trouble and after a long period of sick leave he accepted the offer of a premature retirement on medical grounds.* He left the NTSB in December 1974.

After he had accepted retirement—and therefore, in his own eyes, defeat—Miller put down on paper his reflections about the six years he spent as head of the Bureau of Aviation Safety. In a sardonic epilogue he wrote: "I have learned much about the political arena. A lot of it stinks. Some of it is good. In the final analysis . . . there are political cats and there are technical cats. The job of management at the interface between them is difficult at best. It becomes damn near intolerable if the people on either side do not remain ethical in their conduct. . . ."

It may well be that Miller overestimated the amount of political influence that was exerted against him. But there is no doubt that the NTSB's criticisms of the FAA for failing to take tough enough action against manufacturers was viewed with great disfavor by the Nixon White House. It is also indisputable that on at least one occasion Nixon aides directly interfered with the NTSB's affairs. In its 1972 report to Congress the NTSB recommended that it should be made completely independent of the Department of Transportation, and any other government body. Before the report was sent to Congress, Egil Krogh Jr., then assistant transportation secretary and a former White House aide, warned NTSB Chairman Reed that the Nixon Administration disapproved of the recommendation.† If it was not withdrawn, Krogh said, the Republican members of the NTSB Board would be "disciplined." When the Senate learned of that threat it set up a subcommittee to investigate, and the chairman, Senator Warren Magnuson, left no doubt about the seriousness of the situation:

> This is no matter of bureaucratic nicety. The NTSB was established to insure the optimum safety of the travelling public. To the extent that the Board's voice has been muted, to the extent that the Board has failed to monitor those operating agencies charged with the safety of the travelling public, and to the extent that the board has failed to press for compliance with its recommendations for safety-related reforms, the safety of the public as well as the integrity of the governmental process have been gravely compromised.‡

* Happily, Miller's health has since improved considerably. He is now an aviation safety consultant and spends most of his time lecturing on accident prevention.
† Krogh later resigned after being implicated in the Pentagon Papers burglary of the office of Daniel Ellsberg's psychiatrist.
‡ Oversight Hearings before the Aviation Sub-Committee of the Senate on activities of the National Transportation Safety Board, May 21, 1973.

Some eighteen months later both Houses of Congress passed legislation removing the NTSB from under the umbrella of the Department of Transportation, and it is now a uniquely independent agency, although it is still run by a board of political appointees and Mr. Spears remained general manager until March 1976.*

Had Congress had the foresight to grant that total independence when legislation establishing the NTSB was passed in 1966, it is possible that the consequences of the Windsor incident might have been very different. If, in the summer of 1972, Chuck Miller had not been spending so much time fighting against what he saw as a direct political attack on his Safety Bureau, he might have demanded more loudly that the FAA enforce the radical modifications to the DC-10, which the NTSB had recommended in its report on the Windsor incident. As it was, Miller was too preoccupied even to realize that the NTSB's recommendations about the DC-10 were largely disregarded by the FAA and McDonnell Douglas.

And those men at the FAA who were technically qualified and close enough to the problem to appreciate the threat posed by the DC-10 cargo door were in no better a position to take the kind of action that might have prevented the Paris tragedy. Under the regime of John Shaffer there was little encouragement for men like Arvin Basnight and his staff in the Western Region office to adopt a minatory line dealing with the failings of airplane builders.

ALL THE SAME, Basnight and his men had no choice but to be fairly tough in order to get cooperation from McDonnell Douglas when they started to look into the cargo-door problem. On June 13, the day after the Windsor incident, FAA engineers contacted the Douglas plant at Long Beach to find out if there had been any previous problems with DC-10 cargo doors, or if its failure over Windsor had come quite out of the blue. The company omitted to hand over operating reports filed by the airlines using DC-10s and acknowledged only that there had been a few "minor problems." (Every untoward incident in an airplane's life is reported to the manufacturer, and the airlines have an obvious legal incentive to make their reporting systems accurate. But it is a weakness of the regulatory system that it is left to the manufacturer to decide whether to draw the FAA's attention to any particular report.)

Richard (Dick) Sliff, then head of aircraft engineering under Basnight, is a highly experienced test pilot, well acquainted with the fact that airplane systems nearly always give some kind of warning before they fail. He was "disturbed" by the company's attitude and "raised a fuss" to get the airlines' reports.†

* See *Epilogue*.
† Testimony of Arvin Basnight to the Interstate and Foreign Commerce Committee of the House of Representatives, March 27, 1974.

On examining the records, Sliff found that during the ten months of DC-10 service, there had been approximately one hundred reports of doors failing to close properly and that Douglas had already had to recommend modifications to the system. The trouble was that the electric actuators were not always succeeding in driving the over-center latches fully over the latch spools.* They were sticking partway, requiring extra applications of electric power, and in some cases hand winding. (All DC-10 doors can be opened and closed by a hand crank, in case of power failure.) The Douglas engineers had proposed lubrication of the latch spools when the problem was first reported, but that had not been effective. They had then sent out a service bulletin recommending that the power supply to the electric actuators be rewired in a heavier gauge of wire, thus lessening transmission resistance and increasing the power developed by the actuators. The four operating airlines, United, American, National, and Continental, were still rewiring doors when the Windsor incident occurred, and on Captain McCormick's plane it had not been accomplished.

The Western Division office had certificated the DC-10 (on July 29, 1971), and among many other things that had meant certificating the cargo and passenger doors. Basnight, Sliff, and their colleagues were therefore familiar with the arguments about tension-latch doors and plug doors, and with the choice between C-latch and over-center locking systems. They were also aware, in general terms, of the relationship between door sealing, floor strength, and airworthiness in a plane which, like the DC-10, carries its control cables on the underside of the floor beams.

The NTSB team investigating the Windsor incident were suggesting that the DC-10 cargo door should be modified to make it "physically impossible" for the door to be improperly closed. And the NTSB also recommended that the cabin floor should be modified and strengthened to prevent its collapsing after a sudden decompression. (There was informal contact between NTSB and FAA headquarters throughout June. The NTSB made its formal recommendations to the FAA in writing on July 6.)

But in the immediate aftermath of the Windsor incident, neither the FAA's Western Division office nor anyone at the Douglas plant could see any neat and immediate engineering answer to the problem. And there were DC-10s taking off virtually every hour.

If the plane was not to be grounded—a step with serious economic consequences in the middle of the summer air-travel season—there would have to be an interim "fix": something which could be agreed upon and installed rapidly and which would ameliorate the situation until a proper redesign could be accomplished.

The problem was simple enough to define: Because of the inadequacy of the manual-locking mechanism, a man closing the cargo door could not be sure that the locking pins had actually gone home. Why not, therefore, place a small peephole made of toughened glass in the middle of the metal door

* See Chapter 8 for explanation of over-center principle.

skin over one locking pin? Then the man closing the door would be able to *see* whether the pins were safely home. Nobody in the Western Region office thought this was a complete solution, for the door sill of a DC-10 stands some fifteen feet above the ground, and each locking pin is less than two inches long. To make a proper inspection, each baggage handler would have to wait for the door to come down, and then move his mobile platform along to peer into the one-inch peephole. At night, he would need a flashlight, and it might in any case be difficult to see through glass streaked with oil, dirt, and water. Still, given that all electric actuators were rewired, and given suitable alerting of ground crews, this kind of change would make the DC-10 reasonably safe until something better could be worked out.

The one thing that no one in the Western Region office doubted was that the reworking of the door would have to be enforced by a series of Airworthiness Directives, and by Wednesday, June 15, a certain amount of drafting had been done. On the morning of June 16, four days after McCormick's landing at Detroit, a preliminary text of the first proposed AD was sent by telecopier to the FAA headquarters in Washington, although it did not go beyond giving mandatory force to the wiring changes that Douglas had already recommended. During the day Everitt Pittman of the Western Region office, guided by Sliff and consulting with engineers at the Long Beach plant, worked on a more elaborate text which would specify the size and location of the inspection hole and require warning words and diagrams to be put up alongside them. But although the day had started with the assumption that an AD would be issued against the DC-10, within a few hours, Basnight's staff began to realize that something had gone wrong.

Basnight had started his day early, for the case of the DC-10 door was a considerable crisis on his beat. At 8:50, he received a phone call from Jackson McGowen, president of the Douglas division of McDonnell Douglas. Late the night before, said McGowen, he had spoken on the phone with Jack Shaffer, Basnight's boss, in Washington. They had "reviewed" the work Douglas had done on beefing up the wiring in the DC-10 door system. The Administrator, said McGowen, had been pleased to hear that this work had been going on in cooperation with his own FAA officials—so much so, it appeared, that he saw no need for any Airworthiness Directives to follow from the Windsor incident. The call to Basnight was clearly to inform him that there was an understanding between his boss, John Shaffer, and the builders of the DC-10. As Basnight received the message: "Mr. Shaffer . . . had told Mr. McGowan [*sic*] that the corrective measures could be undertaken as a product of a Gentleman's Agreement thereby not requiring the issuance of an FAA Airworthiness Directive." *

Basnight called in Sliff, who said—as Basnight was well aware himself— that work on an AD was already far advanced. Furthermore, so far from Douglas having cooperated in a handsome manner with the FAA, Sliff

* From Basnight's Memorandum to File, June 20, 1972. This document is the prime source for the events of June 16, 1972. It is reproduced in full as Appendix D.

pointed out that he had been forced to raise a considerable fuss before the company had been ready even to disclose that a serious problem had been identified. Sliff was "disturbed" at the way the company had behaved, and in any case he did not think that rewiring the power supply to the actuators would be enough to deal with the problem.

With Sliff present, Basnight telephoned Washington and spoke to James Rudolph, effectively Number Three in the FAA hierarchy under Shaffer and Deputy Administrator Ken Smith. Basnight gave Rudolph "the background data available to me," and "asked his guidance" as to continuing work on the AD text. This may be taken as public-service language for: I put him in the picture and asked what the hell was going on at his end. Rudolph said that he had not yet seen the proposed AD himself, but he agreed that the work of drafting a refined version should continue. Some time later that day Deputy Administrator Smith called Basnight from Washington. According to Basnight's account, Smith was particularly worried as to why the FAA had been unaware of the cargo-door problems which the airlines had been experiencing. Why hadn't the Maintenance Reliability Reports, which are regularly submitted by airlines, told the Western Region what had been going on? Basnight said that he was still looking into the question of why the FAA had been in the dark and that he would let Washington know as soon as he found out anything. Finally: "Mr. Smith indicated that he concurred with our judgment that an Airworthiness Directive should be issued, and that he would consult with the Administrator."

Throughout the afternoon Pittman, Sliff, and other Western staff members continued working on the text of the AD. But no word came from Washington. Uneasily aware that Washington's day ends three hours sooner than California's, Basnight and Sliff placed another call, fairly late in the afternoon, to Rudolph. They were told that John Shaffer was not in the office and could not be found—from which, it seems, they gathered that there could be no question of rescinding Shaffer's decision against issuing an AD.

Almost immediately Deputy Administrator Smith came on the line with Rudolph, and the two of them explained to Basnight and Sliff that they were trying to set up a telephone conference between Washington, the Douglas plant, and the four airlines then flying DC-10s. The aim of the conference would be to obtain all-around agreement on effecting the changes that would have been demanded in an AD. By this time Basnight must have recognized that he was being decisively cut out of the executive process. And if confirmation of this was needed, it came when, after waiting for some time without information, he and Sliff heard *from the Douglas plant* that the telecon had taken place.

Basnight called Rudolph yet again, to try and find out what exactly had been agreed with the airlines. The purpose of the telecon had been to achieve voluntarily the aims that the Western Office had wanted to achieve with an AD. Had the airlines, then, been asked to make all the changes included in the latest draft, and in particular to install the peepholes?

The answer was no. All that the airlines and Douglas had specifically agreed to do was beef up the electric wiring and install a placard on each cargo door warning baggage handlers not to use more than fifty pounds of force when closing the latching handle: precisely how a baggage handler was supposed to measure his muscle power was not explained. The peephole idea had not been discussed in detail because, of course, at the time of the telecon the semantics were still being worked out at the Western Region office. Well then, could the specification for the peephole now be sent to the airlines? Again the answer was no. Rudolph explained that a telegram, confirming the agreement reached over the phone, had already been sent to the airlines, signed by the Deputy Administrator Smith. Smith had now left the office (it was late on Friday evening in Washington) and it was too late to modify or add to the telegram. "Mr. Rudolph . . . suggested we work the problem on Monday, but indicated he agreed with our proposed amendment."

On Monday, June 19, Basnight did indeed have further dealings with his Washington HQ, but none that produced any more substantial guidance for the airlines using DC-10s. Early in the morning he received a call from Joe Ferrarese, one of Rudolph's senior colleagues, who said that he was instructed to ask Basnight to destroy all but one copy of the teletype message to airlines that the Deputy Administrator had signed for transmission on Friday night. Basnight replied—no doubt rather drily—that he had not yet received any copies of his message, but on doing so he would act as required.

Shortly afterwards, three copies of Smith's message arrived in a sealed envelope. It consisted of a briefly stated request to all airlines operating DC-10s to accomplish Douglas Service Bulletin SB 52-27—the wiring modification—if they had not already done so and install the warning placards. It made no mention of inspection holes.

Basnight called in Sliff, handed him one copy of the message for his records, and destroyed the other two. Next day, feeling no doubt that he was a little exposed, he sat down and wrote a 1,500-word Memorandum to File recording the events of the previous few days. Later, he testified that it was something he had never before felt moved to do during thirty years as a public servant.

IT IS HARDLY surprising that after this crushing defeat the Western office made no further attempts to issue ADs relating to the DC-10 door system. However, Sliff and his colleagues remained in touch with the engineers at the Long Beach plant, who did at least make some effort to honor the gentleman's agreement that Jackson McGowen had made on their behalf.

It was, of course, obvious to any serious engineer that the situation could not be left as it stood on Friday, June 16. In addition to the actuator rewiring, Douglas proposed three more changes to the door, all of which were sent out to the airlines by Service Bulletins from Douglas over the next two months.

First, in SB 52-35 they adopted the peephole idea. This was an "Alert" Service Bulletin, printed on blue paper, rather than white, to show airlines that it concerned a safety problem. Alongside each door frame there was also to be a decal showing diagramatically what a baggage handler would see if he looked in after the locking pin behind the peephole was safely home. This was to be labeled SAFE. There was another decal diagram, giving an idea what might be seen if the pin had for some reason not moved, and this was to be labeled UNSAFE. As Chuck Miller was to point out later, this would demand remarkable devotion from a baggage handler, working on a snow-swept airport in a January night.

Then, in SB 52-37 they went somewhat further, producing for the first time an approximation of a satisfactory design. Tests after Windsor had shown that the essential fault lay in the weakness of the linkages which were intended to drive the locking pins home. When the pins, on that occasion, encountered unclosed latches, then the locking handle outside the door ought to have become immovable in the ground crewman's hand.* Instead, when he pressed down on the handle with his knee, the mechanism merely gave way, enabling him to push the handle right down into its slot. The top of the handle passed through the door skin and was attached to a "torque tube" just inside. This was supposed to revolve with the downward movement of the handle, thereby moving a crank, which in turn would push a long vertical rod going down to the latch assemblies in the lower door. There, the movement of another crank was to translate the vertical thrust into a horizontal sliding movement, and thus drive the bar with all four lock pins mounted on it.

But the torque tube was made to revolve in between bearings which were a long way apart, and the span was too much for its strength. In the Windsor case, the tube had simply bent out of shape when the other parts of the linkage refused to move. Unhappily, though, it did revolve a little—just enough to operate the other linkage working off it, which ran upward to shut the vent door, thus completing the ground crew's illusion that all inside was well.

In SB 52-37, the Douglas engineers therefore recommended the fitting of a "support plate" to hold up the torque tube just beside the handle. It was simply an aluminum bracket fixed to the inner skin of the door, with a half-circle cut out in which the torque tube could rest and revolve. This, it was calculated, would prevent the tube sagging under any conceivable manual pressure which might be applied to the handle. In the same bulletin, they recommended that the linkages be adjusted so as to extend the normal travel of the locking pins by one-quarter of an inch. This would make the mechanism jam all the more conspicuously if unclosed latches were to stand in the path of the locking pins.

* See Appendix B for diagram.

It was not a perfect solution. But if these changes had been properly carried out on every DC-10 built and put into service, then the slaughter of March 3, 1974, might well have been averted. The trouble, of course— as our story later shows—is that they were not properly carried out. (SB 52-37 was sent out as a *routine* service bulletin, printed on white paper with no indication that it was vital to the DC-10's safety. As a result, very few DC-10s were modified with any alacrity. When the bulletin was issued on July 2, 1972, there were thirty-nine DC-10s in service, all of them operated by U.S. domestic airlines. Only five airplanes were modified within ninety days; eighteen were not modified until 1973; and one DC-10, owned by National, was still flying around without a support plate on March 5, 1974— nineteen months after the bulletin was issued.)

Because history is an unrepeatable experiment, we cannot prove that the extra urgency, legal weight, and publicity which go with Airworthiness Directives would necessarily have made the difference. But the crucial point is not so much the issuing or nonissuing of any particular directive, as the general determination on John Shaffer's part that the Douglas Company itself could be left to handle the matter in its own way. And, of course, the very way that the thing was done weakened the authority of Arvin Basnight and his staff. And the secrecy in which the whole business was accomplished was damaging also: Douglas employees later testified that they were simply unaware of the significance of the various things that were supposed to be done to DC-10 doors.

According to Shaffer, the term gentleman's agreement is an unhappy one, conveying the incorrect impression that his transaction with Jackson McGowen amounted to "a handshake behind the barn." * Yet the fact is that at the time he agreed with McGowen that no AD was necessary, it would have been quite impossible for McGowen to have given him any detailed account of what might be done to make the door safe—for no one at the Douglas plant or in the FAA had yet worked out a detailed solution. Shaffer covers this point by saying, simply, that "you have got to have faith in people." But Shaffer should have been aware that the summer of 1972 was a moment when nobody in the Douglas division of McDonnell Douglas could be expected to give an objective judgment on any matter which might involve publicity for actual or potential drawback in the DC-10. Briefly, it was a deeply inconvenient moment for Douglas to have an AD issued against their plane.

IN 1971, IT had seemed that Lockheed and the TriStar were to be destroyed by bankruptcy. Against all the odds, and against the ideological prejudices of both the American and British governments, the TriStar and the RB-211 had survived, with the result that the Lockheed sales team had returned to the field with a new and desperate vigor. The competition had

* Television interview, Thames TV (London) and *The Sunday Times* (London), 1975.

reached the point where the allegiances of most major American airlines had been declared, and the vital need was to capture foreign airlines. Both sales teams had the same philosophy: that what mattered in 1972 more than actual numbers sold was to obtain first orders from certain airlines thought to occupy influential positions. Both Lockheed and McDonnell Douglas were now looking hard at Turkish Airlines, believed to be ripe for a purchase of new equipment and thought to be capable of influencing other Middle Eastern airlines by its example.

The two airbus makers had each set up round-the-world sales odysseys for summer 1972, using in each case a specially prepared airplane which would be laden with executives, engineers, salesmen, and publicity experts of all kinds. Jackson McGowen was proposing to lead the Douglas excursion in person, and the Windsor incident occurred just before the scheduled departure of "Friendship '72" from Long Beach.

The special unpleasantness of the kind of AD that Basnight's men began preparing after the Windsor incident was that it was not unlikely to be clear-cut and swiftly disposable. It could not be dismissed as a once-only problem that had been examined and cured—at least, not within the time-scale of Friendship's tour. Because ADs are distributed through the diplomatic network, the problem of the exploding door might well have pursued Friendship '72 the whole way round the world. (There is little doubt that the Lockheed team, on their own circumnavigation, would have found discreet ways of capitalizing on the issuing of ADs against the DC-10.)

The question of political contributions must be examined because McDonnell Douglas executives emerge as substantial contributors to Richard Nixon's funds for the 1972 Presidential election. "Mister Mac," together with his fellow executives and the company itself, gave just over $90,000 during the first six months of 1972. There is, however, no case for assuming that there was any purchase of particular favors, because nearly all of this money was committed *before* the gentleman's agreement was made—and all of it was publicly declared.*

Herbert Kalmbach, one of Nixon's chief fund-raisers during the 1972 campaign, did visit Jackson McGowen that summer and said: "We have you down for some substantial money . . ." McGowen told him: "I am sorry, but I don't have any money." McGowen is a lifelong Democrat.

IT IS OF course, argued with increasing support that the system of political contributions too often places American Government under unsuitable obligations to particular interests—but even if that case is granted in theory, there was nothing in the McDonnell Douglas case that was unusual by the practical standards of the time. The immediate key to the curiosities

---

* Part of the cash was actually paid *after* the agreement, but this was only because one individual donor, Sanford McDonnell, had to sell some stock to make good his pledge, and it took a few weeks to do this.

of the gentleman's agreement is probably to be found in the personality of John Hixon Shaffer, a man who, in his own words, "likes to think optimistically about the future."

He is a muscular, outgoing man who graduated from West Point during World War II and survived a tour of duty on the horrendous B-26, an airplane widely thought to have inflicted more damage on the U.S. forces than those of the enemy. From an examination of the DC-10 affair, he emerges as an honest man, but one very unwilling to perceive any faults in the industry that had made him prosperous and successful.*

He arrived at the FAA through the introduction of Melvin Laird, then Secretary of Defense, and from the beginning he saw his chief task not in terms of safety administration, but as one of defending the aviation industry against the depredations of the increasingly active environmentalist lobby. (This, he somewhat eccentrically believed, had been devised by H. R. Haldeman and John Ehrlichmann to divert people's attention from the necessary unpleasantnesses of the war in Indochina.) Therefore, to Shaffer the sad failure over the DC-10 appears as a relatively minor incident in his tenure at the FAA. The major achievement, he feels, is that during that time "to the everlasting credit of the USA, we didn't fall for the misbelief that aircraft noise is really so bad." As he points out: "Only seven million people are affected by airport noise in this country." †

Shaffer does mildly regret his period as Administrator, in that it cost him, he estimates, some $400,000 in salary and "missed opportunities" generally. However, it was worth financial sacrifice on a considerable scale, he feels, to have had the "best job in Washington" where he defended an industry for which he has a virtually romantic affection—an attitude which emerged with great clarity when he was asked if he saw any end of the FAA's system of "designated engineering representatives."

This is the system mentioned earlier, under which most of the FAA's supervision of manufacturing is done by the industry's own employees. It would hardly be possible, Shaffer said, to find enough full-time FAA inspectors to do the work. But in any case, to put outside inspectors into aircraft factories would be "like hiring somebody to watch your wife while you're at the office." ‡

After the Paris crash, Shaffer said that he would never have "sat around, fat, dumb and happy" if he had been aware that aircraft were coming out of the Douglas plant with cargo doors that were still unsafe. But that was afterwards: At the time, he expressed nothing but officerlike distaste for

---

* In 1972 Buckley Byers, a fund-raiser for the Committee to Re-Elect the President (CREEP) asked Shaffer for a list of aerospace people who might make good fund-raising targets. Shaffer formed the view—no doubt accurately, given what is now known about the Committee's methods—that CREEP was inviting him to provide a shakedown list. He told Byers that a full list of all aerospace executives could be found in the World Aviation Directory, available at a government discount price of $13.50.
† Authors' interview, July 1974.
‡ Authors' interview, July 1974.

the grubby minutiae of inspection. His own summary of the conversation with McGowen is that "I rang him and said what the hell is wrong with your goddam plane? And he replied: 'Jack, we'll have it fixed by Friday night.' "

Shaffer's FAA career ended with Richard Nixon's 1972 election victory. By tradition all political appointees in Washington formally offer their resignation at the end of each administration's four-year term, and Shaffer duly submitted his in November 1972. He hardly expected it to be accepted for he had just been awarded the Wright Brothers' Memorial Trophy "for aviation": the President had sent him a glowing tribute for "bringing credit not only on yourself but on my administration." Yet, on December 9, while he was having lunch with a magazine correspondent—to celebrate the Wright Brothers' award—Shaffer received a telephone call from his secretary. A message had just come down from the White House saying that his resignation had been accepted and would he mind beefing it up a little? The White House suggested that he could, perhaps, add a few words about what a pleasure it had been to serve. As he cleared his desk Shaffer must have reflected on whether his advice to Buckley Byers of CREEP about the World Aviation Directory had been considered sufficiently responsive.

Shaffer was replaced as FAA Administrator by Alexander Butterfield, then a little-known aide to Bob Haldeman who had been in charge of internal security at the White House. (Just after taking up his new appointment Butterfield let slip that for two years all of the conversations in Nixon's office had been tape-recorded—an admission that gave the Ervin committee and the Watergate Special Prosecutor's office the break they had been waiting for. There is considerable irony in the fact that Shaffer lost his job to the man who was inadvertently responsible for Nixon's final downfall: Shaffer remains a considerable admirer of Nixon and convinced that a "great President" was crucified by the media.)

Butterfield was not a great success as FAA Administrator and in February 1975 he was forced to resign by the Ford Administration. But to his credit he did, two days after the Paris crash, order the establishment of an FAA ad hoc committee to investigate the history of the DC-10 and the role that the agency had played in its development.

Much of the report the ad hoc committee produced on April 19, 1974, was highly technical, but it amounted to a damning indictment of the DC-10 door design and of the regulatory system in the United States that had permitted its certification. On the subject of the DC-10 door, the report said:

> While there is no longer any doubt that [the DC-10 cargo door] is safe, it is an inelegant design worthy of Rube Goldberg. . . . The DC-10 intercompartment structures [partitions, floors, bulkheads] were not designed to cope with or prevent failures that would interfere with the continued safe flight and landing after the sudden release of pressure in any compartment due to the opening of a cargo door. It was therefore incumbent on McDonnell Douglas to show that loss of the cargo door . . . was "extremely remote" for compliance with Federal Airworthiness Regulation 25. . . . It

would appear, in the light of the two accidents, that the level of protection and reliability provided in the cargo door latching, safety locking mechanisms, and the associated warning systems was insufficient to satisfy the requirements of FAR 25. Additionally, it now appears that the possibilities of improper door operation were not given adequate consideration for compliance with FAR 25.

On the subject of the gentleman's agreement the FAA report had this to say:

> Review of the FAA's . . . correction programs associated with the DC-10 airplane has again pointed out that the agency has been lax in taking appropriate Airworthiness Directive action where the need for ADs are clearly indicated. This situation is by no means unique to the DC-10 airplane or to the Western Region. So-called "voluntary compliance" programs have become commonplace on a large number of aeronautical products and in most, if not all, of the [FAA] regions. Voluntary compliance programs have been . . . most commonly implemented on the smaller private and executive aircraft types where the number of ADs issued is most likely to be considered a sales deterrent by the manufacturer.
>
> Many complaints on this issue have been received from a number of foreign airworthiness authorities who have airworthiness responsibility over U.S. manufactured products in their respective countries; and, for the most part, the FAA has ignored those complaints.

Dealing specifically with the Windsor Incident, the report said:

> The agency was not effective in attaining adequate fleet-wide corrective action on a timely basis after problem areas were clearly indicated by the . . . accident. Non-regulatory procedures and agreements were used in lieu of established regulatory AD procedure [and] in the long run proved to be ineffective in correcting design deficiencies . . . to prevent reoccurrence of the [Windsor] accident.

After reading that report and conducting its own review of the FAA and the DC-10 affair, a special subcommittee of the House of Representatives concluded that between June 1972 and the Paris crash of March 1974, "through regulatory nonfeasance, thousands of lives were unjustifiably put at risk." * In defense of the technical experts of the FAA it has to be said that in the thirteen months leading up to the Paris crash they did begin to display some unease about the fundamental hazard (the vulnerability of the DC-10's floor) that the Windsor incident had publicly revealed. In February 1973—after Shaffer had been fired by Nixon, but a couple of weeks before he actually quit his desk—the FAA wrote to McDonnell Douglas, Lockheed, and Boeing expressing concern about the consequences of rapid decompression in wide-bodied jets. The letter said:

---

* Report by the Special Sub-Committee on Investigations of the Committee of Interstate and Foreign Commerce, House of Representatives, January 1975.

We are concerned by the long-range safety implications of a recent incident wherein an in-flight loss of a nonplug type cargo door resulted in failure of the pressure cabin floor and disruption of flight and engine controls. This incident represents a new failure mode involving the wide body jets . . . which tend to make them more critical in this regard than past designs.

In view of the foregoing, we request that you re-assess your jumbo jet designs with regard to the effects on safety . . . at an early date and recommend a course of action.

This sudden concern on the part of the FAA was undoubtedly belated. The Rijksluchtvaartdienst (RLD), the Dutch counterparts of the FAA, had been expressing concern about the vulnerability of the DC-10's floor since April 1971, when KLM decided to buy the plane. As the RLD had pointed out to both the FAA and McDonnell Douglas, there was "a marked difference" between the Douglas DC-10 and the Boeing 747.

In airplanes where flight and engine controls, electrical cables and system lines are attached to the floor, a sudden decompression below or above the floor might not only cause damage to the floor and block the control of the airplane and its engines, but also the vital systems like hydraulics and oxygen lines, and can therefore endanger the safety of the aircraft.

After Windsor the RLD had sent a delegation to Los Angeles to meet with the FAA and McDonnell Douglas. By that time—September 1972—the NTSB had also recommended that the DC-10's floor should be strengthened, but the plurality of opinion seems to have had little immediate impression. McDonnell Douglas pointed out to the Dutch delegation that the DC-10 floor met all of the requirements laid down in the Federal Airworthiness Regulations. Then, said the Dutch, the requirements were inadequate. In a formal reply to the RLD, the FAA said: "We do not concur with RLD views concerning the inadequacy of FAA requirements."

We do not know what happened between September 1972 and February 1973 to bring about the FAA's change of heart. Perhaps John Shaffer's imminent departure from the post of administrator had something to do with it, but in any event by February 1973 the technical experts at the FAA's headquarters in Washington, D.C. had decided there was a need to do something about jumbo-jet floors. In its letter, dated February 2,* the FAA asked the manufacturers to consider two alternative proposals: (1) the vital control systems could be re-routed away from the floor; or (2) the floor itself could be strengthened by reinforcing the supports and by adding vents which, in the event of decompression would allow pressurized air in the passenger cabin to escape before its weight could collapse the floor.

Boeing was plainly displeased at being involved in the aftermath of Windsor.

---

* The copy of the letter sent to Boeing appears to be an afterthought. It was dated February 8.

As the company's reply to the FAA pointed out, the 747's control cables do not run under the floor (although the hydraulic tubes do) and added: "In the decompression cases quoted in your letter, the door locking systems [on the DC-10] were not fail safe in that single failures permitted the pressurization of unlocked doors which led to the opening of the doors in flight."

Lockheed felt equally strongly that it was unfair to include the TriStar in any redesign program that the FAA obviously now contemplated because of Windsor. The company felt there was no question of TriStar doors coming open in flight because they were semiplugs. As for the danger of small bomb explosions (an alternative hazard to floors which the FAA had also raised in its letter), Lockheed argued that jumbo jets were no more vulnerable to this threat than all other pressurized airliners.

For its part, McDonnell Douglas displayed no embarrassment over the fact that it was the DC-10 that had caused all the trouble, insisting that the chances of a DC-10 cargo door opening in flight were "extremely remote." The company's reply to the FAA, dated March 15, 1973, continued: "A reassessment of the DC-10 design, with regard to the effects on safety, for non-plug cargo doors and small bomb explosions shows that the present standard and levels of substantiation [of the floor] are adequate."

To their credit, the FAA's safety experts in Washington were not satisfied with these bland assurances and in June 1973 they asked Arvin Basnight's Western Region office to get from McDonnell Douglas and Lockheed some technical data about jumbo-jet floors. (The FAA's Northern Region office, located in Seattle, was asked to get the same information from Boeing.) The Western Region's reaction to that request is one of the greater ironies of the DC-10 affair. In the squabble with FAA headquarters over the gentleman's agreement the Western Region had advocated a minatory approach: now the Western Region urged restraint. In its reply to FAA headquarters— written on the first anniversary of the Windsor incident—Basnight's office said:

> We are aware of no subsequent service difficulties [since Windsor] which would constitute a basis for recertification [of the floor]. In the light of the above, we feel that an investigation of the detailed nature presented in the memo of 30 May 1973 would be premature.

And there the matter stood for the rest of 1973. The Northern Region did ask Boeing to make a technical survey but the company continued to insist on the 747's basic integrity. The only success FAA headquarters achieved was with a foreign manufacturer over which it had no jurisdiction: informed by the FAA as a courtesy, of the concern about wide-bodied floors, Airbus Industries of France decided to strengthen the floor of its A300B European Airbus which was then about to go into production. In September 1973, the Dutch RLD, having failed to make much impression on either McDonnell Douglas or the FAA, certificated the DC-10. It did, however, place on record its reluctance: "The RLD isn't happy with the present situation, but for this generation of aircraft, we have to live with it." *

It was not until February 5, 1974, that FAA headquarters formally repeated its request to the Western Region that McDonnell Douglas and Lockheed should be asked to carry out a technical study of jumbo-jet floors. This time the Western Region did at least pass on the request. Lockheed replied that the study called for would be expensive and time consuming and might be better done by the aviation industry as a whole, rather than by individual manufacturers. McDonnell Douglas said much the same: "We do not have the manpower available at this time to undertake this study, nor are we in a position to accept this burden alone."

In its reply to the FAA, dated February 25, 1974, McDonnell Douglas also said that if the Government wanted a study made, then Government funds should pay for it. Six days later came the Paris crash.

The irresistible question is: Could the Paris tragedy have been averted if the three manufacturers—and specifically McDonnell Douglas—had taken more notice of the concern over jumbo-jet floors? By February 1974 it was, of course, far too late for a technical study to have altered the fate of Turkish Airlines' DC-10. But in April 1971, when the Dutch RLD first voiced concern, the DC-10 had not yet been certificated and only a handful of models had been built. True, it would have been a highly embarrassing moment—from a commercial viewpoint—to return to the drawing board, as it were, but as McDonnell Douglas has continually stressed to us, safety is *always* the paramount factor. There is no doubt that if McDonnell Douglas had listened to the Dutch, and if the DC-10 floor had been strengthened to tolerate a hole in the fuselage of twenty square feet (as all wide-bodied floors must be able to do under the FAA's regulation introduced in July 1975), then the Paris crash would not have happened. The area of a DC-10 rear cargo door is a little over 14.5 square feet. If the Turkish Airlines' DC-10 floor had been reinforced and fitted with additional vents, the pressurized air in the passenger cabin would have escaped, leaving the floor intact and the control cables undamaged. At most, the passengers and crew would have suffered mild discomfort. However, McDonnell Douglas's decision to disregard the Dutch warnings—and later the warnings of the NTSB—relied on the fact that the DC-10 floor complied with the requirement of the FAA's current airworthiness regulations.

Most of the blame for the failure to do something about the extreme vulnerability of jumbo-jet floors must therefore lie with the FAA. Even after Windsor—when there would still have been time to prevent the Paris crash— the FAA was dilatory. In July 1975, by which time the danger could no longer be denied, the agency did finally order the modification of all DC-10, TriStar, and 747 floors. But the delay undeniably cost the lives of 346 people, and it will take until the end of 1977 before all jumbo jets have received

* Letter from RLD to Western Region of the FAA, dated September 19, 1973.

added protection against a hazard that was clearly identified five years ago.*
However, and in fairness, before the Paris crash the FAA was apparently
confident that, one way or another, the DC-10 cargo-door problem had been
fixed. That confidence might have evaporated if the FAA had known that
the DC-10 designers had chosen to withhold documents which were far more
alarming than the hundred-odd reports of door-latching failure that Dick
Sliff's persistence eventually uncovered in June 1972.

To some of its designers, the faults of the DC-10 had been obvious long
before that. And to judge by their written prophecies, neither Windsor nor
the later tragedy outside Paris could have come as much of a surprise.

---

* By December 1977, the floors will have to be able to withstand the consequences of a
hole up to twenty square feet appearing suddenly in the fuselage. In theory, the FAA's
order affects only U.S. operators but, in practice, all airlines flying jumbo jets are expected to comply. Most DC-10s, TriStars, and 747s built since the beginning of 1976
have been fitted with strengthened floors: older versions are being modified as they are
grounded for routine maintenance overhauls.

10

The Applegate Memorandum

The careful text-books measures
(Let all who build beware!)
The load, the shock, the pressure
Material can bear.
So, when the buckled girder
Lets down the grinding span,
The blame of loss, or murder,
Is laid upon the man.
Not on the Stuff—the Man!
—RUDYARD KIPLING,
Hymn of Breaking Strain

A REGULATORY BODY such as the FAA is, in the end, no better than the information it receives. A safe airplane is the product of a dialectical process between builders and regulators. The builders should bring issues and information to the attention of the regulators, who are supposed to employ an experienced, but impartial, viewpoint in assessing the design problems. Without this dialectic—as we suggested earlier on page 164—the Federal Airworthiness Regulations are worth no more than the paper they are printed on.

The process requires, of course, alertness and intellectual rigor on the part of the regulators. But even more, it requires cooperation and openness on the part of the aircraft manufacturers. Even when directed by a temperament more skeptical than John Shaffer's, the FAA's resemblance to a police force is strictly limited. It is a body of engineers and administrators, not of detectives, and it has no institutional aptitudes, such as even a good newspaper may have, for detecting that which has been concealed. Briefly, if the regulators have less than complete access to design information, then their role becomes meaningless.

This is axiomatic among aeronautical engineers. Yet the Douglas and Convair design teams responsible for the DC-10 did not submit to the FAA material about the DC-10 which was even more disturbing than the records of the one hundred or so cargo-door malfunctions that emerged in the week after the Windsor incident. It is hard to believe that if the FAA officials had been fully aware of the design history of the DC-10 cargo-door system, they would have certificated the aircraft as readily as they did. And it is almost impossible to believe that the adjustments made under the gentleman's agreement would have been taken as sufficient.

In our view, the root cause of the catastrophe, on March 3, 1974, was a failure of communication. No doubt some part of this was caused by John Shaffer's conception of the role of the FAA. Some part of the blame must also lie with the major subcontractor for the DC-10, the Convair division of General Dynamics. But the central responsibility, at least in terms of morality, must lie with McDonnell Douglas and in particular with its Douglas division. Douglas's claim was, and remains, that the DC-10 is "the product of a forty-year tradition of quality," in which the consideration of safety has been uppermost at all times. As one of McDonnell Douglas's current advertising campaigns says:

> Airlines recognized our heritage of quality.
> DC transports began writing aviation history more than forty years ago. Since then, we've built more than 3,000 commercial airliners bearing that famous McDonnell Douglas trademark.
> The growth of modern air transport began in 1933 with the DC-1: followed by the world renowned DC-2 and DC-3 which established DC as a mark of performance and dependability.
> The DC-10 wide-cabin jetliner is the crowning achievement in the distinguished DC line, upholding this tradition of service to airliner and air travellers around the world.

Douglas's behavior in the case of the DC-10 is the more remarkable because the company's own records must include the best possible account of a classic precedent: Perhaps the clearest single case in which a failure of communication between an aircraft manufacturer and its regulatory agency resulted in avoidable slaughter. This precedent is not well known in public, partly because of the curious way in which it was investigated and the lengthy legal proceedings that surrounded it. But it would be remarkable indeed if it left no impression upon the corporate consciousness of the Douglas Company, particularly as one of those intimately concerned was Donald Douglas Jr.

The airplane involved was the DC-6, a piston-engined, pressurized airliner, which was produced immediately after the war in fierce competition with the Lockheed Constellation and the Boeing Stratocruiser. After a United Airlines DC-6 had crashed near Bryce Canyon, Utah, on October 24, 1947, killing fifty-two people on board and an American Airlines DC-6 had landed with flames streaming from it at Gallup, New Mexico, on November 11 of the same year, all DC 6s in service in the United States were grounded. It turned out that in the DC-6 fuel could spill when being transferred in flight from tank to tank, and that fuel vapor could then make contact with the electric coils of the cabin-heating system, causing inevitable fire.

In those days the Civil Aeronautics Board was responsible for inquiring into airline disasters and the official investigation by the CAB into the United crash accused both Douglas and the CAA (the predecessor of the FAA) of carelessness in building and certificating the plane.* What followed, however, was worse.

Douglas was told that, in addition to preventing fuel spills, the in-flight fire-fighting capability of the DC-6 would have to be radically improved, and this was achieved by fitting banks of carbon dioxide (CO_2) cylinders in several parts of the aircraft, including the cargo compartments under the cabin floor. Then, in order to get the plane recertificated, it was necessary, among other things, to show that discharging the CO_2 would not incapacitate the flight crew—for carbon dioxide stops fires by replacing oxygen in the immediate atmosphere, and human beings need oxygen just as much as fires do.

Douglas ran a series of flight tests in January and February 1948, with alarming results. On January 16, test pilot Lawrence Peyton discharged the extinguishers in the central baggage compartment, to fight a fictional fire, whereupon he and his copilot suffered "extreme physical discomfort" and almost lost control of the plane.† In a second test, the same day, the pilots were equipped with rebreathing oxygen masks but the same trouble occurred: the pilot "was almost completely out . . . [he] just sat there . . . [he] seemed to be excited and was staring straight ahead." Captain Peyton's conclusion was that *pure* oxygen masks must be available for any occasion when the extinguishers were used. Douglas called in Dr. Clayton White of the Lovelace Clinic, Albuquerque, New Mexico. On February 9, Dr. White, a noted aero-medical expert, produced a devastating report.

Douglas had approached the problem on the proposition that CO_2, although it could cause damage to human beings by replacing oxygen, was not in itself toxic. But as White pointed out, the gas can be actively lethal, and everything depends upon the concentration in which it is present. The problem, he wrote, was "of utmost importance in aviation now that carbon dioxide fire-fighting gear is so widely used, and this is particularly true in pressurized aircraft in flight."

The doctor said that the tests that Douglas had carried out to measure CO_2 concentrations in the DC-6 cabin had been "inadequate" and he urged that the testing methods should be refined and the experiments rerun. "This . . . might avoid some embarrassment at a future date." Finally, he made a number of recommendations. The first one said:

> Maximum allowable CO_2 concentrations for habited compartments in aircraft be established in commercial aviation, *and that the relevant material in this report be made available to the CAA, the aircraft industry, the Airline Medical Directors' Association and the Air Transport Association as an aid*

* The CAB, a U.S. Government agency, is now exclusively concerned with regulating airlines, deciding which routes they can fly, the fares they can charge, and so on.
† This and all other references in this section are from *DeVito* v. *United et al.*: Civ. No. 9555, U.S. District Court, Eastern District of New York, May 1951.

to establishing and accepting industry-wide regulations in this regard. (Emphasis added.)

Douglas did exactly the opposite. Captain Robert Bush, the company's chief test pilot, wrote on the face of the White report: "Do not discuss with the CAA." And it wasn't.

But the minimizing of the dangers of carbon dioxide did not stop there. The CAA officials suggested some minatory language in the sections of the DC-6 operating manual to deal with the new firefighting procedures. The DC-6 was fitted with pressure-relief valves and the premise was that if the valves were opened before the CO_2 extinguishers were discharged, the diminution of pressure would mitigate the level of carbon dioxide in the cockpit by allowing some of it to escape. The CAA wanted the manual to say: "WARNING: Failure to open valve may result in lethal concentrations of CO_2 in cockpit and cabin." Douglas persuaded the CAA to change *lethal* to *excessive.*

In March 1948, the DC-6 was recertificated as airworthy and in June United Airlines began using its fleet of DC-6s again. With the DC-6s back in service, and with the aviation industry in general largely ignorant of the danger, an accident was now almost inevitable. However, there was one more chance to avert catastrophe. On May 13, a TWA Constellation had almost crashed near Chillicothe, Missouri, when the flight crew had been overcome by oxygen deficiency in the cockpit, after a false fire warning had prompted them into discharging CO_2 extinguishers. Subsequent tests left TWA in no doubt as to how serious the hazard was and in June it alerted the Air Transport Association, which frequently acts as a clearinghouse for information about aircraft safety. Immediately, the ATA sent a telegram to all member airlines warning them of the danger and passing on TWA's recommendation that flight crews should be told to depressurize and open cabin windows *before* using CO_2. The telegram also recommended that flight crews should have available pure oxygen masks and, finally said: "It has been recommended to us by TWA that a *similar situation may exist on DC-6 aircraft.* Therefore, suggest if possible you carrying out necessary tests to determine if such can occur to DC-6." (Emphasis added.) Douglas's response was to send a telegram of its own demanding that the ATA contact the airlines and withdraw all references to the DC-6. Douglas said that the CO_2 problem had already been dealt with in the DC-6, on the basis of "extensive tests made under varied conditions *carrying CAA approval.*" (Emphasis added.)

Whether the CAA would have approved those tests had it known that Douglas's own consultant, Dr. White, had specifically rejected them as inadequate seems highly doubtful. But Douglas had concealed Dr. White's report, and on June 15, the ATA meekly informed all DC-6 operators of Douglas's advice that the CO_2 problem had been dealt with adequately.

Two days later, on June 17, Flight 624, a United Airlines DC-6 carrying thirty-nine passengers and four crew, took off from Chicago bound for New York. On board were one or two celebrities—Earl Carroll, the Broadway showman, and his girlfriend, Beryl Wallace; Henry Jackson, one of the founders of *Esquire* magazine—and Anthony DeVito, a young businessman whose family eventually managed to expose the reasons why Flight 624 never reached its destination. (DeVito had just been made head of the Chicago division of the company he worked for and was heading for New York to tell his wife and children.)

At 12:27 over Pennsylvania, Flight 624 routinely acknowledged New York's clearance to descend for approach to LaGuardia Airport. Four minutes later there came a loud garbled message which New York could not read. However, Captain William Bach, flying a United DC-3 somewhat behind Flight 624, made out part of it: *"New York New York* [unintelligible] *. . . this is an emergency descent."* This came over, according to Bach, in a scream.* Near Mount Carmel, Pennsylvania, witnesses on the ground saw the plane swoop, climb, and pass two hundred feet over the mountaintop in a tightening righthand turn. Three miles northeast—at 12:41, seven or eight minutes after the last radio message—it crashed through a 66,000-volt power line and exploded on a hillside. A few minutes later Captain Bach in his DC-3 saw the smoke column and reported to LaGuardia, which was still calling for Flight 624: *"Flight 624 stands in no further need of clearance— wrecked at Mount Carmel."* There were no survivors. The crash investigation concluded that the crew had discharged CO_2 extinguishers in response to a false fire alarm and had been overcome by carbon dioxide. Just how many people on board were conscious during the DC-6's last minutes of eccentric flight will, of course, never be known.

In 1951 Anthony DeVito's family sued both Douglas and United and United filed a cross-claim against Douglas. The precise extent of United's knowledge before the crash about the dangers of CO_2 was one of the most bitterly contested issues of the litigation. Donald Douglas Jr.—then a senior Douglas executive—claimed that he had discussed the White report with a Mr. Christenson, United's flight safety engineer, but Christenson flatly denied that he had ever heard anything about the report from Douglas or anyone else. After the issue had been fully investigated, U.S. District Court Judge Clarence Galston decided: "It is clear that Dr. White's report was never transmitted to United. . . ." The judge said that even if Douglas's evidence was "taken at face value . . . the jury might well have questioned whether

* Later, after a tape recording of the message had been electronically cleaned, it was possible for an expert to make out other parts of the transmission:

> Flight 624: *"624 . . . 624 . . . Fire. Discharging* [unintelligible] *forward bag-gage pit."*
> New York: *"Can't hear, try again, please."*
> Flight 624: *"624. Emergency."*
> New York: *"Can't hear, try VHF"* [Very High Frequency radio band].
> Flight 624: *"624. Coming down. I . . . ah . . . ah."*

this so-called discussion constituted knowledge on United's part of the complete contents of Dr. White's report." After a seven-week trial in New York the jury awarded $300,000 against United and Douglas and dismissed United's cross-claim against Douglas. However, Judge Galston reduced the damages to $160,000 (still the highest award in a negligence case in the United States at that time) and set aside the verdict against United because of the "misconduct" of Douglas's chief defense lawyer, Lasher Gallagher, who, said the judge, had deliberately misled the jury. Among other things, one of the defense attorneys had erased the damning words, "Do not discuss with the CAA," from the cover of the White Report before producing it in evidence. Judge Galston was particularly critical of Douglas's "deliberate concealment" of that report, saying:

> It is difficult to dismiss the thought that it was human failure—negligence —that caused the death of forty-three people; and that had Douglas distributed the White report to those in the aviation industry and particularly to the CAA and United, and followed its recommendations, the catastrophe might never have happened.

It does seem astonishing that such a horrifying episode would not have left some irremovable trace in the Douglas corporate memory, particularly in view of the characteristic Douglas insistence on the importance of "tradition" in their approach to aeronautical engineering, and the role of their long-standing experience in procuring safety.

McDonnell Douglas does not see it that way. When we invited the corporation to comment on the DC-6 affair, its spokesman replied:

> Not only do I see little point in you and I attempting to retry a court case of twenty-five years ago, but I suggest to you that it is a very long reach for copy, and patently unfair, to include this kind of ancient history—involving a different company, different management, and a vastly different technology —in a discussion of the Turkish Airlines accident. Rather like questioning the Concorde on grounds of the Comet 1 disasters, or attacking the QE2 because of the Titanic.*

That is, of course, a rather different story from the one told in the corporation's advertisements: "DC transports began writing aviation history more than forty years ago. Since then, we've built more than 3,000 commercial airliners bearing that famous McDonnell Douglas trademark. . . . The DC-10 wide-cabin jetliner is the crowning achievement. . . ." There is nothing there about "different company, different management and a vastly different technology."

But, that apart, it also has to be said that the DC-6 affair was not Douglas's only lesson of what can happen when a manufacturer, and its regulatory agency, ignore manifest danger signals. Some of the later versions of the

---

* Letter to *The Sunday Times* (London) from McDonnel Douglas, dated February 13, 1976.

McDonnell Douglas DC-8 incorporated, through dubious design, a serious safety hazard. When the fault was revealed, McDonnell Douglas's reaction was to play down the danger and protest against the issuance of an Airworthiness Directive.

And this episode can hardly be called ancient history. It began less than two years before the DC-10 lost its cargo door over Windsor.

AIR CANADA'S FLIGHT 621 operates between Montreal and Los Angeles with a brief stopover en route at Toronto. On July 5, 1970, the first leg of the flight was in every way normal until the aircraft, a stretched model of the DC-8, was just sixty feet above the runway at Toronto International Airport. Without warning, the plane fell out of the sky. The starboard wing hit the tarmac first and the outboard engine was ripped from its pylon. Aviation fuel spewed out of the wing tanks and caught fire.

The pilot, Captain Peter Hamilton elected to get the plane airborne again. He opened the throttles to their maximum and pulled back on the stick. The DC-8 climbed and Hamilton prepared for a second-landing approach. Neither he nor the rest of the flight crew knew about the loss of the engine or the fire. When the airplane was at 3,100 feet, there was a series of three explosions. Most of the right wing fell off and the DC-8 plummeted to the ground. All of the one hundred passengers and nine crew on board were killed.

The cause of the disaster was straightforward. Like most modern jets, the DC-8 is fitted with ground spoilers—flat metal plates hinged to the upper surface of each wing—which, as their name suggests, are designed to spoil or disrupt the airflow over the wing and, consequently, reduce lift. As their name also suggests, they should be used only *after the aircraft has touched down on the runway:* during the last few seconds of flight, when the plane is very close to its stalling speed, it needs all the lift it can get.

At the time of the Air Canada disaster, the DC-8 offered its pilot two methods of deploying the spoilers. Either he could wait until touchdown and do the job manually by *pulling* a lever, or, during the final approach, *lift* the same lever to "arm" the spoiler mechanism. If he chose the latter course, the spoilers would automatically deploy when the landing gear contacted the runway.

What happened at Toronto, on July 5, 1970, was that the copilot inadvertently pulled the lever instead of lifting it; in other words, he deployed the spoilers instead of arming them. The cockpit voice recorder, recovered from the wreckage of the DC-8, proved that the Captain had immediately realized the mistake and yelled: *"No, no, no."* The copilot said: *"Sorry—Oh! Sorry, Pete."* Simultaneously, the right wing hit the ground.

McDonnell Douglas's first reaction to the disaster was to send a cable to all DC-8 operators, on August 7, 1970, warning that ground spoilers should not be used in flight. The cable made no mention of the Air Canada crash. Three months later the FAA Western Region decided to issue an Airworthiness Directive against the DC-8 in an attempt to guard against a repeat of the Air Canada tragedy. In the circumstances, what the Western Region had in mind was, to say the least, not very demanding: the AD would require a phrase to be added to the DC-8 flight manual saying: *"Do not extend ground spoilers during flight,"* and a placard to be installed in each cockpit alongside the spoiler lever, saying: *"Deployment in Flight Prohibited."*

It was an absurd proposal. The hazard revealed by the Air Canada crash was *inadvertent* deployment of the spoilers—the copilot certainly did not extend the spoilers deliberately. As one Canadian aviation official later pointed out, the warning placard might just as usefully have said: "Do not crash this plane." *

McDonnell Douglas realized that the warning placard would serve no purpose and told the FAA so. But the corporation went further. It did not know, it said, of a single instance of inadvertent extension of the ground spoilers in flight (the Air Canada example apart, presumably) and did not see the need for any corrective action. The Air Transportation Association, representing U.S. airlines, also objected to the proposed AD on the grounds it was "unnecessary" and Eastern Airlines protested that the installation of the warning placard in the cockpit would be "too expensive."

Unabashed the Western Region went ahead and issued its AD on December 5, 1970. For its part McDonnell Douglas sent out an "Alert" Service Bulletin to DC-8 operators telling them how to install the placard. Both the directive and the bulletin were mute about the Air Canada crash. Indeed, the AD didn't mention ground spoilers until the third paragraph, and then it said only that the aim was to prevent their "mis-use." In hindsight it seems inevitable that the consequences of this lethargy would be another DC-8 crash, caused by the inadvertent extension of the ground spoilers.

That is precisely what happened—twice and possibly three times. On October 12, 1971, the International Air Transport Association (IATA) reported that a second DC-8 had fallen out of the sky after the captain had accidentally pulled the spoiler lever. Fortunately, the aircraft was only ten or fifteen feet from the runway when the mistake was made, and although the DC-8 was damaged, the crew were not hurt. (This was a training flight and there were no passengers on board.) IATA is an association of the world's major airlines and in reporting incidents such as this does not identify the airline involved or go into very much detail. The purpose of Safety Information Exchange Bulletins—as these reports are called—is to alert other members to a potential hazard and, rightly or wrongly, IATA feels that it needs to preserve anonymity: if it didn't, airlines might be reluctant to report incidents which they found embarrassing.

Although this report of a second DC-8 accident caused by ground spoilers

---

* This section relies on the report of the Committee of Interstate and Foreign Commerce of the House of Representatives which in 1974–75 reviewed the FAA's performance in the light of the DC-10 crash. The report (No. 44-731 0—U.S. Government Printing Office) does not name the Canadian official.

named no names, it is difficult to believe that both McDonnell Douglas and the FAA would not have learned about it. If they did, the information provoked no discernible ripples, and certainly no practical action by either. It took tragedy, which cost the lives of sixty-one people, to focus serious attention on the problem. On November 28, 1972, a Japan Air Lines DC-8 crashed a few seconds after becoming airborne from Sheremetyevo Airport, Moscow. The Russian investigation concluded that the accident was caused either by engine failure or by the "mistaken" deployment of the ground spoilers.

McDonnell Douglas did at this point begin "an extensive design study" to find ways of making the spoiler system inoperable in flight, but that effort came too late to prevent one more crash. On June 23, 1973, a Loftleidir Icelandic Airlines DC-8 was forty feet above the runway at John F. Kennedy Airport, New York, when the copilot, Arni Sigurbergsson, accidentally extended the spoilers. The airplane fell to the ground, one engine was torn off, and fire broke out in one wing, but the captain, Olaf Olsen, managed to keep control. Of the 128 people on board, thirty-eight were injured—eight of them seriously.

The McDonnell Douglas design study was speeded up and, on October 5, 1973—three years and three months after the Air Canada crash—the company sent out a *routine* Service Bulletin telling airlines how to fit a lock on the spoiler mechanism. The SB said that the necessary parts would be supplied free of charge. Even now there was no mention of the Air Canada, Loftleidir, and Japanese accidents and the modification was described as a "design improvement."

The Western Region of the FAA remained confident that no action on its part was necessary. On September 6, 1973, it sent a letter to the FAA's Eastern Region office in New York (the Loftleidir crash had occurred in the Eastern Region's territory) which contained this bland assurance: "We consider that limitation, warning and information contained in the cockpit placards and airplane flight manuals . . . are adequate for the safe operation of the ground spoiler system of the DC-8 aircraft."

The NTSB did not agree. After investigating the Loftleidir accident, it recommended that the entire DC-8 fleet (about five hundred aircraft) should be fitted with ground spoiler locks saying, "the hazard potential well justifies such action." The Western Region of the FAA did finally issue an Airworthiness Directive on January 29, 1974, requiring all DC-8s to be fitted with the locking system, but there was still no sign that the agency considered the matter to be urgent. The AD said that each DC-8 must be fitted with spoiler locks within the next 3,600 hours of passenger service; for the average American airline that meant within the next one-and-a-half years.

When the Interstate and Foreign Commerce Committee of the House of Representatives reviewed the DC-8 affair in late 1974, it concluded that most of the blame lay with the Western Region of the FAA. The Airworthiness Directive issued against the DC-8 after the Air Canada crash was "almost totally worthless," the Committee said. It remained a matter of speculation as to whether the Japan Air Lines crash in Moscow could have been averted because the cause had not been positively determined. But there was no doubt, the Committee said, that the Loftleidir crash could have been prevented by tougher action against the DC-8. The Committee added:

> In summary, a pattern emerges from the Ground Spoiler matter which is not unlike that noted in the DC-10 cargo door problem. The Icelandic plane which was involved in a near disastrous accident had the placard warning "Deployment in Flight Prohibited" mounted on the instrument panel. The Turkish plane, which was involved in aviation's worst disaster, had the "Peephole" in the aft cargo door. Both provisions made no allowance for human error, and were manifestly inadequate.

WE HAVE ALREADY shown that there were inherent weaknesses in the design of the DC-10 and that given the vulnerability of the cabin floor, and its crucial role in carrying the flight controls, the cargo-door locking system was quite inadequate. Often, such perceptions are available only in hindsight. But what the FAA seemingly did not know was that some well-qualified members of the DC-10 design teams were aware almost from the first that the cargo-door system was suspect.

The detail design of the DC-10 fuselage and its doors was largely done by engineers from the Convair division of General Dynamics in San Diego, California. This was for financial as well as industrial reasons: McDonnell Douglas was naturally looking for subcontractors capable of sharing the enormous load of financing a program which could not run into profit for many years. (This is standard procedure in the U.S. aerospace industry. Subcontractors pay their own start-up and engineering costs, getting the money back bit by bit over a predetermined period as they deliver units to the main contractor.)

Convair and its unsuccessful rivals for the contract (North American Rockwell, Rohr Aircraft, and Aerfer) were given, early in 1968, a weighty Subcontractor Bid Document which set out Douglas's requirements for the DC-10. The passenger doors were to be plugs, but the lower cargo doors were to be outward-hinging tension-latch doors. They were to have over-center latches driven by *hydraulic cylinders,* a system already used on some DC-8 and DC-9 doors. In addition to hydraulic latches, each cargo door was to have a manual locking system "designed so that the handle or latch lever cannot be stowed unless the door is properly closed and latched." The bid document was insistent that weight should be saved wherever possible (paint was to be used "to a minimum"), and the subcontractors were told that one pound of weight should be thought of as costing $100.

On August 7, 1968, McDonnell Douglas signed a subcontract with Convair. William Gross of Douglas later said that the reason for choosing Convair, in addition to the financial strength of the parent company, General Dynamics, was the excellent reputation Convair had for structural design. If that was so, it is surprising that Douglas did not take more notice of Convair's reservations about the DC-10 cargo-door design. These began to emerge in November 1968, when Douglas told the San Diego engineers that instead of hydraulic cylinders they must use *electric* actuators to drive the cargo-door latches.

The principles are very different. To continue the discussion begun in Chapter 8, a hydraulic actuator uses high-pressure fluid to drive a piston along a cylinder. An electric actuator, on the other hand, works through gears, and is driven by an electric motor.

Douglas gave two reasons for changing from hydraulic actuators: to do so would save twenty-eight pounds of weight per door and also would "conform to airline practice." The fact was that American Airlines had asked for electric actuators, saying that as they would have fewer moving parts they would be easier to maintain. And in the competitive conditions of the airbus market, requests of this kind from airlines were not likely to be rejected.

Some Convair engineers, and in particular their Director of Product Engineering, F. D. "Dan" Applegate, were never fully reconciled to the change. The hydraulic system was, in Applegate's judgment at least, better because it was more "positive." When engineers say that one system is more positive than another, they are making a value judgment rather than a precise, mathematical statement just as they do when they say that one system is "simpler" than another. Nonetheless, the Convair argument has much substance to it.

Airplane hydraulic systems make power—pressure—available *continuously* to the mechanisms they serve. Fluid pressure from a hydraulic system is pumped through valves into an actuator cylinder. It then pushes the piston through its travel and continues to exert pressure against it. Piston travel is reversed by adjusting the valves so as to allow fluid to flow in from the other end of the cylinder, and thus press upon the other surface of the piston. The point is that whatever the position of the piston, hydraulic fluid is always working on it.

Electric power, on the other hand, is not used continuously. It is switched on to produce movement, and then switched off again. As there is no continuous pressure to maintain the position that an actuator has reached, some kind of mechanical device must be used to prevent the electrically driven piston from slipping back. Most electric actuators, and certainly the type fitted to the DC-10 doors, are therefore *irreversible* in action. Once they have achieved the maximum travel that they can achieve in any particular direction, they remain fixed until the travel is specifically reversed by the application of an opposite electric impulse.

Therefore, an electric latch will behave very differently from an hydraulic one. An hydraulic latch, though positive, is *not* irreversible. If it fails to go over-center, it will in the nature of things "stall" at a point where the pressure inside the cylinder has reached equilibrium with the friction which is obstructing the travel of the latch. Thereafter, quite a small opposite pressure will move it in the reverse direction—and what this means in a pressure-hull door is that if the latches have not gone quite "over," they will slide open quite smoothly as soon as a little pressure develops inside the hull and starts pushing at the inside of the door. Thus, they will slide back and the door will open, well before the pressure inside the aircraft hull is high enough to cause a dangerous decompression. The door will undoubtedly be ripped from its hinges by the force of the slip stream but, at low altitude, that poses no threat of structural damage to the plane and no danger to its passengers. The crew will immediately become aware of the problem, because the aircraft cannot be pressurized and can simply return to the airport.

However, if an irreversible electric latch fails to go over-center, the result will usually be quite different. Once current is switched off, the attitude of the latch is fixed, and if it has gone quite a long way over the spool, there will be considerable frictional forces between the two metal surfaces, holding the latch in place. Pressure building up inside the door cannot *slide* the latches open. It can only force the fixed, part-closed latches off their spools. This, typically, will happen in a swift and violent movement, occurring only when pressure inside the airplane has built up to a level when sudden depressurization will be structurally dangerous.

Hydraulic latches would have been intrinsically safer with even a mediocre manual-lock system to back them up. But once the changeover was made to electric power, it became essential to provide a totally foolproof checking-and-locking backup.

In the summer of 1969 Douglas asked Convair to start drafting a Failure Mode and Effects Analysis (FMEA) for the lower cargo-door system of the DC-10. The purpose of the FMEA is, as the name suggests, to assess the likelihood of failure in a particular system, and the consequences of failure should it occur. Before an airplane can be certificated, the FAA must be given an FMEA for those major systems which are critical to safety.

Convair submitted a draft FMEA for the door system in August 1969. The design examined was an early one, in which the back-up locking system consisted simply of spring-loaded locking pins (later, of course, the spring-loaded pins were replaced by the manual locking handle and its complex linkages). Convair apologized for having taken two months over the work, due to the fact that their engineers had been working lately on military programs and were not familiar with current FMEA procedures for civil aircraft. In spite of this, and the fact that the door design analyzed was a relatively early one, Convair produced a document which accurately foresaw the deadly consequences of cargo-door failure.

Among the "ground rules and assumptions" of the FMEA, those dealing with failure-warning systems stand up especially well to hindsight. First, said Convair, no great reliance was to be given to warning lights on the flight deck because "failures in the indicator circuit, which result in incorrect indication (i.e., 'lights out') of door locked and/or closed, may not be discovered during the checkout prior to take-off."

Convair claimed that even less reliance should be placed on warning systems which relied on the alertness of ground crews. In this early design, the only way of telling from the outside whether the latches had gone home was to look at the "manual override" handle provided to wind them shut by hand in case of electrical failure. If it had moved through its full travel, the latches must be safe. The Convair FMEA found:

> . . . That the ground crew requirement to visually check the angular position of the manual override handle, to detect an "unlocked" condition, to be subject to human error. It is assured that routine handling of repetitive aircraft could result in the omission of this check or visual error due to the location of the handle on a curved surface under the lower fuselage.

The substance of the FMEA then went on to show that there were nine possible failure-sequences which could lead to a "Class IV hazard"—that is, a hazard involving danger to life. Five of these involved danger to ground crew by doors falling suddenly shut or coming open with undue violence, and these do not concern our narrative. But there were four sequences shown as capable of producing sudden depressurization in flight: also a "Class IV hazard," and meaning in this context the likely loss of the airplane. One of these sequences was, in principle, remarkably similar to what actually occurred over Windsor and later outside Paris.

The starting point was seen as a failure of the locking-pin system, due to the jamming of the locking tube or of one or more of the locking pins. In that case, said the FMEA: "Door will close and latch, but will not safety lock." There should of course be a warning against this, and if it works properly: "Indicator light [in the cockpit] will indicate door is unlocked and/or open." But one of the ground rules of the FMEA was that circuit failures in the indicator system might well go undetected, in which case: "Indicator light will indicate normal position." If that happened, malfunction of the electric latch actuators could produce a situation in which the "door will open in flight—resulting in sudden depressurization and possibly: structural failure of floor; also damage to empennage * by expelled cargo and/or detached door. *Class IV hazard* in flight."

The difference between this and what actually happened over Windsor is: first, the FMEA envisaged failure in a spring-loaded, rather than a hand-driven, locking-pins system; second, the failure mentioned was one of inadvertent electrical reversal of the latches rather than a failure of the latches to go "over-center" in the first place. But it was a powerful demonstration that the door design was potentially dangerous without a totally reliable fail-safe locking system.

The other three depressurization sequences were rather different in that they envisaged total latch failure due to electrical faults, with the door being held shut until danger point by just the electric system which closes the door. These were not relevant to the accidents which ultimately occurred; nonethe-

---

* Tail flying surfaces.

less, together with the FMEA's general skepticism about warning lights and ground-crew assessments, they helped to produce a document that spelled out very clearly the terrible consequences that could follow from ill-thought-out door design.

But neither this FMEA draft nor anything seriously resembling it was shown to the FAA by Douglas, who, as lead manufacturers, made themselves entirely responsible for certification of the airplane. (Indeed, under the terms of the subcontract, General Dynamics was forbidden from contacting the FAA about the DC-10.) Our evidence is drawn from documents produced by Douglas and testimony given in the complex of compensation lawsuits which resulted from the Paris crash.* Evidence given by J. B. Hurt, Convair's DC-10 support program manager during the litigation, was that Douglas never replied to the Convair FMEA.

FMEAs submitted by Douglas to the FAA, leading up to certification of the DC-10, do not mention the possibility of Class IV hazards arising from malfunction of lower cargo doors.

THE DOCUMENTARY WARNING of the dangers of depressurization were followed in 1970 by a physical manifestation at Long Beach. But even this it seems could not dent the self-assurance of the Douglas design team.

By May 1970, the first DC-10 (Ship 1) had been assembled at Long Beach and was going through ground tests to prepare for the maiden flight scheduled for August. On May 29, outside Building 54 the air-conditioning system was being tested, which involved building up a pressure-differential inside the hull of four to five pounds per square inch. Suddenly the forward lower cargo door blew open. Inside, a large section of the cabin floor collapsed into the hold.

The Douglas response to this foreshadowed the company's response to later and more serious accidents. It was simply blamed on the "human failure" of the mechanic who had closed the door. This explanation was still adhered to by William Gross when he gave evidence during the Paris crash litigation in 1974–75. He gave no sign of thinking that there might be something basically dubious about a system that could become dangerous simply because one man, fairly low down in the engineering hierarchy, failed to perform exactly to plan.

However, Douglas did at the time acknowledge that some modification of the door was required before presenting the whole system for FAA certification. It had already been decided before the Ship One accident that the spring-loaded locking-pin system should be replaced by hand-driven linkages, and now it was decided to try and build some extra safeguards into that system. Ship One took off on its maiden flight on schedule with an unmodified door. But by the autumn of 1970 the "vent door" concept had been adumbrated.

The essence of this system has already been described in Chapter 8.

---

* *Hope* v. *McDonnell Douglas et al.:* Civ. No. 17631, Federal District Court, Los Angeles, California.

The miniature plug door was let into the main door above and to the right of the locking handle. This was supposed to stand conspicuously open until the main door was latched, and the locking handle was pulled down to drive the locking pins home.

In truth, it added little or nothing to the safety of the system. Such a vent door can only provide a check on the position of locking pins if its closure is a *consequence* of the pins having gone home. This is the case with the Boeing 747 tension latch cargo door, which was already flying by 1970. But in the Douglas scheme, the closure of the vent door was merely *coincidental* to the action of the locking pins. If the rod transmitting the locking-handle's movement were to break, or to be absent altogether, then the Douglas vent door would still close.

On the face of it it seems especially remarkable that such an obvious flaw should have been overlooked. The certification process which the FAA imposes on every new airplane is supposed to identify and reject the offspring of dubious design philosophy. But the process, long and exhaustive though it undoubtedly is, suffers from a fundamental weakness: Although it is carried out in the name of the FAA, much of the work involved is actually done by the manufacturers themselves.

The FAA says it has neither the manpower nor, in some instances, the specialized expertise to inspect every one of the thousands of parts and systems that go to make up a modern airliner. It therefore appoints at every plant Designated Engineering Representatives (DERs)—company men, paid by the manufacturer, who spend part of their working lives wearing, as it were, an FAA hat.* Their job during the certification process is to carry out "conformity inspections" of the plane's bits and pieces to insure that they comply with the Federal Airworthiness Regulations. In the case of the DC-10, there were 42,950 inspections. Only 11,055 were carried out by FAA personnel. The rest were done by McDonnell Douglas DERs.

Designated representatives are chosen by the FAA with a careful eye to their experience and integrity, but inevitably conflicts of interest can arise when manufacturers are called upon, in effect, to police themselves. The system also reduces the chance of mistakes being spotted, and the DC-10 vent door system stands as a classic illustration of what can happen. Before certification the vent door system was submitted to a series of tests by McDonnell Douglas and the results were approved, on behalf of the company, by a senior engineer. *Later, this time wearing his DER hat, the same engineer approved the report of the tests as acceptable documentation for showing that the DC-10 cargo door complied with the airworthiness regulations.*

There were other faults in the design. The linkages had not been stressed correctly, so that they later turned out to be capable of flexing out of shape when submitted to pressure. And their various degrees of travel were all

---

* The FAA also appoints Designated Manufacturing Inspection Representatives (DMIRs) to assist the agency in monitoring production.

Captain Bryce McCormick and flight engineer Clayton Burke, explaining how they saved a DC-10 over Windsor.

John Hixon Shaffer (previous page) was appointed administrator of the FAA by Richard Nixon in 1969 and soon demonstrated his faith in the aviation industry's ability to regulate itself. On June 12, 1972, the DC-10 revealed the lethal flaw that had been built into every aircraft when an American Airlines plane lost a cargo door over Windsor, Ontario. Only luck and the skill of Bryce McCormick and his crew prevented disaster. The FAA's technical experts wanted to issue an Airworthiness Directive against the DC-10 to prevent reoccurrence, but Shaffer intervened and made a gentlemen's agreement with McDonnell Douglas, relying on the company's promise to get "their goddammed plane" fixed. The problem was not fixed in at least one DC-10, Ship 29, which crashed near Paris twenty months later. After the tragedy Shaffer said he would not have sat around "fat, dumb and happy" if he had known about the consequences of his action. But at least one FAA executive, Arvin Basnight, was always very dubious about Shaffer's intervention: a few days after the Windsor incident Basnight wrote a secret memorandum to the files, recording the events. It was the first time in his thirty-year career that Basnight felt the need to protect himself and his staff.

**The DC-10 cargo hold and the remains
of the door after Windsor**

REPORT FOR THE FILES

On Friday, 16 June 1972, at 8:50 AM, I received a phone call from
Mr. Jack McGowan, President, Douglas Aircraft Company, who
indicated that late on Thursday, 15 June, he had received a call
from Mr. Shaffer asking what the Company had found out about the
problem about the cargo door that caused American Airlines to have
an explosive decompression.

Mr. McGowan said he had reviewed with the Administrator the facts
developed which included the need to beef up the electrical wiring
and related factors that had been developed by the Douglas Company
working with FAA.

He indicated that Mr. Shaffer had expressed pleasure in the finding
of reasonable corrective actions and had told Mr. McGowan that
the corrective measures could be undertaken as a product of a
Gentleman's Agreement thereby not requiring the issuance of an
FAA Airworthiness Directive.

In light of this data, I consulted with Dick Sliff as we were already
preparing an airworthiness directive and Mr. Sliff advised that several

**Arvin Basnight and part of his secret
memorandum**

The arrival of Turkish Airlines' first DC-10 at Istanbul in December 1972 was a matter of great national pride. Turkish Government officials were taken on special demonstration flights, and John Brizendine, now Douglas president, was on hand to join in the celebrations. The sale of three DC-10s to Turkey had followed a classic high-pressure sales campaign for which the entertainer Danny Kaye was recruited to add his weight. (The photograph below shows the flight deck of one of THY's two surviving DC-10s. The crew member in the foreground is the flight engineer.)

Edward Evans

Henry Noriega

Ray Bates, William Gross, and Bill Glasgow,
three of the senior Douglas executives responsible for development
of the DC-10.

During the sales negotiations with THY Jackson McGowen promised that the DC-10 cargo doors would be modified. The door of Ship 29 was not modified properly yet McDonnell Douglas's records falsely showed that the work had been done and inspected. Edward Evans could not explain how his stamp appeared on those records: Henry Noriega believes that he must have loaned his stamp to somebody else— but he cannot remember to whom. McDonnell Douglas's executives blamed human failure for the breakdown in the inspection system. (Below, the DC-10 assembly line at Long Beach, California.)

Jackson McGowen and John Brizendine, testifying to the U.S. Senate about the missing support plate

The rear cargo door of the DC-10. The first photograph (left) shows the height of the door above the ground and the need for the operator to use steps in order to close and inspect it. The locking handle is situated in the upper lefthand side of the door alongside the vent door. The peephole and the warning placard are situated in the lower lefthand corner. The third photograph (below) shows the inside of the door containing the vent door cage, the latches, and the spools around which they are supposed to grip. (Diagrams and a full explanation of the DC-10 door mechanism can be found in Appendix B.)

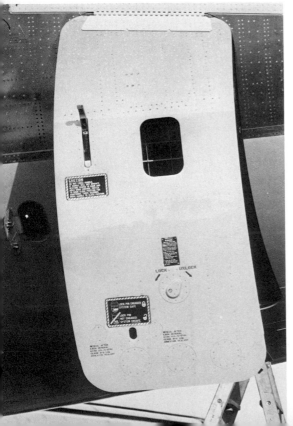

adjustable, so that the whole system was equivocal. Its validity as a check on the function of the latches depended upon whether the linkages in any given door happened to be correctly rigged.

Conceivably such design faults resulted from inexperience with doors of this kind. (Incidentally, if Douglas needed proof of the efficacy of the hydraulic system they had abandoned in the DC-10 cargo doors, it was abundantly available during 1970. There were five examples that year of hydraulic tension-latch doors in DC-8s and DC-9s opening in flight *before* pressure had built up to a dangerous level. All the aircraft landed safely.) But inexperience is no excuse, because in safety matters most airplane designers are willing to render assistance across competitive commercial boundaries.

Indeed, during November 1970, Convair was able to obtain via American Airlines considerable detail on Boeing vent systems for tension-latch doors. Not only did they discover that the two vent doors in the 747's cargo door were driven off the locking tube itself: they also learned that the locking-system consisted of nonadjustable, and so unequivocal, linkages.

Not that everyone at Douglas's Long Beach plant thought that things were going the right way. In November a Convair engineer named H. B. ("Spud") Riggs, who was attached to the Douglas design team, wrote an internal memo headed: "Approaches to Eliminate Possibility of Cabin Pressurization with Door Unsafe." Riggs wrote that the design conception of the vent door was so far "less than desirable." He canvassed other ways of dealing with the problem: going back to hydraulic actuation; adding redundant electric circuits as back-up on the electric actuators; interlinking the door-closing system with the pressurization system; increasing the floor strength sufficiently to make the floor resist any possible pressure differential after a door blowout; providing vent space in the floor to enable high-pressure air to escape without doing damage.

None of these possibilities was incorporated in the DC-10—nor was Riggs's memo given to the FAA. By the following month Douglas as a corporation seemed to be more concerned about the financial consequences of the May blowout than the engineering ones. As with the doors, Convair was responsible for the detailed design of the DC-10 floor—albeit to Douglas's specifications. On December 4 an internal Convair memorandum recorded a negotiation with Douglas officials:

> Douglas indicated that there was nothing defective in the door but that the passenger floor, having failed in the pressure test, was defective, and Convair owed Douglas a new floor. However, since it was not practical to change the floor Douglas wanted Convair to install the blow out door [vent door] in the 300 aircraft to satisfy its obligation on the floor.

There had always been a case for making the floor of the DC-10, and all other wide-bodied jets, strong enough to withstand full pressure differential. To do so would have cost some three thousand pounds in the DC-10 and the TriStar: say, one dozen passengers and their luggage, which in terms of present load factors might seem quite tolerable.

But that had been rejected—by Lockheed and Boeing, as well as Douglas —and the floor which Convair had built was by this time exactly as strong as Douglas had specified. Now Douglas was trying to say that the floor was not really strong enough ("defective"), so the door needed to be more reliable; therefore, the insertion of the vent door should be paid for by Convair.

On December 15, 1970, another Convair memo noted that Douglas had decided that in all cargo versions of the DC-10 the *upper* cargo doors would not have electric actuators and vents, but would go back to hydraulic actuation. M. R. Yale, Convair's Manager of DC-10 Engineering, wrote: "[We] asked Douglas why this approach would not be a better solution to lower cargo doors than vent doors. Only answer received was that Douglas had considered these factors and concluded that the vent door was the appropriate solution."

In the New Year of 1971, with seven months to go before the scheduled date for the certification of the DC-10 as a commercial airliner, Douglas formally directed the installation of vent doors on all lower cargo doors, and although they were delayed by tooling problems and shortages of parts, deliveries were getting under way in June.

On July 6, 1971, a Convair memo gave a résumé of the situation:

> (1) Design criteria and design features of operating, latching, and locking mechanisms were specified by Douglas for Convair.
> (2) Basic design work was done by Convair engineers working at Long Beach under supervision of Douglas Engineering Department. Douglas retained total responsibility for obtaining FAA approval of DC-10 and prohibited Convair from discussing any design feature with FAA.
> (3) After the 1970 incident Douglas unilaterally directed incorporation of the vent door, even though there were in Convair's opinion several simpler, less costly alternative methods of making the failure more remote.

On July 29, 1971, the DC-10 was certificated by the FAA. Less than one year later came the blowout over Windsor, Ontario.

FIFTEEN DAYS AFTER Windsor, after the gentleman's agreement had been struck, and after Jack Shaffer had relaxed in the knowledge that Jack McGowen had fixed "his goddam airplane," Dan Applegate of Convair wrote a remarkable memorandum, which demands to be quoted in full. It expresses, with a vehemence not commonly found in engineering documents, all the doubts and fears that some of the Convair team felt about the airplane they were working on.

27 June 1972
Subject: DC-10 Future Accident Liability.

The potential for long-term Convair liability on the DC-10 has caused me increasing concern for several reasons.
  1. The fundamental safety of the cargo door latching system has been progressively degraded since the program began in 1968.
  2. The airplane demonstrated an inherent susceptibility to catastrophic failure when exposed to explosive decompression of the cargo compartment in 1970 ground tests.
  3. Douglas has taken an increasingly "hard-line" with regards to the relative division of design responsibility between Douglas and Convair during change cost negotiations.
  4. The growing "consumerism" environment indicates increasing Convair exposure to accident liability claims in the years ahead.
Let me expand my thoughts in more detail. At the beginning of the DC-10 program it was Douglas' declared intention to design the DC-10 cargo doors and door latch systems much like the DC-8s and -9s. Documentation in April 1968 said that they would be hydraulically operated. In October and November of 1968 they changed to electrical actuation which is fundamentally less positive.

At that time we discussed internally the wisdom of this change and recognized the degradation of safety. However, we also recognized that it was Douglas' prerogative to make such conceptual system design decisions whereas it was our responsibility as a sub-contractor to carry out the detail design within the framework of their decision. It never occurred to us at that point that Douglas would attempt to shift the responsibility for these kinds of conceptual system decisions to Convair as they appear to be now doing in our change negotiations, since we did not then nor at any later date have any voice in such decisions. The lines of authority and responsibility between Douglas and Convair engineering were clearly defined and understood by both of us at that time.

In July 1970 DC-10 Number Two * was being pressure-tested in the "hangar" by Douglas, on the second shift, without electrical power in the airplane. This meant that the electrically powered cargo door actuators and latch position warning switches were inoperative. The "green" second shift test crew manually cranked the latching system closed but failed to fully engage the latches on the forward door. They also failed to note that the external latch "lock" position indicator showed that the latches were not fully engaged. Subsequently, when the increasing cabin pressure reached about 3 psi (pounds per square inch) the forward door blew open. The resulting explosive decompression failed the cabin floor downward rendering tail controls, plumbing, wiring, etc. which passed through the floor, inoperative. This inherent failure mode is catastrophic, since it results in the loss of control of the horizontal and vertical tail and the aft center engine. We informally studied and discussed with Douglas alternative corrective actions including blow out panels in the cabin floor which would provide a predictable cabin

---

* We have been unable to establish whether Applegate's reference to an accident involving Ship Two, in July 1970, is a mistake on his part or whether there were *two* blowout incidents. Certainly Ship One was damaged on May 29, 1970, in circumstances very similar to those described by Applegate.

floor failure mode which would accommodate the "explosive" loss of cargo compartment pressure without loss of tail surface and aft center engine control. It seemed to us then prudent that such a change was indicated since "Murphy's Law" * being what it is, cargo doors will come open sometime during the twenty years of use ahead for the DC-10.

Douglas concurrently studied alternative corrective actions, inhouse, and made a unilateral decision to incorporate vent doors in the cargo doors. This "bandaid fix" not only failed to correct the inherent DC-10 catastrophic failure mode of cabin floor collapse, but the detail design of the vent door change further degraded the safety of the original door latch system by replacing the direct, short-coupled and stiff latch "lock" indicator system with a complex and relatively flexible linkage. (This change was accomplished entirely by Douglas with the exception of the assistance of one Convair engineer who was sent to Long Beach at their request to help their vent door system design team.)

This progressive degradation of the fundamental safety of the cargo door latch system since 1968 has exposed us to increasing liability claims. On June 12, 1972 in Detroit, the cargo door latch electrical actuator system in DC-10 number 5 failed to fully engage the latches of the left rear cargo door and the complex and relatively flexible latch "lock" system failed to make it impossible to close the vent door. When the door blew open before the DC-10 reached 12,000 feet altitude the cabin floor collapsed disabling most of the control to the tail surfaces and aft center engine. It is only chance that the airplane was not lost. Douglas has again studied alternative corrective actions and appears to be applying more "band-aids." So far they have directed us to install small one-inch diameter, transparent inspection windows through which you can view latch "lock-pin" position, they are revising the rigging instructions to increase "lock-pin" engagement and they plan to reinforce and stiffen the flexible linkage.

It might well be asked why not make the cargo door latch system really "fool-proof" and leave the cabin floor alone. Assuming it is possible to make the latch "fool-proof" this doesn't solve the fundamental deficiency in the airplane. A cargo compartment can experience explosive decompression from a number of causes such as: sabotage, mid-air collision, explosion of combustibles in the compartment and perhaps others, any one of which may result in damage which would not be fatal to the DC-10 were it not for the tendency of the cabin floor to collapse. The responsibility for primary damage from these kinds of causes would clearly not be our responsibility, however, we might very well be held responsible for the secondary damage, that is the floor collapse which could cause the loss of the aircraft. It might be asked why we did not originally detail design the cabin floor to withstand the loads of cargo compartment explosive decompression or design blow out panels in the cabin floors to fail in a safe and predictable way.

I can only say that our contract with Douglas provided that Douglas would furnish all design criteria and loads (which in fact they did) and that we

* Murphy's Law: "If it can happen, it will." Also known sometimes as the totalitarian or Hegelian law of physics, from Hegel's view that all that is rational is real: that is, if you can think of it, it must exist. This is a case where implausible philosophy makes for good engineering.

would design to satisfy these design criteria and loads (which in fact we did).* There is nothing in our experience history which would have led us to expect that the DC-10 cabin floor would be inherently susceptible to catastrophic failure when exposed to explosive decompression of the cargo compartment, and I must presume that there is nothing in Douglas's experience history which would have led them to expect that the airplane would have this inherent characteristic or they would have provided for this in their loads and criteria which they furnished to us.

My only criticism of Douglas in this regard is that once this inherent weakness was demonstrated by the July 1970 test failure, they did not take immediate steps to correct it. It seems to me inevitable that, in the twenty years ahead of us, DC-10 cargo doors will come open and I would expect this to usually result in the loss of the airplane. (Emphasis added.) This fundamental failure mode has been discussed in the past and is being discussed again in the bowels of both the Douglas and Convair organizations. It appears however that Douglas is waiting and hoping for government direction or regulations in the hope of passing costs on to us or their customers.

If you can judge from Douglas's position during ongoing contract change negotiations they may feel that any liability incurred in the meantime for loss of life, property and equipment may be legally passed on to us.

It is recommended that overtures be made at the highest management level to persuade Douglas to immediately make a decision to incorporate changes in the DC-10 which will correct the fundamental cabin floor catastrophic failure mode. Correction will take a good bit of time, hopefully there is time before the National Transportation Safety Board (NTSB) or the FAA ground the airplane which would have disastrous effects upon sales and production both near and long term. This corrective action becomes more expensive than the cost of damages resulting from the loss of one plane load of people.
F. D. Applegate
Director of Product Engineering.

Although this was not a formally set-out safety analysis, like the Convair-drafted FMEA which Douglas did not give to the FAA, the Applegate memorandum was in some respects an even more disturbing document. Yet not only did its contents not reach the FAA, where surely they would have eroded even John Shaffer's durable complacency, they were not even put by Convair to Douglas.

The immediate fault in this must, of course, lie with Convair. But something must also be said about motivation: Briefly, Convair's experience over the previous three years had led to the belief that to raise such major safety questions with Douglas was chiefly to give away points in an ongoing financial contest. Certainly the record appeared to be one in which Douglas had been unimpressed by the safety propositions argued by Convair in its draft FMEA, and by the middle of 1972, it seems clear the position of the two

---

* Douglas's design criteria called for the floor to withstand a pressure of 3 psi, and it eventually did so—although not until Douglas had challenged Convair's original stress analysis, and a stronger kind of aluminum alloy had been introduced to the floor beams.

companies was essentially an adversary one. Five days after Applegate wrote his memorandum his immediate superior, Mr. Hurt, wrote an equally revealing comment upon it:

3 July 1972.

From: J. B. Hurt.
Subject: DC-10 Future Accident Liability.

Reference: F. D. Applegate's Memo, same subject, date 27 June 1972.

I do not take issue with the facts or the concern expressed in the referenced memo. However, we should look at the "other side of the coin" in considering the subject. Other considerations include:
  1. We did not take exception to the design philosophy established originally by Douglas and by not taking exception, we, in effect, agreed that a proper and safe philosophy was to incorporate inherent and proper safety and reliability in the cargo doors in lieu of designing the floor structure for decompression or providing pressure relief structure for decompressions or providing pressure relief provisions in the floor. The Reliance clause in our contract obligates us in essence to take exception to design philosophy that we know or feel is incorrect or improper and if we do not express such concern, we have in effect shared with Douglas the responsibility for the design philosophy.

  2. In the opinion of our Engineering and FAA experts, this design philosophy and the cargo door structures and its original latch mechanism design satisfied FAA requirements and therefore the airplane was theoretically safe and certifiable.

  3. In redesigning the cargo door latch mechanism as a result of the first "blowout" experience, Douglas unilaterally considered and rejected the installation of venting provisions in the floor in favor of a "safer" latch mechanism.* Convair engineers did discuss the possibility of floor relief provisions with Douglas shortly after the incident, but were told in effect, "We will decide and tell you what changes we feel are necessary and you are to await our directions on redesign." This same attitude is being applied by Douglas today and they are again making unilateral decisions on required corrections as a result of the AAL Detroit incident.†

  4. We have been informally advised that while Douglas is making near-term corrections to the door mechanism, they are reconsidering the desirability of following-up with venting provisions in the floor.
I have considered recommending to Douglas Major Subcontracts the serious consideration of floor venting provisions based on the concern aptly described by the referenced memo, but have not because:
  1. I am sure Douglas would immediately interpret such recommendation as a tacit admission on Convair's part that the original concurrence by Convair of the design philosophy was in error and that therefore Convair

---

* Vents built into the floor would allow pressurized air to escape into the hold without buckling the floor.
† American Airlines. Detroit is where the DC-10 landed after the Windsor incident.

was liable for all problems and corrections that have subsequently occurred.

2. Introducing such expression at this time while the negotiations of
SECP 297 and discussion on its contractual justification are being conducted would introduce confusion and negate any progress that had been
made by Convair in establishing a position on the subject.* I am not
sure that discussion on this subject at the "highest management level"
recommended by the referenced memo would produce a different reaction
from the one anticipated above. We have an interesting legal and moral
problem, and I feel that any direct conversation on this subject with
Douglas should be based on the assumption that as a result Convair may
subsequently find itself in a position where it must assume all or a significant portion of the costs that are involved.

J. B. Hurt
Program Manager, DC-10 Support Program.

On July 5, 1972, Mr. M. C. Curtis, the Convair vice-president who was
in overall charge of the DC-10 project, called a meeting to decide corporate
policy in the light of the Applegate memorandum. (Hurt's memo was chiefly
a briefing to Curtis for the meeting.) Convair's chief counsel, director of
operations and director of contracts, attended along with Applegate and Hurt.
(It seems that one person not consulted was David Lewis, the one-time heir
apparent of "Mister Mac" who became president of General Dynamics in
1970. Ironically, it was Lewis as president of the Douglas division who, in
1968, negotiated the financial details of the contract with Convair.)

It was acknowledged that Applegate was closer than Hurt to the engineering of the DC-10 and had a better knowledge of the safety factors involved.
But Mr. Curtis and his colleagues preferred the reasoning of Hurt's memo
and resolved the "interesting legal and moral problem" by deciding that Convair must not risk an approach to Douglas. According to Hurt's testimony,
two-and-a-half years later in Hope versus McDonnell Douglas, the meeting
came up with a rationalization which, though touched with cynicism, had
justification of a sort. "After all," said Hurt, "most of the statements made
by Applegate were considered to be well-known to Douglas and there was
nothing new in them that was not known to Douglas." And it is certainly
hard to believe that the Douglas design team could have claimed, after three
years of close collaboration with Convair, that the arguments of the Applegate
memorandum were unknown to them.

Both to Douglas and to Convair the dangers were, or should have been,
obvious. The determination of Convair, as subcontractors, was not to take
upon itself the duty of pointing them out.

And so, because of the interrelated failures of McDonnell Douglas, Convair, and the Federal Aviation Administration, a fundamentally defective
airplane continued on its way through the "stream of commerce" (as plaintiff
lawyers like to call it). Unhappily, some examples of the DC-10 were soon

---

* A reference to negotiations over the cost of fitting three vent doors in each of 300
projected aircraft: a total cost of $3 million.

to find their way into places where that stream flowed in a somewhat troubled fashion. Just as the aviation industry likes to pretend that all airplanes are created equal, so the same superstition is attached, at least in public, to the airlines that operate them. But, as a little investigation will show, the fact is that some airlines are quite strikingly less equal than others.

11

The Only Way to Fly

I wonder if we could contrive . . . some magnificent myth that
would in itself carry conviction to our whole community.
　　　　　　　　　　　　　　　　　　　　　　—PLATO,
　　　　　　　　　　　　　　　　　　　　　　Republic

ACCORDING TO JOHN BRIZENDINE, who in 1973 succeeded Jackson
McGowen as president of the Douglas division of McDonnell Douglas, airplane makers have a dread of telephone calls in the middle of the night. Like
doctors and newspapermen and the man who holds the keys to the store,
aviation executives invariably get called out when disaster strikes. Brizendine
told us that a midnight call to his home has him "nine feet up in the air."

The dread is understandable. Airplane manufacturers remain intimately
involved with their products long after they have left the plant through the
need for spare parts, overhauls, and modifications. When things go dramatically wrong with a plane, the first thing an airline does is call the people who
made it, and that is especially true when there has been a crash.

The consequences of the midnight call (catastrophes have a habit of occurring at unsocial hours) are always unpleasant. It is immensely distressing for
a plane maker to learn that one of his machines has become an instrument
for dozens or even hundreds of deaths. And an airliner crash is also inevitably
bad for business, particularly if the failure that caused it was mechanical
rather than human. There will be massive publicity, recrimination, maybe the
need for expensive modification and, increasingly nowadays, costly litigation.

"I've lived with this all my working life," Brizendine added, "and I can tell
you that I've had sleepless nights and I've worried through the years about
things."

To a large degree that concern—which Brizendine's counterparts in the
industry undoubtedly share—is born out of impotence. The quality and integrity of the airplanes they produce must bear a direct relationship to the
degree of safety that the mass transport system can attain. Yet, once an airplane is sold and delivered, the manufacturer rarely has any control over the
way in which it is operated and maintained. And, whatever the official fiction
may say about all airlines being equal, every manufacturer knows that some

189

airlines are far more safety conscious than others. Hence, the fear of midnight calls.

The sanctions that any one manufacturer can employ against less safe airlines are obviously limited. In the last resort he can refuse to sell his airplanes and, by his own account, Brizendine once did that. "I can remember sitting here in this chair telling an airline operator if you will do this, this, and this, *then* we will let you operate our planes." *

But that sanction is not foolproof because other manufacturers might not feel similarly constrained and, anyway, there are always secondhand planes on the market. A better remedy might be to help unsophisticated airlines become safer by providing them with the experience and the technical knowledge they lack and which is so often the root cause of their deficiencies. Again there are problems. A great many international airlines owe their existence to national pride as much as, and perhaps more than, anything else. In many a developing country, an airline is *the* symbol of progress, as potent as the head of state's Rolls-Royce. And where pride and national machismo are rampant, offers of help, which may well smack of paternalism, are not always welcome. As Brizendine put it: "We have to accept that they don't necessarily accept their ineptness as you state it or we might see it. . . ."

This moral and commercial dilemma that confronts airplane manufacturers is not new but it has, in this decade, been greatly exacerbated. It was one thing to introduce into the stream of commerce the relatively simple jets of the 1960s and quite another to float the vast and enormously complex machines that constitute the jumbo-jet generation. Those machines—747s, DC-10s, and TriStars—are now being operated by some sixty-five airlines. Officially, of course, all of them are regarded as fine airlines, fully paid-up and respected members of the international airline industry. In fact, some of them have safety records and labor under handicaps which might suggest that they could cause the plane makers many sleepless nights.

In 1972 that group had to include Turkish Airlines. Turkey's reasons for running an airline are not concerned primarily with national chauvinism—they are economic and social, and soundly based. Nevertheless, the airline did suffer from a marked deficiency of technical sophistication and experience.

Those deficiencies were evident before THY joined the jumbo-jet club and long before the Paris crash in that its safety record was, to put it mildly, moderate. That did not prevent the airline from being extravagantly courted by both McDonnell Douglas and Lockheed during the airbus stakes. If McDonnell Douglas was not aware of the problem before it won that particular sales competition, it certainly was afterward, as our narrative will show.

But, in more general terms, it is unrealistic to hoist all the blame for the premature promotion of airlines exclusively onto manufacturers. Less safe airlines could not operate without the acquiescence of the entire aviation community: the insurers, the international bodies that are supposed to regulate and monitor the industry, and the airlines and their own industrial

* Authors' interview, August 1975.

organizations, who have done more than anyone to promote the myth of universal excellence.

FOR AN INDUSTRY that was so recently in its infancy, the scheduled airline business has come a long way. It is just sixty years since the St. Petersburg–Tampa Air Boat Line inaugurated the world's first sustained passenger service. Before the service was suspended—because a subsidy from the local chamber of commerce didn't cover costs—1,205 intrepid passengers had paid $5 each for the twenty-one-mile flight across Tampa Bay, Florida, in the open cockpit of a Benoist 14 flying boat. Today U.S. scheduled airlines carry, on average, half a million passengers daily. That is not, perhaps, too many compared to the New York subway system (3,250,000 a day) or the Paris Metro (3,350,000 a day), but when it comes to long distance travel within America (a hundred miles or more) the airlines have captured well over 80 percent of the public transportation business. In the foreign travel market they have done even better: in 1950 about 40 percent of those Americans who went abroad traveled by sea; in 1973 over 99 percent went by air.

While no other country—except perhaps the USSR—has been able to match the U.S. boom in domestic air travel, international airlines have flourished at a remarkable rate throughout the world. There are now more than two hundred scheduled airlines flying between them 514 million passengers a year, and there is scarcely a nation, no matter how poor, that does not boast a national, flag-carrying airline.

Whether humanity—or indeed, the airline industry itself—benefits from such an abundance is open to question. It does, for instance, seem a little excessive to have fourteen scheduled airlines competing for passengers on routes across the Pacific and twenty-four fighting it out in Central and South America. It seems nothing short of absurd that, on the North Atlantic routes, there should be *twenty-seven* scheduled carriers (not to mention twenty-one charter and supplemental airlines) offering, each year, millions more seats than there are passengers to fill them.

But, commercial idiosyncrasies apart, there is no denying that in its short history the airline industry has made spectacular progress. Within the space of a couple of decades it has transformed international travel from an uncertain adventure into a routine daily affair. In the process it has overcome some mighty barriers—technical and logistical ones, of course, but frequently political barriers too: Pan Am and Aeroflot were, for example, commuting between Moscow and New York long before U.S.–U.S.S.R. détente became fashionable. There have also been the financial ups and downs to cope with, not least the Arab oil embargo of 1973 and the subsequent quadrupling, overnight, of the price of jet fuel.

And then there has been a more prosaic but also a more stubborn obstacle for airlines to overcome. Very large numbers of airline customers, and potential customers, are more or less afraid of flying. This is an area of imperfect research and reliable statistics are hard to come by but in 1968 the Behavioral Science Corporation of Los Angeles carried out a survey into fear of flying, on behalf of U.S. airlines, and estimated that eighteen million Americans suffer from the phobia. Half of those simply refuse to fly and the other half feel obliged to but only with varying degrees of apprehension. On top of that, there are, perhaps, another seven million person who fly fearfully but who are ashamed to publicly admit it.* Given that only thirty-four million adult Americans fly regularly—at least once a year—it is a sobering thought that almost half of them are probably reluctant customers.

Fear of flying is a largely irrational phobia that cannot easily be cured by facts (for instance, 99.9999 percent of U.S. air travelers arrive at their destination safely) no matter how compelling. On the other hand, the fear can easily be inflamed: Every airline knows that on the day after a well-publicized crash it should expect a rash of passenger cancellations.

The airline's dilemma is therefore a considerable one. If no amount of education will cure the phobia, what is the point of even talking, publicly, about safety? Yet it is manifestly in an airline's interest to persuade those potential customers who are scared that in its planes they have a better chance of getting from, say, Chicago to New York, intact. By and large the compromise that airlines as individual entities have struck is to pretend in all of their promotion that none of the customers has even considered the possibility of the plane crashing. "Safety" is the *verbum non gratum* of the airline-advertising copywriters' vocabulary. However, the ads do deal out euphemisms which the nervous can readily interpret into a promise of safe deliverance. United Airlines flies through "friendly skies"; TWA will make you "happier" than any other airline; British Airways "takes more care of you"; and, at Iberia, "only the plane gets more attention than you."

The understanding that airlines should not compete for passengers by boasting of a better safety record is an unwritten one, but it has rarely if ever been breached. Instead, traditionally seats have been filled and revenue won or lost for more ephemeral reasons—the convenience of schedules, the flavor of the cuisine, and the quantity of the booze and, more recently, the star of the in-flight movie. Sex or at least the illusion of sex has also played a part. Over the past few years airline passengers have been tempted by the subtle allure of a variety of Far Eastern maidens and, at the other end of the scale, by the brash promise that Continental Airlines' women flight attendants will really move their tail for you. Of course, they *really* mean the airplane's tail. . . . The most famous—or, in the eyes of feminists, most infamous—campaign of this genre, has been National Airlines' invitation to "fly me," me being Linda/Cathy/Mary, etc. (Long before that campaign was dreamed up *The New Yorker* magazine doubted the wisdom of employing

* The seven million estimate was made by the late Nat Coote, a New York engineer whose own fear of flying prompted him to begin group-therapy sessions for fellow sufferers. The Behavioral Science Corporation dismissed Coote's claim as "sheer speculation."

as flight attendants women who were far too young and pretty not to be convinced of their own immortality:

> The stewardess we have always longed for at such moments most resembles the ample, white-haired, cushiony, unhurrying elder we relied on when we were about five years old and were beset by a sudden anxiety—in the course of a noisy country thunderstorm, say. In our anxious flying dreams, this lady appears from up forward at the first whisper of Trouble; she is dressed in plain black silk and is wearing sensible, ground-gripping (*air-* and ground-gripping) black shoes. She is carrying an orange tabby cat under one arm, and she is holding a bit of crochetwork. As she passes us, on her way to close a window somewhere, she shoots us a quick, shrewd glance over the top of her steel-rimmed spectacles and murmurs, "Land sakes, child, don't fret yourself about *that*. That's just clouds bumping together." *

Some years later in the midst of its "Fly Me" campaign, National acknowledged that *The New Yorker* had a point, or at least that the airline was aware of the art of self-parody: for a spell Linda/Cathy/Mary, etc., were replaced by a "Mrs. Goldblum." Mrs. Goldblum was everything that the magazine had ever dreamed of.)

All of this is, of course, harmless enough. So long as we select the airlines we fly with on the basis of convenience or some in-flight triviality we can hardly expect those airlines—who are after all in business to make money or, in some cases, to massage national pride—to devote their promotion to discussions of safety, especially since to do so would scare as many as it would attract. It is undoubtedly true that we get the airline advertising we deserve: we, the passengers, have created as much as anyone the myth that there are only Fine Airlines and Very Fine Airlines. (Very Fine Airlines generally have bigger advertising budgets.) Where that myth ceases to be either harmless or reasonable is where it is endorsed by official bodies who should, and do know better.

As only a small amount of investigation will reveal, there are very safe airlines and not so safe airlines. (See Appendix C.) That is to say, there are airlines that have the means and the ability to transport their passengers in conditions of great safety: Delta and American Airlines in the United States; the Scandinavian airline, SAS, the Portuguese airline, TAP, Qantas and TAA of Australia, Japan Air Lines, and Iran Air are some of those that deserve special appreciation. On the other hand, there are other airlines including TAROM of Rumania, VIASA of Venezuela, ALIA of Jordan, Nigeria Airways, and Egyptair, who have established decidedly inferior safety records. Between those two extremes lie the rest of the world's leading airlines, many of whom would have difficulty in supporting their Very-Fine-Airline status if safety were to be the only criterion.

The airlines are aware of the disparities and so too are aviation experts like Peter Martin, who is an English lawyer, a partner in the distinguished

* From Notes and Comment in *The New Yorker* issue of May 4, 1963. © 1963 The New Yorker Magazine, Inc.

firm of Beaumont & Sons, which among other matters represents airlines that have obtained insurance coverage for their companies from Lloyd's of London. In times of trouble Martin acts as a negotiator between the relatives of people killed in an air crash and the airline/or Lloyd's. His clients include Turkish Airlines, and in the early part of 1975, we visited him to discuss the aftermath of the Paris crash. During the course of the interview Martin— who is naturally very knowledgeable about airlines—said: "There are airlines [as an example he named Iran Air] that I fly with happily. There are other airlines that I would not fly with if I was paid in heavy gold."

The trouble, of course, is that Martin and others like him feel themselves unable to share their knowledge publicly. And neither are official organizations that exist to monitor and regulate the aviation industry normally willing —or, in some cases, able—to broadcast the deficiencies of airlines.

In the United States the FAA, NTSB, and the Civil Aeronautics Board (CAB), all government agencies, must be aware that while U.S. airlines generally share an impressive safety record, there exists, among individual carriers, varying degrees of excellence: American Airlines, for example, has achieved over the years a safety record that is twice as good as the record of Braniff which, in turn has twice as good a record as Allegheny. However, none of the agencies has a mandate to retail such information. And the U.S. airlines' own body, the Air Transport Association, has no intention of judging its own members publicly.

Similarly, IATA (the International Air Transport Association) only concerns itself, at least publicly, with the safety record of its members as a whole. IATA is perfectly content to compare the performance of scheduled airlines, which make up its membership, with the safety record of charter airlines. But it will not acknowledge that there is any validity in comparing the records of, say, British Airways and British Caledonian, both IATA members. When we began conducting a survey of over seventy major airlines in an attempt to establish how Fine was Fine, IATA advised its members not to cooperate in what it feared could only be a "negative" exercise.*

ICAO, the International Civil Aviation Organization, based in Montreal, is, in theory, the most able body to monitor the performance of individual airlines because it receives data on over two hundred of them. But ICAO is made up of 126 contracting states and many of the airlines are owned by member governments who might not take kindly to comparative studies: ICAO's published safety statistics are therefore naturally concerned with worldwide safety trends, not the nitty gritty of whether it has been safer to travel with Alitalia or Air France. (The answer, according to our survey, is Alitalia. The detailed results of the survey are given in Appendix C.)

As things stand, those air travelers not blessed with inside knowledge have little choice but to select the airlines they fly with for reasons that are often every bit as irrational as the fear of flying. Ironically, it is the fear of flying

---

* We are not suggesting that IATA is complacent about aviation safety, but it does go to considerable lengths to protect the reputation of its individual members.

phobia that, to a large extent, inhibits those airlines with worthy records from initiating the kind of debate about safety that might bury, once and for all, the myth of universal excellence. And, as a result, many airline passengers are unknowingly exposing themselves to greater risk than they need.

IN 1960 UNITED Airlines, easily the biggest carrier in the United States, took delivery of twenty Caravelle jet airliners, manufactured by the French aerospace company, Sud Aviation. Twelve years later United sold the last of its twenty Caravelles on the secondhand market having avoided seriously damaging any of them. It was a remarkable achievement for the Caravelle was not technically a happy airplane and—in the words of Alan Hunter, a London aviation insurance assessor—other airlines crashed Caravelles with "monotonous regularity." *

For men like Hunter, who of course have a vital interest in the comparative safety of airlines, the Caravelle experience confirmed that United is a very remarkable enterprise with the corporate ability to take even a troubled airplane and operate it safely. United, and the other three big U.S. domestic carriers, American, Eastern, and Delta, have, over the last twenty-five years, established safety records that are remarkable. That is not to say that any of them is perfect. Running an airline the size of United, which has 383 planes, 49,000 employees, and flies to 114 cities every day, demands, besides anything else, eternal vigilance. Very occasionally the vigilance has been lacking: Perhaps United's most notorious failure on record was in giving command of a Boeing 727 to a pilot who was (in the opinion of the CAB) "indifferent" about following normal landing procedures. After his initial training as a jet pilot—on a DC-8 in 1960—the man was sent back to flying DC-6s because, according to United's instructors, he had shown "unsatisfactory performance in the areas of command, judgment, standard operating procedures, landing technique and smoothness and coordination." However, two years later the man was promoted to 727s, although two FAA inspectors reported that during the second training program he had "deviated from accepted procedures." On November 11, 1965, during a landing approach to Salt Lake City, Utah, the pilot descended his 727 at 2,300 feet per minute—nearly three times the recommended descent rate. The airplane landed 335 feet short of the runway, slid on its belly for more than half a mile and caught fire. Forty-three passengers were cremated. There were other factors involved in the accident, but the official investigation into the crash concluded, not surprisingly, that airlines should take more care in insuring that their pilots not only have the necessary technical skills but "qualities of prudence, judgment and care as well." †

---

* Between 1960 and 1972 there were twelve major Caravelle crashes, seven of them involving European airlines, in which a total of 739 people were killed.
† This and all other references to the Salt Lake crash from Civil Aeronautics Board Aircraft Accident Report, File No. 1-0032, June 7, 1966.

American, too, has had at least one 727 pilot who disobeyed the rules, with awful consequences. On November 6, 1965, American's Flight 383 from LaGuardia crashed two miles short of the Greater Cincinnati Airport near Constance, Kentucky, while trying to land during a thunderstorm when the visibility was deteriorating rapidly. The official investigation report said: ". . . There can be no valid excuse for such a gross deviation [from normal operating procedures] as was presented in this accident." * Fifty-eight people died.

Eastern Airlines has the unfortunate distinction of being the first—and, to date, the only—airline to have crashed a TriStar. On December 29, 1972, while most of the flight crew was trying to work out whether the airplane's undercarriage was down for landing at Miami (the annunciator light said it wasn't) Eastern's Flight 401 from New York flew itself into the swamps of the Florida Everglades. The crew had set the autopilot to maintain the TriStar's altitude at two thousand feet but had then, unknowingly, disengaged it by bumping one of the control columns. They should have been automatically alerted but the two computers—one for the pilot and one for the co-pilot—which controlled the autopilot had been accidentally mismatched by Eastern and gave no warning. It was not until five seconds before impact that the co-pilot realized something was wrong.

Co-pilot: *"We did something to the altitude."*
Captain: *"What?"*
Co-pilot: *"We're still at two thousand, right?"*
Captain: *"Hey, what's happening here?"*

Of the 176 passengers and crew on board, 101 were killed. Many of the victims' relatives sued Eastern and Lockheed for negligence: both corporations settled out of court.†

For its part, Delta has never been held negligent for a crash,‡ but on two occasions, in 1972 and 1974, it carried improperly packed radioactive materials in the cargo compartments of passenger jets. Some of the passengers involved in the 1974 incident who allege that they were exposed to radiation are now suing the airline for negligence.

But occasional lapses such as these apart, United, American, Eastern, and Delta have demonstrated that they are, indeed, Very Fine airlines. Over the last twenty-five years they have, among them, made some forty-eight million flights: thirty-five of those crashed. Even the most tremulous passenger should be able to draw some comfort from that.

Few, if any, of the world's other two-hundred-odd scheduled airlines can

---

* Civil Aeronautics Board Accident Report, File No. 1-0031, October 7, 1966.
† All references to the Everglades crash are taken from the pretrial hearings in the case of *Mandell et al. v. Eastern et al.*, U.S. District Court, Florida, Civ. No. 17931, 1974.
‡ The NTSB investigation into the crash of a Delta DC-9 at Boston on July 31, 1973, which killed eighty-nine people, concluded that the accident was due in part to errors by the flight crew. However, Delta claims that the pilot was given wrong information by an FAA air traffic controller and is currently suing the U.S. Government in an attempt to clear the crew of blame.

claim to have matched this achievement. For one thing only four other carriers have even come close to enjoying the massive patronage (a total of 1,150 million passengers between 1950 and 1974) that has enabled United, American, Eastern, and Delta to evolve into such immensely sophisticated concerns. But neither should it be imagined that the remainder sit in comfortable equilibrium, just below the pinnacle.

The disparities that exist between the leading (in terms of size) airlines are both large and alarming. The survey which we conducted into the safety records of seventy-four airlines showed, as might have been expected, that U.S. carriers have the best overall record. But, within that boundary, some have proved to be less safe than others. TWA and Pan Am, for example, both lag some way behind the big four U.S. domestic carriers. This may well be due, in some measure, to the fact that both airlines are predominantly overseas carriers (Pan Am exclusively so). As such, they often have to cope with some inadequate airports, the imperfect English of some foreign air-traffic controllers, and time-zone changes which inevitably increase the strain on pilots. But inadequate airports are to be found not only in the more remote corners of the world: according to a pilots' survey Los Angeles International, which handles four hundred *domestic* departures a day is sadly deficient. And no amount of rationalization about the difficulties of international flight can entirely explain why, after five accident-free years, Pan Am should in less than nine months have suffered three major disasters, or why eleven of TWA's fifteen accidents since 1950 should have occurred *inside* the U.S.*

It is equally difficult to find an obvious explanation for the disparity between the records of National and Continental, both established U.S. scheduled carriers of approximately the same size. Judged by world standards National has a very impressive record—since 1950 it has suffered only six accidents in which 139 passengers died. But that record cannot compare with Continental which in the same period has had only two accidents (forty-two passengers killed), one of which was caused by a bomb. (In justice it should be said that only one National passenger has died since 1960. He was sucked out of the window of a DC-10 in November 1973 after an engine fan-blade disintegrated and shrapnel punctured the fuselage. Continental has been accident-free since May 1963 when a bomb destroyed one of its 707s over Unionville, Missouri.)

However, the differences between airlines really begin to show when one moves outside the United States. Among European carriers, Lufthansa, which has carried twice as many people as KLM, has had half as many casualties. SAS, the Scandinavian airline, is better than both of them and so is TAP of Portugal. Among the big league carriers, British Airways has shown itself

---

* The Pan Am crashes all occurred in the South Pacific between July 1973 and April 1974, and killed a total of 283 people. After the third accident, the FAA ordered an in-depth investigation into Pan Am's worldwide operations. The results of the investigation have not been made public, at the request of the airline.

to be twice as good at carrying people safely as Air France. (Air France, once derisively known in the business as Air Chance has improved dramatically since 1968—but then it has one hell of a reputation to live down, with 812 of its passengers killed since 1950. In May 1974, one of its pilots did little to help the airline's public relations when he refused to land his Boeing 707 after it had encountered severe turbulence over Nebraska. Nine passengers were significantly injured—one sixty-four-year-old man had his knee broken in six places—but the pilot, Captain Robert Espes, insisted on continuing the seven-hour flight to Paris.)

All of the major South American airlines—Aerolineas Argentinas, LAN-Chile, AVIANCA of Colombia, VARIG and Cruzeiro of Brazil and, especially, VIASA of Venezuela—have undistinguished safety records. More surprisingly, perhaps, so too does the Philippine airline, PAL, which used to boast that it carried more passengers than BOAC. PAL's international record is fine: on domestic routes, it has suffered fifteen crashes in which 306 people were killed.

At the other end of the scale, both of Australia's airlines, Qantas and Trans-Australia, have superb records. Contrary to aviation legend, Qantas passengers *have* been killed—but its last accident was twenty-four years ago on July 16, 1951, and only seven people died. Since then the airline has carried nearly 10 million people 62 billion passenger-kilometers.*

Such an achievement has not come about through pure chance. It would be disingenuous to entirely attribute Qantas's record—and TAA's—to Australia's hospitable climate: Qantas does most of its business on overseas routes and, anyway, the weather in parts of Australia, particularly the southeast and the north, is often far from clement.

A great deal of the credit undoubtedly belongs to the Air Transport Group —Australia's equivalent of the FAA—which has imposed, and enforces, exceptionally tough air-safety regulations. But the attitude of the airlines' management must also have played a considerable part. In 1974 Captain Ron Gillman, a former Master of the British Guild of Air Pilots and Navigators, visited Qantas on behalf of the influential aviation magazine *Flight International* to investigate the airline's modus operandi. He came to the conclusion that Qantas owed its impressive record to a management "which is not prepared to narrow safety margins in the cause of economic expediency. . . ." †
Captain Gillman says that the attitude is not universal throughout the airline industry: "I once heard an executive say that his airline could afford to have one fatal accident every seven years."

There is one other factor that might have contributed to Qantas's record, although in sympathy for the feminist movement we regret mentioning it. Qantas discriminates against female flight attendants, preferring in the main male flight attendants on long-distance flights. This factor may lessen the

---

* A passenger-kilometer is the distance flown, multiplied by the number of passengers carried on each flight.
† "Safety Is No Accident," *Flight International,* London, October 3, 1974.

danger of a weak flight crew becoming distracted. Those who would dismiss that proposition as male chauvinism might bear in mind the misadventure that overtook British Airways Flight 910 on December 4, 1974. The airplane, a VC-10, was operating a scheduled service between Hong Kong and Tokyo and when it had reached its cruising altitude of 37,000 feet, the flight engineer began the routine task of redistributing fuel among the wing tanks to balance the plane. Unfortunately, in the middle of the transfer, the engineer was "distracted" by a stewardess and one of the fuel tanks—the one that was feeding all four engines—ran dry and the pilot suddenly found himself without engine power. He was forced to put the VC-10 into a dive to prevent it from stalling. As the engines died, one by one, the plane lost all electrical power and went into a Dutch roll, with the wings oscillating up and down. It took three minutes, nine thousand feet, and a great deal of piloting skill to get the plane back under control and the engines restarted.*

IN GENERAL TERMS our survey shows there are perhaps thirty airlines who deserve the accolade Very Fine. It also shows, encouragingly, that the overall record of the scheduled airline industry is getting better, although with over 1,800 deaths to its debit in 1974, there is no room for complacency.

One standard method of measuring airline safety is to divide the number of passenger-kilometers flown by the number of passengers killed. Using that method, the International Civil Aviation Organization has calculated that in the five years between 1969 and 1974, less than 0.2 people were killed for every 1 billion passenger-kilometers flown by the world's scheduled airlines as a whole.

According to our figures for that period, thirty-six airlines did better than average, mostly by not having any accidents at all. Of the remaining airlines we surveyed, only four returned death rates of less than one per billion and twenty-nine had figures ranging between one and 11.37 deaths per billion. The death rates achieved by the other four airlines deserve to be quoted precisely: TAROM (Rumania) 31.75; Turkish Airlines 60; AVIACO (Spain) 60; ALIA (Jordan) 139.34. In other words, based on this record, the death rate was 150 times greater than the average on TAROM, 300 times greater on THY and AVIACO, and 700 times greater on ALIA. Also between 1965 and 1969 only thirty-four airlines beat the industry average; twenty-one carriers had figures of twenty deaths per billion or worse. The pattern was the same for every five-year period we examined. (Though, statistically, it would be surprising if the *distribution* of air safety performance varied greatly from period to period.) What these figures underline is that huge variations exist between the safety records of the Fine and the Not So Very Fine airlines. These variations are carefully shielded from public view though are well known inside the industry.

It is clear that the industry's reputation for safety has been consistently

* British Airways internal operations report, dated January 1975.

propped up by a *minority* of its members. And the support is not only statistical. The premiums that airlines pay to insure their aircraft against loss or damage—and their passengers against death or injury—are based in some measure on their performance. Men like Alan Hunter, the London assessor we quoted earlier in this chapter, are employed by insurance syndicates such as those at Lloyd's to run an experienced eye over each airline's operation and estimate the risk. But, too often, if the premium demanded were to fully reflect Hunter's opinion, airlines would be driven out of business and that, most assuredly, is not the aim of the game.

Rather than do that the insurance men employ a compromise. From past experience they assume that from every type of airliner in commercial service one example will crash, and all on board will perish, once every few-hundred-thousand hours of flying time. (The estimated interval varies from aircraft type to type. For the 707, for example, it was assumed that there would be a crash once every 200,000 hours of flying time. With the 747, the anticipated interval was 500,000 hours; in fact, the worldwide fleet of 747s notched up two-and-a-half million hours of flying time before Lufthansa crashed the first example in Nairobi in November 1974, and even then the death toll was fifty-nine and not, as had been expected, several hundred.)

The premium that an airline pays is primarily based on the size and make-up of its fleet and the number of flying hours that it performs in a year. It is not a very sophisticated actuarial basis but by and large it assures that the insurance men will collect enough premiums to pay for those crashes which occur and still be in pocket.

The problem from an air safety point of view is that the burden falls heaviest on the bigger airlines who are usually, though not always, safer. From a business point of view, of course, it makes perfectly good sense: the bigger airlines can more readily afford the premiums.

None of this is meant to imply that insurers are complacent about air safety. Undoubtedly, they want to see the record improved just as much as the airline industry and the aircraft manufacturers. After all, numerous crashes would be bad for all of their businesses.

But, also undoubtedly, the self-interests of each group sometimes may lead to situations that are horribly dangerous. Just such an event was the premature and hasty promotion to the jumbo-jet league of Turk Hava Yollari—Turkish Airlines.

12

Friendship in Turkey

Difficult indeed it is for those to emerge from obscurity whose
noble qualities are cramped by narrow means at home.
—JUVENAL,
Satires

THE ARRIVAL OF Turk Hava Yollari's first DC-10 at Istanbul, on December 10, 1972, was a matter of some national pride. Before the end of
that month THY was able to proclaim itself the first airline outside the
United States to operate DC-10s on scheduled services. "This," noted a
Douglas engineer, "improved the image of the Turks throughout Europe."
The achievement was only possible, naturally, with the support and encouragement of McDonnell Douglas—and when revenue flights began, THY
was still dependent on Douglas engineers for the performance of many vital
tasks.

Whether it was right that a small, inexperienced airline should have set
out to bring a huge and complex new plane into service in about one-eighth
the time thought appropriate by airlines like Swissair and KLM was not
asked publicly at the time. In private, however, some of McDonnell Douglas's
own staff expressed a good many reservations about the technical quality of
the operation they were supporting.

For McDonnell Douglas in 1972, the deal with THY meant that after a
long and energetic sales campaign, three out of six unwanted DC-10s had at
last been disposed of. These particular airplanes had been ordered in 1970
by a Japanese financial conglomerate named Mitsui Bussan (Trading), which
expected to pass them on profitably to the Japanese internal carrier All
Nippon Airlines. Mitsui's confidence was based on the fact that in 1970 the
TriStar development program was falling behind schedule, largely due to the
problems which Rolls-Royce were experiencing with the RB-211 engine. It
was well known that All Nippon was about to decide which airbus it wanted.
In the circumstances, the DC-10 was the odds-on favorite.

However, Mitsui had not reckoned on the fierce determination of Lockheed to sell the TriStar to All Nippon—or on the very special influence of a
little-known Japanese businessman, Yoshio Kodama. On paper Kodama was
nothing more than the chairman of an insignificant colliery company and an

adviser to a number of industrial concerns. In fact, he was one of the most powerful men in Japan. His patronage was responsible for the election of three, if not four postwar Prime Ministers; he provided most of the funds for the foundation and support of the ruling Liberal Democratic party; and, through a carefully constructed network bound together since 1948 by skeins of personal friendship and mutual obligation, he had an immense, although hidden, influence in some of the most important industrial concerns in Japan. From 1958 Kodama was also an agent, a secret one, for Lockheed.

Just at the moment in 1970 when All Nippon was due to announce its choice of airbus, the airline acquired—through a boardroom coup—a new president, Tokuji Wakasa, the former deputy minister of transportation. Immediately afterwards, All Nippon postponed its airbus decision. It was unclear, at the time this manuscript was completed, as to what precise role Kodama played in the affair but, according to Lockheed's account,* he was primarily responsible for achieving that delay. His reward, paid over a two-year period, was more than $2.5 million.

The significance of this episode was that it left McDonnell Douglas with six unwanted DC-10s. By the time All Nippon had decided to delay its choice of airbus, the production of the six Mitsui planes had already begun and it was too late to cancel. McDonnell Douglas and Mitsui decided to abandon hope of selling the planes to All Nippon—wisely, as it turned out, for the airline eventually ordered TriStars. But selling them elsewhere was not an easy task for these were Series 10 DC-10s, the medium-range version, which were never in great demand outside of the United States: most foreign potential DC-10 customers were interested in the longer-range version, the Series 30. By the end of 1971 it looked as though the unwanted planes might overhang the market indefinitely. Their only real sales advantage was that, having been ordered by Mitsui in 1970, they could be delivered comparatively quickly. A buyer who could find employment for a short-to-medium-haul jumbo would not have to wait for eighteen months or more, as would be the case on ordering the TriStar or the European A-300B Airbus or, indeed, the longer range DC-10. In 1971, Turkish Airlines emerged as just such a potential buyer, and so became a major target for Jackson McGowen's salesmen.

An airline in a country as poor as Turkey does not generally have one mass-transit market. But THY is remarkable in having two. First, there is the flow of Turkish *gastarbeiter*—migrant workers—who provide much of the manual labor for the mighty economy of West Germany. As their families normally remain in Turkey, the *gastarbeiter* travel back and forth as frequently as they can afford to do so. Second, there is the pilgrim trade to Saudi Arabia. Many Turks are devout Moslems and feel an obligation to undertake the *Hajj*, the sacred journey to Mecca. The demands of the slowly mod-

---

* In evidence given in 1975–76 to the Senate Sub-Committee on Multi-National Corporations, Lockheed admitted paying $22.5 million in bribes to foreign government officials between 1969 and 1974. Of that amount, Lockheed said, $2 million was paid to Japanese Government officials to influence, among other things, the sale of TriStars to All Nippon.

ernizing Turkish economy now make the journey by foot and by mule impractical for many people. But Islam is not an obscurantist religion, and the aerial *Hajj* is an acceptable substitute.

In the early seventies when the European economies were climbing into their simultaneous boom, THY's business seemed likely to expand rapidly beyond the capacity of its small, mixed fleet of Boeing 707s, DC-9s, and Fokker F-28s. McDonnell Douglas hoped at first to sell four of the Mitsui DC-10s to them. Early in March 1972, just before the first Mitsui aircraft were due to emerge from the production line, McDonnell Douglas applied to the U.S. Government's Export-Import Bank for financial support on the sale to THY of four DC-10/10s at $75 million dollars, about 10 percent off what was by then—through inflation—the normal price. There was a kind of hopeful domino theory about this, also, for both Douglas and Lockheed subscribed to the belief that the firm which first gained a sales beachhead in the Middle East might take over the whole territory.*

The Export-Import Bank was not encouraging. Although the bank exists to help American industry by financing exports, it is a government agency and it was advised, via the State Department, that the U.S. Embassy in Ankara was not altogether happy at the prospect of THY buying four DC-10s. The embassy thought that THY would be overextended by the burden of training and maintenance and would not be able to "maintain a viable operation." It suggested that two DC-10s would be enough.

Those reservations would seem to have been well founded. The strengths, but also the considerable weaknesses, of THY derive from the fact that it is even more closely associated with the state than is usual in the national airline of a developing country. The fact that it is largely owned and controlled by the government gives it automatic control of the lucrative *gastarbeiter* and *Hajj* trades, and a good measure of financial support. But state control has been extended so far as to make THY something very like a branch of the Turkish Air Force, with most of its senior management posts held by ex-military flyers. (At one point in 1972 the top four men in the operations-management hierarchy were friends who had flown together in the Turkish Air Force aerobatics team.) Operating an airline, which has modern, complex jets like a DC-10 is a very different business from military flying, and this is particularly the case with the Turkish Air Force, which does not possess any large jet transports. (The fighters and small turboprop transports which the Turks do have come, of course, from the United States, which is anxious that its lesser NATO allies should be able to resist aerial attack and suppress internal disorder, but is understandably chary of providing them with the capacity to airlift troops into each other's territory.) Another consequence of the military connection is simple instability of management. The

---

* This has not worked out in practice. Only three other Middle Eastern airlines have, so far, gone into the jumbo-jet business since THY bought DC-10s: Middle East Airlines chose a short-range version of the 747; Saudi Arabian Airlines and Gulf Air bought the TriStar.

armed forces play a major political role in Turkey, with the result that air force and airline executive posts are susceptible to sudden change. In the five years up to 1972, THY had five different chief pilots.

However, despite the fact that acquiring a DC-10 fleet would require drastic expansion of its small corps of ex-air force pilots and flight engineers, THY showed great interest in the prospect, though agreeing with the U.S. embassy in Ankara and the EX-IM bank that four airplanes would be too many. Negotiations with McDonnell Douglas continued throughout the first half of 1972, but not at a lively enough pace to satisfy Jackson McGowen, who therefore resolved to make a strong personal pitch when Friendship '72 arrived at Istanbul in August. Friendship '72 also had Danny Kaye, the entertainer, on board: whether the show-biz side of the expedition impressed the Turks is hard to say, but McGowen probably made some impact by promising to cut yet another million dollars off the price if THY would order three DC-10s. This meant that THY would get them for $17.2 million each, against the 1972 starting price of $21.6 million.

Still THY hesitated, and in September McGowen sent a reproachful telex message from California to the THY board: "We do not see how you can deprive THY of the opportunity to make this great saving and operate DC-10s 18 months in advance of when you would otherwise be able to do so." He added a promise that any aircraft sold to Turkey would have all the required modifications made to their cargo doors before being dispatched from California. (Although the gentleman's agreement had moderated awareness of the June Windsor incident—and the THY pilots did not receive copies of the NTSB report of that incident until 1974, shortly before the Paris crash—it seems clear that the THY board had some anxieties about it.)

In the end THY succumbed. The airline was only asked to put up 10 percent of the price in cash, the rest being loaned by the EX-IM Bank and the Bank of America at 6 percent interest, a rate that was generous four years ago and now looks almost nominal.* In addition, McDonnell Douglas agreed to give $400,000 worth of crew training free of charge.

From THY's side a remarkable condition was imposed, without apparent demur from Douglas. It was stipulated that at least two of the three airplanes must be in service before Christmas 1972. This meant that flight crews, maintenance crews, and cabin staff would have to be trained in just under twelve weeks, starting virtually from scratch.

Highly experienced airlines like KLM and Lufthansa spent two years preparing for the adoption of the DC-10 into their fleets. Robert H. La Combe, who led the Douglas technical support team to THY's HQ in Istanbul, said

---

* The other three Mitsui airplanes were sold to the British charter airlines, Laker Airways. Freddie Laker, the airline's founder, is a remarkable entrepreneur, and although he is not anxious to reveal all of the details, he has acknowledged that he also obtained very generous terms from McDonnell Douglas, and Mitsui which still had financial obligation for the six airplanes so long as they remained unsold.

that meeting Turkey's twelve-week deadline meant setting a record because "nobody has ever tried to do it that fast before."

THY's board minutes, disclosed in the lawsuits which followed the Paris crash show that the airline was anxious to get the DC-10s in time to catch the Christmas *gastarbeiter* trade between Turkey and Western Europe. (The Turkish Government pays 40 percent of the fare of a *gastarbeiter* who uses the Christian holiday to refurbish the bonds of Moslem family life.) The revenue involved was indeed substantial: Douglas estimated that in their first two weeks of service, two DC-10s shuttled fifteen thousand people between Europe and Istanbul, which, if correct, means that each plane did the round trip twenty-one times in fourteen days.

But there was, it seems, another and more intriguing reason for THY's desire to get into the jumbo-jet business with great haste. The minutes of the board meeting at which the decision to buy the DC-10 was made show that the directors discussed the comparative merits of the DC-10 and the TriStar. It was acknowledged that the DC-10 had a cargo-door problem (the Windsor incident), but McDonnell Douglas had promised to fix that. And the great attraction of the DC-10, according to the minutes, was that it was available almost immediately, whereas a decision to buy the TriStar would involve a wait of at least eighteen months. The next passage of the minutes was deleted, before the document was disclosed in the litigation "for reasons concerning the security of the Turkish state." The subsequent surviving passage simply records the board's decision to opt for the DC-10.

In the autumn of 1972 the talks between the Turkish and Greek communities in Cyprus were becoming steadily more angry and less productive: it was the beginning of the escalation that led in 1974 to open war and the Turkish invasion. It is a fair guess that the military men on the THY board were attracted by the idea of rapidly acquiring an aircraft that could, in a pinch, remedy most of the Turkish Army's transportation problems. The runway at Nicosia, Cyprus's main airport, is long enough for DC-10s to shuttle in with 350 armed troops on every trip. (The first Turkish thrust when they invaded was made with paratroops against Nicosia Airport; but in spite of prolonged attacks, they never gained control.) The THY board could hardly help realizing, when they signed up with McDonnell Douglas, that the U.S. Government's reluctance to supply large military transports might be canceled out by the anxiety of American airplane builders to sell large airliners.

THE OVERSEAS SALES offensive that McDonnell Douglas conducted on behalf of the DC-10 was, without doubt, aggressive. The corporation has admitted that, like Lockheed, it made payments to foreign government officials in attempts to sell planes, and although none of that money went to Turkish officials, McDonnell Douglas's activities in Turkey do raise considerable questions about business ethics. The wisdom of pressing huge and complicated aircraft on an airline such as THY has to be challenged—the more so since McDonnell Douglas now chooses to blame the shortcomings of THY's maintenance operation for the Paris disaster. In the words of J. M. McCabe, Douglas's manager of flight crew training, THY had not long made "the first giant step to jet operations. . . ." THY felt confident that, with *Douglas Aircraft Company assistance,* the second step into wide-bodied tri-jet was feasible." (Emphasis added.) In fact, the report that McCabe wrote after several weeks' working with THY suggested that the airline had not even mastered the first step fully.

It is necessary to be frank, but also tactful and cautious, when comparing an airline like THY with American or West European standards. In the previous chapter, the absurd pretense of airline equality was demolished. The myth only exists, of course, because virtually every state in the world considers that it should have a national airline. And aviation authorities in the developed countries must cooperate with this ambition, because commercially vital landing rights are only granted by governments. If Turkish Airlines may not land in London, then British Airways will not be allowed to land in Istanbul.

Pilots, engineers, and airline administrators know perfectly well that this is the reality of the case, and sometimes there is a tendency to compensate for a degree of public hypocrisy by technological racism expressed in private. Disagreeable jokes circulate about traffic controllers with inadequate command of English, or ground engineers who fail to understand the uses of tools. It therefore needs to be said with emphasis that the racial make-up of an individual is not of the slightest predictive value in estimating his or her potential ability to learn complex skills and handle sophisticated machinery. The problems we are dealing with arise from the fact that during the last few decades of history, people in most Western countries have had more opportunity to engage in the costly and time-consuming process of accumulating technological education. (How many impatient Western engineers giving instruction in Moslem countries reflect that the mathematics upon which their discipline depends might not be available had not Arab civilization preserved and expanded the Greek learning that Europe chose for a long time to neglect?)

The THY flying teams which the Douglas pilots and engineers encountered in 1972 and 1973 were very much the people their background might be expected to produce. The pilots were in general excellent aircraft handlers, as jet interceptor pilots usually are. However, their common pedigree in the sense that they were *all* experienced military flyers, had produced an absurd command structure: McCabe, Douglas's manager of flight training, found in early 1973 that of 160 pilots employed by THY, 153 were captains, with the remaining seven about to be upgraded from the rank of first officer. McCabe reported back to Long Beach that this meant that responsibilities on the flight deck were not clearly divided between pilot and co-pilot, and that this lack of authority led to even greater difficulties with flight engineers, who in some cases were "refusing to obey the captain's commands."

McCabe submitted his written report to McDonnell Douglas in April 1973; Leo Hazell, a Douglas flight-engineering specialist wrote the report of his experiences in Istanbul in February 1973. Both documents were stamped "sensitive." * They were disclosed in the litigation in California in 1974, and together with the testimony of Douglas's Robert La Combe (in the same lawsuit: Hope versus McDonnell Douglas) about THY's ground facilities at Istanbul, they give a detailed inside view of the "first airline in Europe with the DC-10." (To be fair, it is necessary to enter two caveats. First, the documents and La Combe's statements do not relate to the present operations of THY. Turkish flight-crew training is now conducted by the U.S. airline United which is widely regarded as being about the best there is. Second, nobody from Turkish Airlines has, at the time of writing, given evidence in the Los Angeles litigation and had the opportunity of refuting the allegations.)

McCabe, in his report, appears to have been extremely sympathetic to THY's difficulties, and his words have none of the impatience which occasionally emerges in Hazell and La Combe. "Perhaps the picture of Turkish Airlines painted in this . . . report is too severe, but it's important that Douglas management understand reality in that part of the world," he wrote. "The Turks are a proud people and a curious mixture of Eastern and Western cultures. They know they need help and yet are reluctant to admit this to someone that might look down on them as inferior people. However, if they sense a genuine interest in the individual or company attempting to assist them, they are most receptive and accept advice willingly." THY's flight-operations management "wants very much to run a safe, efficient and profitable airline, but they are dealing with bureaucracy at every level, unwillingness to accept responsibility, and personnel that are technologically deficient."

McCabe was somewhat astonished by THY's concept of passenger relations. For some time the airline could not provide undercarriage locking pins at all the airports it flew to, and these pins are essential to prevent the undercarriage from buckling when the aircraft stands at rest. THY crews were quite happy to take off with the long, oily pins stowed in the aisles of the passenger cabins. Worse, McCabe noted, that on overnight delays there was sometimes no provision for passengers, who found themselves sleeping in the airport terminal or, on one occasion, in the unheated aircraft standing on the tarmac. After one all-night vigil at Frankfurt, it became clear that the plane was overbooked, so the infuriated *gastarbeiters* hurled their luggage into the engine intakes and sat down on the runway in front of the wheels. THY's customers were removed from the tarmac with a fire hose by the German authorities. These were, however, cosmetic points. THY's most serious weakness was the lack of properly trained flight engineers.

THY had gained most of its jet experience with the DC-9, which is a relatively small twin-engined jet and was designed specifically by Douglas to be flown by two pilots without a flight engineer. There was never any

* The reports are produced in full in Appendix E.

doubt, however, from the very first design stages of all the wide-bodied jets that because of their huge and sophisticated engines, their proliferation of hydraulic, electrical, and pneumatic systems, because of the scale of their air conditioning and pressurizing equipment—and, apart from anything else, because of the number and size of their doors—three pairs of hands would be required to fly them. The flight engineer in an American or European airline is not a mere mechanic. Usually he is a qualified commercial pilot himself, but in any event, he must hold a separate and specialized certificate. His task is to calculate the aircraft's takeoff weight and its readiness for flight. He monitors and controls the engines and their fuel supply during takeoff, transit, and landing, so that the pilots are free to handle the flying controls and to navigate. He controls all the airplane's life-support systems, including those which are meant to insure safe working conditions on the flight deck.

In order to put their DC-10s into service, THY had to find some fifteen flight engineers. There were obvious candidates available because at that time THY leased three Boeing 707s which also require flight engineers, but the airline's management decided against transferring any of the experienced men to the DC-10. Those who did make the switch had, at best, two or three hundred hours experience on the 707 and the rest of the flight engineers needed for the DC-10 were recruited directly from the interceptor and ground-attack squadrons of the Turkish Air Force. As Leo Hazell said in his written report:

> [Their] training background is not commensurate with other world air carriers. . . . The typical THY F/E [flight engineer] . . . has had no training in F/E theory similar to that which is necessary to pass the U.S. license requirement. His background would even disqualify him from taking the U.S. F/E exam.

It is not surprising that Hazell should have been disturbed. Most American flight engineers are fully qualified commercial pilots or have a degree in aeronautical, electrical, or mechanical engineering. And the implications of THY's policy became alarmingly clear to Hazell when he traveled as an observer on the delivery flight of the first DC-10 sent out to Turkey. After takeoff from Bangor, Maine, Hazell found that the THY engineer was about to set off across the Atlantic without a fuel log on board, an act of almost suicidal overconfidence.

On reaching Turkey, he discovered how this man and his colleagues had received their certificates. The Turkish equivalent of the FAA, operating from Ankara, five hundred miles away (La Combe testified that he did not see one of its officials in Istanbul during his eight months' stay there), had delegated the power of certificating flight engineers . . . to THY itself.

It was not, perhaps, surprising that the engineers "failed to give sufficient weight to their own jobs," as Hazell put it. The average THY engineer "regarded the flight engineer's task as an interim stepping stone to becoming a captain," and could not "fall back upon a solid background of flight engineering experience."

The normal practice of having the engineer handle the engine throttles on takeoff understandably went by the board in most THY aircraft. The senior captain generally delegated the task to the co-pilot, leaving the engineer staring at the complex array of engine instruments in the hope that he would pass on to the pilots any relevant information. Frequently, Hazell wrote, the engineers did nothing else but stare, "regardless of how high, low or staggered" the performance of the three engines might be. As long as that happened, he said, "THY airplanes and engines will be in jeopardy."

Failure to monitor engine performance on takeoff is supremely dangerous in a big jet. And as often as not, in addition THY engineers were calculating airplane takeoff weight incorrectly: The power-settings cannot even be properly selected without knowledge of the correct weight. Sometimes, Hazell reported, the errors were as much as 40,000 pounds—nearly eighteen tons, or the weight of 250 passengers. It appears that few of the THY engineers had even heard of the weight-and-balance theory before they went to Long Beach for brief DC-10 familiarization-training sessions and not many of them, Hazell thought, had fully understood it when they left to go back to Turkey. In any case, he said, THY's commercial policy would have made really accurate calculation difficult, because one of the airline's selling points was that it allowed *unlimited* hand baggage.

To add to the confusion, THY had its weight-and-balance forms calibrated in kilograms, although the DC-10's instruments were calibrated in pounds. As a result, airplanes were taking off without proper fuel reserves, which could have led to grave emergencies in the case of flight diversion.

Both Hazell and McCabe commented on the fact that engineers were apt to leave the cockpit without permission during flight. McCabe thought that often they did not know how to operate oxygen masks, monitor flight deck atmosphere, and operate aircraft doors. They seemed prepared to take off in an aircraft without any real knowledge of its state of airworthiness: few of them had the flight bags of reference books, manuals, and tools that all properly trained flight engineers carry with them. And they generally did not have the pocket calculators which are now essential equipment in their complex trade.

Apart from managing the fuel supply, the engineer in flight is supposed to make a log of "squawks": malfunctions of the airplane which need to be reported to the maintenance crew after landing. The main feature of the flight engineer's station in a jumbo jet is the imposing array of annunciator lights which signal faults in the plane's systems. The engineer's task is to put the light out, if possible, by juggling the systems. In any case he should note it. On one occasion, a Douglas engineer on a THY flight logged fifteen "squawks" while the THY man logged none. Many of the THY engineers,

Hazell thought, lacked the necessary experience to "differentiate between a normal and an abnormal functioning airplane."

Hazell thought he had managed to persuade the THY engineers of the importance of logging "squawks," but La Combe, who remained in Turkey after Hazell's visit, testified in 1974 that his maintenance support staff could not rely upon log books compiled by THY crews. Douglas flight engineers riding with THY to provide on-the-job training reported to La Combe that the Turkish engineers often did not know how to interpret a "squawk" signal.

> . . . A light would come on and my man sitting on the fourth seat [the Douglas man] would look over and see it and the flight engineer would just be sitting looking at it . . . hoping it would go away, I guess. . . . Sitting in the fourth seat and seeing a light on is sometimes rather a stimulating experience. And you would want to get it out.

The Turkish flight engineers were not helped by the fact that the courses they went through at the Douglas plant were abbreviated. Anxious to meet the twelve-week DC-10 adoption deadline, THY cut Douglas's recommended thirty-two hours of flight-simulator training to twenty-four hours per crew. (Possibly in the early stages of the DC-10 program, THY did not fully appreciate the value of ground simulators, as against actual flight, when dealing with an airplane whose flying costs run to $1,775 per hour. Not only is it far cheaper, but with the danger of a crash eliminated, far more risks and difficulties can be explored.) Out of the first batch of prospective crews who went to California in late 1972, Douglas failed three pilots and two engineers, in spite of the pressure of THY's deadline.

Hazell left no doubt in his report to McDonnell Douglas of how serious he considered the problem to be. On page three he wrote:

> The THY F/Es have no direct supervisor or Chief Flight Engineer. . . . With this policy the F/Es are left to drift without direction or leadership. A typical case involving the lack of supervision resulted in a non-qualified THY F/E being sent solo on a DC-10 to London and back to Istanbul without a [Douglas] or THY check F/E on board the aircraft. Fortunately the aircraft got back to Istanbul without disaster, but the number one hydraulic system was contaminated.

On page nine he reported:

> The THY F/Es have a habit of getting up and leaving the cockpit without advising the pilot. Once again, this is something the F/E should already know, but due to their limited training and background, they do not know they are doing anything wrong. If an emergency should arise in the cockpit during the F/E's absence, obviously the safety of the DC-10 is in jeopardy to a greater degree. . . .

Summarizing the situation Hazell wrote:

> One could say conservatively that (the flight engineer training program) has a long way to go. One of the most baffling aspects of the program is the

fact that the THY F/E's ask few or no questions. Whether this is due to shyness or indifference, the writer cannot tell. . . . The flight engineer situation at THY should be resolved immediately. As insignificant as this job may seem to some people in the THY organization, it should be noted that one incompetent DC-10 flight engineer could trigger an incident and possibly jeopardize future sales by Douglas.

Broadly speaking, McCabe concurred, although by the time he wrote his own report, a couple of months after Hazell, he thought the airline was "smoothing out" and that "significant progress can be noted especially by one that is intimate with the whole situation from its beginning." He added: "Nevertheless, it is this writer's opinion that the Douglas influence and support either contractually or non-contractually should continue."

THE DOUGLAS SUPPORT team did not report that the THY ground engineers were infected with the same casual spirit that was sometimes found among the flight engineers. The maintenance men were not recruits from military service but trained mechanics, and they did not regard their job as a stepping stone to something more prestigious. To judge by the testimony of Robert La Combe, the Douglas executive who ran the support program in Istanbul, they had enough background in airline maintenance work to appreciate the magnitude of the task they were being given when the DC-10 operations began. This, no doubt, is why many of them expressed a degree of unease.

THY actually began running DC-10 passenger flights before any of its own ground engineers had completed their DC-10 conversion courses. The aircraft which were delivered to THY were identified by their production-line numbers as Ships 29, 33, and 78: as it happened, Ship 33 was the first to be delivered on December 10, 1972, and it began service on December 16. Ship 29, arriving three days later, brought home the first of the THY maintenance crews that had completed their DC-10 courses at the Douglas training school in Long Beach.

However, La Combe, together with ten Douglas ground engineers, had established himself in Istanbul some time before Ship 33's arrival and was joined shortly afterward by five Douglas flight engineers (crew chiefs). Together with other THY mechanics, he and his team got the Christmas airlift under way. "It was a hodge-podge sort of a deal of who worked on it," he testified later.

Before leaving Long Beach, the THY mechanics were asked to make written evaluations of the three-week training program. Most of them said that the course had been "too fast," considering that they were trying to understand an airplane far more complex than any they had seen before. Many of them said that they did not feel their English was adequate to deal with the particular course they had undergone—accelerated by seven days in order to meet the service deadline. About three-quarters of them said that they would have to carry out "further extensive research" in their own time. Most of the systems in the plane, wrote one engineer, "are working by electronic circuits, but since we are not familiar with electronic circuits the duration of a course must be a little longer and the speed a little slower and OJT [on-job-training] must be longer."

As La Combe already knew, these men were coming back to perform maintenance work under extremely adverse conditions. To begin with, THY's largest hangar still left the rear fuselage and tail surfaces of a DC-10 protruding, which meant that a good deal of maintenance work had to be done in the open during the appalling Turkish winter. The only usable runway at Istanbul's Yesilkoy Airport was in such disrepair that it was often necessary to send a jeep out to remove debris capable of damaging wing flaps and undercarriage parts. There was a new runway, but it was unserviceable for lack of a lighting system.

Without the presence of La Combe's team, and in particular the five Douglas flying crew chiefs, who, among them traveled on nearly every flight, it seems unlikely that THY could have maintained its services during that first bitter winter of 1972–73. The problems were reduced by the fact that Ships 29 and 33 did not come straight off the production line. Since their completion in March they had done some test flights and demonstration flights (Ship 33 had been to Turkey in the summer as Friendship '72). As La Combe said, some of the "infant mortality" had been worked out of them.

But although it was successful enough in enabling THY to get its operations started, the Douglas-support operation ran into trouble in almost every other respect. In order to supplement the brief courses given to the ground engineers at Long Beach, there was supposed to be a recurrent training program in Istanbul, and there were three instructors attached to La Combe's team for that purpose.

Circumstances in Istanbul soon reduced these brave hopes; first, by way of a battle, quite outside La Combe's control, between the operational and the training divisions of THY. The airline's training establishment was in downtown Istanbul, forty-five minutes away from the airport: perhaps, understandably, in view of the schedules he was expected to keep, the airport maintenance boss said that his men could not spare the traveling time. It took six weeks, during which time the Douglas ground-engineering instructors remained inactive, before permission was given for training at the airport, where, it was discovered, no proper facilities existed. "Although the terminal complex is in a continual state of construction," wrote La Combe, "there are no plans to build a training center. . . ."

Even more serious was the fact that virtually all the training material for ground crews went astray. Some 3,500 color slides, together with tape cassettes and film strips, utterly vanished. And, of sixty sets of *Flight Crew Operating Manuals, Flight Engineer Reference Manuals,* and *Flight Study Guides,* only a single volume of the *Flight Crew Operating Manual* was ever found while the Douglas team were in Istanbul. (The THY training department blamed Turkish Customs officials for the nonarrival.) J. M. McCabe, making his tour of inspection in spring 1973, declared that the missing materials were "the tools of the Training Department. Without them recurrent training will be sadly deficient."

On La Combe's evidence, the first non-American scheduled airline to operate DC-10s gave serious indications of being overtaxed by the accomplishment. On one occasion early in 1973, a DC-10 took off from Istanbul with both fire detection loops out of action. On the advice of the Douglas crew chief on the flight deck, the plane returned and the system was allegedly fixed. But after takeoff, it became apparent that the system was still not working. The THY pilot elected to ignore the Minimum Equipment List (MEL), which stipulates that every system in the plane must have one fire-detection loop in working order, and continued the flight to Munich.

The plane returned to Istanbul with the two loops still inoperative, and it was not fixed at the next departure from Turkey. This time the Douglas flight engineer, on La Combe's instruction, refused to fly. It emerged that the reason for the aircraft breaching the MEL rules was that there were no spares to repair the fire-detection loops; one of them had been placarded "inoperative" for a considerable time. At a formal Douglas–THY meeting held in March to try and thrash out the problems of the DC-10 program, the Turkish pilots complained that there were numerous placards in their cockpits saying that systems and instruments were "inoperative." This was happening, said La Combe in testimony, not only because spare parts were often unavailable, but because sometimes the maintenance people did not have time to make all the necessary repairs before the plane had to be dispatched.

Organizing spare parts and maintenance for a jet fleet is no small task, and La Combe's observations suggest that THY found that handling DC-10s in addition to DC-9s and Boeing 707s was often too demanding. Broken parts were sometimes kept lying in Istanbul for weeks before being shipped to the manufacturers for repair: spares and service-bulletin kits were lost or held up by customs, and even those which survived were not always installed, because there was no workable system for dealing with modifications. As there was a lack of coordination between the different departments of engineering, supply control, aircraft scheduling, and maintenance, it was difficult to insure that airplanes would be available for modification or repair when the parts arrived or that enough time had been allowed for installing them. And in all cases, said La Combe, it was difficult to insure that what changes were done had been properly recorded.

It seems clear that the relationship between the proud, sensitive, but inexperienced Turks and the highly qualified Americans did not develop fruitfully. By the end of January 1973, La Combe, who already knew that the maintenance logs of THY flight engineers were grossly inadequate, found that the Turkish ground engineers would not work from the logs prepared by the Douglas crew chiefs. "I found out that they just were not working our squawks off," he said. Their attitude seemed to be: "Douglas wrote it, Douglas fix it."

Not that the Douglas ground engineers were really allowed to fix anything: After the first "hodge-podge" week, the Turkish unions had insisted that none of the Douglas engineers handle tools themselves. The solution adopted was for the Douglas crew chief on each flight to give his flight sheet to THY's chief inspector, who would write the unrecorded "squawks" into the THY logbooks, thus creating a work sheet that was acceptable to the maintenance crews. The Douglas ground engineers were then permitted—within the limitations of the language barrier—to advise the THY ground engineers on how to clear the various "squawks."

But soon afterwards the Douglas crew chiefs found themselves in difficulty with the THY flight crews. With increasing frequency, they were asked to travel in the passenger cabin because the "fourth seat" in the cockpit was required by a new THY trainee, and the THY operations department formally asked La Combe's men to reduce their flying hours. Their argument was that if the Douglas men were present on every flight, it would not be possible to train new Turkish personnel in sufficient numbers. The view of the Turkish flight crews was that they no longer needed Douglas supervision—which, if the reports of Hazell and McCabe are even roughly correct, could only have been the bravura of inexperience.

Meanwhile, the support operation was costing THY dearly. Originally, it had been intended to run for twelve months, during which period there would have been a Douglas flight engineer on every THY flight. But this would have cost THY $800,000—nearly half of its annual profits—and when confronted by this reality, THY found it clearly unattractive. As early as April 1973, THY began to suggest that the cost of the support program was making it difficult to continue, and plans for extra flight crew training in California were dropped when McDonnell Douglas declined to cut its simulator price from $350 per hour.

There were, of course, two ways of looking at things from Douglas's viewpoint. La Combe, struggling with day-to-day frustrations in Istanbul, felt that there was no point in staying on if the Turks were not prepared to use him and his team. In terms of what it was allowed to do, the Douglas operation in Istanbul was in his view "overstaffed." His testimony suggests that he felt incensed at the exclusion of his crew chiefs from the flight decks of the Turkish DC-10s.

On the other hand, J. M. McCabe, the Douglas manager of flight training, had urged a broader view. There were obviously temperamental difficulties between the Turks and the Americans, which McCabe believed could be overcome. But the chief recommendation of his devastating report of April 1973 was that Douglas support should not be withdrawn from THY. If financial concessions were necessary to keep the support contract going, then concessions must be made. Discussing the question of recurrent training for flight crews, McCabe said that two U.S. airlines were known to be interested in bidding for the contract. "Knowing the Turks they will probably choose the cheapest program," he wrote, "it is suggested that our own management seriously consider a major reduction in our [flight] simulator prices to capture this airline, so that we can continue to exercise a level of control over their operation."

It was not the view that prevailed within McDonnell Douglas. In March 1973, La Combe had received a long letter from Long Beach, signed by Douglas's vice-president for product support Harold Bayer, which made it clear that THY could only have what they were prepared to pay for. The decision, and the responsibility, were entirely THY's, in Mr. Bayer's view, and if the Turks were not ready to pay, there would have to be a serious reduction in the support program. Clearly, McDonnell Douglas felt that every possible economic concession had been made to THY: the company was not prepared to subsidize a state-owned airline any further.

Six months after THY began flying DC-10s, the support operation began to run down. On July 1, the Douglas crew chiefs wound up their reduced program of check flights with THY, and on July 18 La Combe himself departed. Others left at intervals during the second six months of 1973, and by the beginning of March 1974 there were only three of the team left in Istanbul, largely occupied with the supply of spare parts.

La Combe's intention was to insure before leaving that the THY flight engineers were precisely aware of the tasks they had to take over from his men. But like so many other aspects of the support program, this resolution was only partially implemented. One task that La Combe had assigned to his Douglas crew chiefs was that of checking that cargo was properly loaded into the holds and that the cargo doors were properly sealed and locked. La Combe was well aware of what had happened at Windsor, and it was for this reason, he said, that he warned his men to "be very, very conscious of closing the doors." Yet he was unable to say which member of the THY flight crew had been delegated with this task after his engineers left Istanbul.

Granted the difficulties under which they were working, THY's ground engineers and flight crews accomplished no small feat by keeping their three DC-10s flying for fifteen months without a major mishap. What the Turks did not know was that one of those airplanes, Ship 29, was dangerous in a way which went beyond the questions of engineering principle that F. D. Applegate had raised in relation to the design of every DC-10. In spite of the gentleman's agreement made with Jackson McGowen, Ship 29 had a "lie" in it, in exactly the fatal sense that Rudyard Kipling meant by the words quoted on page 2 of this book. To inquire into the mystery of how it got there, we must examine the ancestry of this particular airplane.

13

The Invisible Airplane

The individual technician who assembles some component for
a plane . . . knows that his work calls for the utmost care and
exactitude and that the lives of many may well depend upon its
reliability. Beauty, so runs the ancient definition . . . is the
proper conformity of the parts to one another, and to the whole,
and this requirement must also be satisfied in a good aircraft.
—WERNER HEISENBERG,
The Meaning of Beauty in the Exact Sciences

THE FINAL ASSEMBLY of the DC-10, sections of which are built in
the Far East, in Europe, and in many parts of the North American con-
tinent takes place in the desolate industrial suburbia at the southern tip of
the Los Angeles megalopolis. The 455-acre Long Beach plant, where the
production lines can roll out one DC-10 each week, lies fifty miles south
of Lockheed's Burbank plant, where amid equally unappealing landscape
much of the rival TriStar is built.

Monotonous rows of corrugated sheds surround the Long Beach plant's
modest seven-story administrative block, the only multistory building visible
in the horizontal townscape. Lakewood Boulevard, a busy major highway,
cuts through the plant, and once every week or so Lakewood closes between
midnight and 4 A.M. to allow an assembled DC-10 to cross from the manufacturing facility to the section bordering on Long Beach Airport where it
will be prepared for delivery.

Sometimes months, sometimes weeks, sometimes only days before the
crossing of Lakewood Boulevard the engines, landing wheels, refrigeration
equipment, navigational systems, the wings, doors, flaps, and lavatories will
have separately arrived at Long Beach to meet a minutely arranged schedule.
In all, contracts worth more than a billion dollars have so far been placed
to obtain the subassemblies and components for the DC-10, the airplane
which bears the name of McDonnell Douglas.

The largest subcontractor is General Dynamics, whose San Diego plant is
one hundred miles down the freeway from Long Beach. At San Diego the
fuselage some fifteen tons of structure is built and fitted with its floors and
doors. The wings come from the McDonnell Douglas subsidiary in Canada,

and most of the tail from the old Chance-Vought concern, in Dallas, Texas, now part of the LTV conglomerate. Huge machine tools, automatic pressing, and casting systems—inhuman machinery of every possible description—is used in fabricating the parts of a big airplane. McDonnell Douglas's job at Long Beach is to do what American industry is best at: orchestrating a fast-moving, cost-effective production line. And the business still depends, as much as it ever did, and perhaps more, on individual but highly organized human skills. It depends upon the behavior of human beings toward each other, for production is first a social phenomenon, and only second a mechanical one.

Naturally, therefore, the glueing, welding, soldering, bolting, and riveting together of each DC-10 generates a vast quantity of documents, in which the relationships of all the people who build the airplane are supposed to be recorded. We have encountered "paper airplanes" before; that is, the set of concepts which map out the general characteristics of a prospective airplane type. Now, in production, we see that each metal airplane is accompanied by another kind of "paper airplane," which is the set of records showing that this particular example has been built and inspected to the standards laid down by the design teams.

Quite obviously, the practical safety of an airplane depends in the end on the quality of this monitoring process. The company's own inspection system is generally supervised by the Federal Aviation Administration, and normally the progress of each individual airplane is closely followed by the customer airline. As positions in the delivery sequence are matters of great importance, most airplanes are destined for one particular airline from the moment the first metal is cut, and that airline keeps engineers at the manufacturing plant watching each stage in its progress.

As these are the people who will have most to do with the airplane when it goes to work, theirs may well be the most rigorous check of all. It happens that the airplane with which we are most concerned—Ship 29—was an "orphan" for lengthy periods during its construction, and there cannot be much doubt that this made a difference to its fate. Nonetheless, the real responsibility for making an airplane measure up has always been ascribed to the builder, and it was said of the Douglas Aircraft Company for many years that it had "never built a bad airplane."

The Douglas quality-control system was built upon the idea of unambiguous personal responsibility. Every action in the history of an aircraft should be the responsibility of a particular person, and recorded as such. This system was supposed to be strengthened, if anything, under the new McDonnell regime. Certainly James McDonnell believed himself to be running a system in which "with military-like precision all projects are signed off, and individual responsibility is accepted for everything, no matter how routine."

Ship 29, however, was undoubtedly a "bad airplane"—and not alone in its badness, as was discovered in the inquisition which followed the world's most devastating air crash. The plane's history reveals that the "responsibility" system at the Douglas plant had become virtually meaningless. Responsibility had been so subdivided as to drift away like sand between the fingers. At times, as we examine the strange history of Ship 29, the Douglas plant appears as a place where everybody is responsible for something particular—when he can remember, amid the lattice-work, exactly what it is—but in which nobody at any level in the company would accept any general responsibility for anything. One wonders where the will and purpose that summoned the DC-10 into existence was located, for the Douglas engineers and executives seem to portray themselves as mere cells in a nervous system whose center is somehow always elsewhere.

Our evidence about the building of Ship 29 is drawn from lengthy depositions taken in Los Angeles by lawyers working for the relatives of the people killed in Paris in March 1974. Witnesses from McDonnell Douglas, from General Dynamics, and the Federal Aviation Administration were questioned in considerable detail; however, a compensation settlement was reached before the case could go to trial, and before any witnesses from Turkish Airlines reached the stand. In addition to the oral depositions, the defendants had to produce more than 50,000 pages of documentation upon the development of the DC-10 and the building of the twenty-ninth example. In the nature of pretrial investigation, the story that emerges is not complete: there are mysteries, uncertainties, and clashes of evidence which further investigation and trial might have been expected to resolve. Even so, this body of testimony and documentation perhaps gives more detail about the creation and destruction of one particular airplane than anything which has previously been available.

It is necessary to say that the depositions were taken in secret. Possibly this accounts for the tenor of the evidence given by senior McDonnell Douglas executives like John Brizendine and William Gross. They certainly seemed at little pains to preserve the image of their company as one in which executives and officers thought hard before putting their decisions on the line.

IT IS CONVENTIONAL to identify aircraft in production by a ship or fuselage number. Thus the very first DC-10 to be built and flown was known as Ship One. Ship 29 emerged on April 5, 1972, a few days after the delivery date originally agreed with Mitsui, the Japanese finance house which during 1970 had ordered this and five other DC-10s. But soon after production began Mitsui's hopes of selling or leasing the planes to All Nippon had collapsed.* Of course, a complex production system cannot be stopped simply because a particular buyer has dropped out, and as Ship 29 and the other Mitsui airplanes moved along the line, Jackson McGowen and his colleagues tried hard to find new customers for them.

In spite of their close pursuit of Turkish Airlines, April 1972 came without a customer having been found. Ship 29 was certificated as airworthy ac-

* See Chapter 12.

cording to the state of knowledge at the time and was kept for the next nine months in the "Rework for Delivery" area adjoining Long Beach Airport. During that period, which lasted until December 1972 (when Ship 29 took off at last for Turkey via Bangor, Maine), it flew an average of a couple of hours each week as a general test airplane. When the much-delayed deal was completed with Turkish Airlines, the seats were fitted, interior decor applied in the appropriate style, and the fuselage painted in THY's colors.

In the period between completion and sale, some three hundred modifications, known as "service changes," were made in the DC-10. These were the results of the operating experience of U.S. airlines which had the plane in service, and for the most part they were trivial enough, ranging down to little matters of cabin-lighting circuits and ways of dealing with problems in the galley area.

Amid this flood of minutiae, there were the changes listed in Service Bulletin 52-37 which were necessary to honor the gentleman's agreement made between the FAA and Douglas to prevent the rear cargo door from coming open as it had in the Windsor accident. According to the bulletin, three things were to be done to Ship 29,* viz.:

> The lock pins of the cargo door were to be adjusted for greater travel, so that in the "closed" position they would overlap the latches by one-quarter of an inch, instead of one thirty-second of an inch.
> A support plate was required to prevent the upper torque tube bending under pressure as it did at Detroit, causing the Windsor incident.
> The tube carrying the locking pins was to be modified ["scarfed"] so that its new and longer travel would not make it give a false signal through the micro-switch system.

Of these three modifications, only the third one was certainly accomplished, and on its own it was meaningless. But the aircraft's records at the plant said that all three jobs had been done, inspected, and passed as adequate. Had Ship 29 crashed, as it might well have done, into the sea, or had its wreckage been irrecoverable for some other reason, its catastrophic end would have been quite impossible to understand. Because the pieces were found, the mechanical failure was not too hard to grasp. But the nature of the human failure that created the false records of Ship 29—and also, as it turns out, of Ship 47 †—is something which cannot be understood so readily.

WHEN MR. McGOWEN and Mr. Shaffer agreed, in gentlemanly fashion, that there was no need for an Airworthiness Directive in order to get the

---

* See Chapter 9.
† Ship 47, another Mitsui orphan, was sold to the British charter airline, Laker Airways. After the Paris crash Laker discovered that the support plate was also missing from the rear cargo door of its plane although the paperwork for Ship 47 claimed that the job had been done. However the other two modifications had been correctly carried out by McDonnell Douglas.

DC-10 cargo doors fixed, Ship 29 was still lying unpurchased. Therefore, it was not listed on Service Bulletin 52-37, which was addressed to the airlines that had purchased DC-10s, and which did eventually persuade most of them to make the improvements which it described.

Ship 29 was not at that time owned by an airline, and it appeared only on an internal list of airplanes requiring modification within the factory. The modifications required by the internal paperwork were just the same as those sent to the airlines in Service Bulletin 52-37. In both cases, there was no indication that safety might be particularly involved.

When we say that the Long Beach records state—falsely—that the work was done, we mean that microfilm photographs exist of three stiff cardboard forms. These forms are of a standard type devised and written in the Planning Department of the Douglas plant, and known as Advanced Assembly Orders (AAOs). There is an AAO for each discrete manufacturing operation on each individual aircraft, and it is made of card rather than paper so that it can follow the aircraft through the plant.

Each of these three particular AAOs describes one of the critical gentleman's agreement tasks—pin travel adjustment, support plate fitting, lock tube "scarfing"—in a few terse words, gives numbers to identify the bits and pieces needed for the work, and provides space for the personal stamps of plant inspectors to be applied in verification of the work.

The quality of the microfilm reproductions produced to the investigating lawyers was not high. Even so, it is possible to read that on July 18, 1972, three inspectors—identifiable as Edward M. Evans, Henry C. Noriega, and Shelby G. Newton—seemingly applied stamps verifying that the support plate had been installed, and the lock tube had been "scarfed." These three men were brought forward in Los Angeles and examined on oath. It emerged that not one of these three could recall having worked on the cargo door of any DC-10 at any time in 1972. Nor could they recall any occasion whatever on which they had worked together—and they were certain they had not done so on July 18 of that year.

And according to the microfilms, a few days later two other inspectors "stamped off" the rerigging of the locking system on Ship 29's door, verifying that the locking-pin travel had been increased and the requirements of the gentleman's agreement completed. These two inspectors had not been identified by McDonnell Douglas or produced by the time that the legal settlement was made. But on the evidence produced, the work which they purported to inspect was certified as complete even before there were instructions available to say how it should be done.

McDonnell Douglas conceded twenty-one days after the Paris crash that the support plate was *not* fitted—and physical examination of the recovered metal of Ship 29's door makes it impossible to say otherwise. But the company has never made an equivalent concession on the equally important question of locking-pin travel adjustment, and as this depends not upon the fixing of one object physically to another, but upon the reversible setting of rigging screws, it is possible to say that Ship 29 might have left Long Beach with one setting and somehow acquired another while in the care of Turkish Airlines.

An inspector's stamp appears as a set of letters and digits enclosed by an elliptical border: this remains the same on each application, and identifies (to McDonnell Douglas) the inspector personally. Alongside is a rolling date, of the sort used on many office stamps, so that a stamp can never be used without a date appearing. The inspectors' stamps purport to have been applied for the travel adjustment work on August 2, 1972, fifteen days after the support-plate work.

At that time, the orders for that work were still incomplete. According to the records of the Planning Department, there were blueprints in existence by that date which showed the cargo-door locking pins protruding past the latch assemblies, and clearly this blueprint would have suggested to anyone familiar with the system that a greater degree of travel should be applied. But it was not until August 11 that a written instruction was added, saying that the extra protrusion required should be set at one-quarter (.250) of an inch. Clearly, the August 2 stamp constitutes a false record, but in the absence of testimony there is little more that can be said about it. There is at least testimony on the other entries, which casts some light on the question of how false entries might occur in the records at Long Beach.

A CHECKING SYSTEM dependent upon "personal responsibility" is only meaningful if the issuing and using of stamps is carefully controlled. This, at the Douglas plant, is part of the work of the Badge Control department, issuing two kinds of stamps: "Q" (quality) stamps issued to selected supervisory workers, such as foremen, and "A" (assurance) stamps, issued only to inspectors who spend all their time checking work which has Q-stamps already applied to it.

The form dealing with the addition of the support plate to Ship 29 bore the impressions of two Q-stamps (those of Henry Noriega and Shelby Newton) and one A-stamp, that of Edward Evans. Each appeared several times, to relate to different stages of the process: there being, for instance, both Q- and A-stamps for the entry "O.K. to Install" (meaning preparation had been done correctly) and also for the entry "Installation O.K." signifying the completion of the work. There is not one simple process, but rather a small network of human actions: William Gross, one of the company's senior engineering executives, now manager of the DC-10 program, suggested throughout his testimony that "human error" must account for the falsity of the records in his plant, and of course in a certain inclusive sense this is obviously so. But the phrase often suggests the idea of a single, isolated slipup. In this case, we are dealing with a set of false moves related intimately with each other, yet apparently committed at different times by different people.

If the July 18 date is to be accepted, we must begin with the mild curiosity that Ship 29 was then one of the first—if not the first—DC-10 to be worked on after the gentleman's agreement. Such urgency seems odd, in that the airplane still had no purchaser. The engineers in the Rework for Delivery section would surely have had other and more pressing tasks.

Even more curious is that McDonnell Douglas itself seems to have been unaware at the time of its own promptness. William Starloff, the company's liaison officer with the FAA, had the task of informing the FAA Western Region office of the progress being made with putting the door modifications into force. On August 7, he wrote to the FAA in these terms: "No airplanes complete as of this date. Kits have been shipped to UAL and AAL.* You will be provided periodic reports as this bulletin is accomplished."

Mr. Starloff's account was physically accurate, in that Ship 29 had certainly *not* been properly modified, and never was. However, if the dates on a microfilm produced to the lawyers have any validity, then at the time he wrote, the plant's records should have indicated that one airplane had actually been completed. Still more curious is the fact that a month later the company still appeared to be unaware of the work supposedly done in July and August. On September 25, 1972, just before Turkish Airlines signed for the purchase of three DC-10s, Hal Bayer, a McDonnell Douglas executive, wrote to the airline in terms which suggest that the work still remained to be done: "As Jack McGowen stated, at the time of delivery of the DC-10 . . . to THY . . . the revision of the cargo door latching system will have been accomplished in accordance with the FAA approved service bulletin."

Given the intensity of the sales effort that Jackson McGowen and his colleagues were mounting, one might have expected them to make something of the fact that the work on Ship 29 was apparently complete. Speed of delivery, after all, was the item about which the Turkish buyers were more concerned than anything else.

ON JULY 18, 1972 Ship 29, along with Ship One and another ex-Mitsui plane, Ship 47, were parked together on the ramp outside Building 54 in the Rework for Delivery area. Airplane workers, unlike men on the line in auto factories, can usually identify the machines on which they have worked, and this makes it the more unusual that Edward Evans, like Henry Noriega, has no memory whatever of having set eyes on Ship 29. Shelby Newton, who gave the most precise, but in some ways the most puzzling evidence, actually remembers Ship 29, but remembers with equal clarity that he did not work on it in July 1972.

However, Edward Evans's case differs slightly from the other two, in that the plant records do actually show that on the relevant day he could have been working on the airplane. Evans is the only one of the three who was a full-time inspector. He is a slightly-stooped, bespectacled man, then aged twenty-nine, and taking classes four nights a week in Quality Control at

* United and American Airlines.

Long Beach City College. Evans comes from Pennsylvania, which he still regards as "home," and moved west out of a fascination with high technology, which made him want to work on airplanes. He joined Douglas in 1965 after four years in the Navy as a sonar technician third class. One year after joining, however, he suffered a brain tumor and was on sick leave for a whole year, recovering from the operation.

He was first assigned to DC-10 work in March 1972. He worked long days that summer, because he left home each day at 6:30 A.M., traveled twenty minutes to the plant, and after work went straight to the college, returning to his wife and small child at 10:30 in the evening. That particular day in July 1972, a Tuesday, he followed the usual routine, so far as he could recollect.

Each day at seven he would get an assignment and collect the relevant documents for that assignment, kept in each case in a small portable office alongside Building 54. On July 18, Ship 29 was parked outside Building 54, along with other DC-10s and it is possible that he could have been assigned to it. He would have collected records which already had Q-stamps on them, and gone on from there to make his own inspection. He explained his job:

> Our purpose, where we demand a Q-stamp on that piece of paper before we go on the job, we want to make sure that someone else has looked at the job generally to make sure it was done. Their Q-stamp, to me personally, doesn't mean an awful lot. I mean I am not going to buy it on the weight of their stamp, but I want the stamp on there. . . . I believe they receive the same type of . . . little documents we do when we receive the stamp, that you are responsible for any work that your stamp appears on the record for accepting. And I would like to hold them to it. It's double assurance that nothing has been overlooked in a double reliability factor.

Evans recognized his own A-stamp as being on the microfilm record in a position suggesting that he had checked the installation of the support plate. As the stamp was there, he said, he accepted that he must have applied it. Yet he had no recollection of ever having seen Ship 29. Nor did he have any recollection of having even seen a torque-tube support plate before the Paris crash of 1974. He certainly could not recall that he had ever inspected any work on such an item, and indeed his evidence suggested that he would not have known how to do so.

He was asked whether he had ever seen an inspector perform the sort of inspection that the records purported to show that he had performed himself:

A. No.

Q. Do you not recall ever having one of these inspections, is that right?

A. No, I haven't.

His stamp, he said, had always remained in his pocket when not in use, and he had never allowed anyone else to use it.

Pressed to account for the presence of his stamp on the documents, he produced two possible theories:

Q. What explanation do you have, if any, for the appearance of your stamp on these documents?

A. I have no explanation, exact explanation for what happened. I have several possibilities.

Q. What are these?

A. O.K. Number one, out in the back of that building it was high summer. . . .

Q. It was what, sir?

Mr. Fitzsimons (for McDonnell Douglas). High summer.

Evans was suggesting that perhaps he had been dazzled by working in strong sunlight, and thus become confused between airplanes and parts of airplanes.

A. Number two is that our attention could have been focused mainly on this lock tube and function of the lock tube. I didn't pay proper attention to some of these other parts. That's all I could think of at the moment.

Evans was here suggesting that he was so closely occupied with a piece of work which did get done (though not one which he remembered), namely the "scarfing" of the lock-pin tube, that he failed to look at anything else.

Q. Is it possible that you just missed the ship completely? Just in the press of business, just didn't get to Ship 29? Have you considered that?

A. No.

Q. You know you did something with Ship 29?

A. The papers say I did.

Q. True, the papers also say a couple of things that apparently weren't on there. I mean, have you considered that as a possibility?

A. No.

Q. Are you sure 29 was there at that time?

A. I have no idea.

Q. I see.

A. Everything you have shown me says it was there.

After the Paris crash, Evans was taken off inspection work. But in the summer of 1974, feeling restless and anxious to get back to the work he felt he knew best, Evans had an interview with the president of the Douglas division, John Brizendine, and asked if he could have his stamp back. Apparently, Evans did not feel inhibited in this request by his own belief that to stamp off a piece of work without having personally checked it would "constitute incompetence almost to the point of ignorance." His account of the meeting with Brizendine suggests that it did somehow resolve his doubts about what he had done (even though it did not result in his going back to inspection work).

A. I told him [Brizendine] I have no idea how I bought that job off without those parts being installed, *if that is what I did.* (Emphasis added.)

Q. Is there a doubt in your mind that you did?

A. There was.

Q. And there still is?

A. No.

Q. No?

A. No.

Q. You believe now that you actually stamped that off with those parts not here?

A. I have no reason to misdoubt it, I am sorry.

Insofar as a clear line emerged from Evans's evidence, there was a suggestion that somehow he was working in conditions where the application of his stamp was an act that might mean nothing to him. Henry Noriega's memory was in certain respects more distinct, but this meant if anything that his account was even more puzzling.

Noriega, who is of Mexican extraction, is the oldest of the three men questioned, having worked at Douglas for twenty-four years. He is an electrician, and during the whole of his career with the firm has worked only on electrical, never on mechanical installations. Yet his Q-stamp appears in the sequence for stamps validating the nonexistent *mechanical* work on Ship 29's door.

Noriega recalled in his evidence exactly what he was doing on July 18. He was working on Ship One, which had just come back from a flight test, and required some rewiring in the nose, and he was doing this with a crew of some twelve men. This was one passage in a month of work concerned exclusively with Ship One. Although Noriega must presumably have been aware of the physical presence of Ships 29 and 47, not so far away, he made it quite clear that he did not identify either of those airplanes in his mind, or have anything to do at any time with either of them.

How, then, did his stamp appear on Ship 29's documents? Noriega could not suggest that he might have been dazzled, or that he might somehow have stamped two pieces of work while attending to one. The presence of Evans's stamp might have been explained by some such accident, because it was actually possible that Evans had at some point worked upon Ship 29. Noriega was always elsewhere, and, as he conceded, his stamp could only have appeared on the support-plate work through someone else having put it there.

All he could suggest was that perhaps a worker superior to him—one who for some reason did not have a stamp and needed one to clear up a job —had borrowed his stamp. Such a thing would be rare, said Noriega, but not entirely unheard-of, even though it was against company rules (and would make nonsense of the principles on which the checking system was erected).

It emerged that for Noriega to use his Q-stamp himself was in itself fairly rare, because most of the work done under his charge was in practice stamped off by his lead woman, using her own stamp. For a superior-ranked worker to appear from another part of the plant and take Noriega's stamp away for use elsewhere, on Noriega's account of things, would have been virtually unique. He could not recall any such thing having occurred on July 18, 1972, or on any other day up to September of that year, when he was moved to another kind of work and returned his stamp to the company through Badge Control.

Both Evans and Noriega, in somewhat different ways, conveyed to the lawyers interrogating them that they were puzzled men, distressed and under strain. Evans, indeed, came close to tears at several points. Both were men of good character but neither appeared as men with much intellectual self-confidence.

Shelby Newton, the third witness, made a sharp contrast. He is a young, articulate black man, precise in speech and determined not to waste words. Newton, possessing college qualifications in airframe engineering and power-plant engineering, had taken his strictly aeronautical education somewhat further than Evans, and considerably further than Noriega.

On July 18, 1972, Newton was working as an assistant foreman with twenty men under him, installing electrical wiring in instrument panels and updating hydraulic systems. He was doing this for Department 549, located in Building 12, which was several hundred yards away from Building 54, where Ship 29 was parked.

Newton testified that during July 1972 his duties might have taken him through the ramp area of Building 54. But he was quite certain that he had never identified the airplane or worked upon it in anyway. Nor had anyone under his charge done so. Neither he, nor any of his crew, had at any time done any work whatever upon a cargo door on any DC-10. Newton said that he had never seen anyone working on a DC-10 door and did not in July 1972 have any idea how such work might have been done. He would not under any circumstances whatever have stamped off work which he had not seen done with his own eyes.

He was asked to examine the Q-stamp outlines appearing on the support-plate records for Ship 29, and say if one of them was his:

A. It's the stamp that was issued to me, but I can't say the mystery of how it got there.

Q. Did you place it on there?

A. No.

Q. Did you ever lend your stamp to anyone?

A. No.

Edward Evans had testified earlier that he kept his stamp in his pocket. Newton stated, with even more emphasis, that he adopted the same practice, and that his stamp was attached to his key ring which also had the key to his car and his house.

Q. You know that you never let go of your stamp on July 18, 1972. Is that correct?

A. Yes.

Q. It was always in your personal possession?

A. Yes.

Q. You never lent it to anyone?

A. No.

After service in the Navy, Newton had spent two years at Los Angeles College. He joined Douglas straight from college in 1965, and began work as a structural assembler on DC-8s and DC-9s. He remembered Ship 29 very clearly, he said, because in October 1972, he had been promoted from assistant foreman to mechanical supervisor, and his first job was directing the installation of seats, cabin panelling and the oxygen system in Ship 29, which had by then been sold to Turkish Airlines and was in the delivery center. But in his new job he did not need to use a Q-stamp, and so he turned his in before he started any work on Ship 29.

Newton's account of the meaning of a Q-stamp was very positive and clear. It was "a number issued only to you . . . it's like a confidential piece of equipment." The purpose was to insure the quality of the product by looking at the work, and the procedure was spelled out in detail by the Douglas Quality Standards manual, a document which "you would need a wheelbarrow to carry around." Before using a Q-stamp, its owner would have to go to the area where a piece of work was being done, and personally verify it.

Q. That is the correct way to use a Q-stamp?

A. It is the only way to use a Q-stamp.

Newton showed no eagerness to be drawn, however, when the interrogating lawyers suggested that his evidence could only be explained if his stamp had been placed on Ship 29's records by forgery—and when they tried to do so, the McDonnell Douglas attorneys moved rapidly to head them off:

Q. Does the appearance of 2Q690 [Newton's stamp number] on this AAO [Advance Assembly Order] in view of the fact that you did not place it on this paper, violate the rules of McDonnell Douglas with respect to the use of these stamps, as you understand the rules?

A. I don't know.

Mr. Fitzsimons (for McDonnell Douglas). If you are having trouble with his question . . .

A. Right, I am having trouble with his question . . . I am having trouble with the question because to me you are asking the question in one context and looking for something else in another context, which I can't give you.

Q. If you didn't put 2Q690 on it, and it was your stamp . . .

A. Who did.

Q. If someone else did, he was breaking the rules of McDonnell Douglas with respect to the use of Q-stamps, as you understand the rules?

A. I see what you are saying, basically, but I have no answer.

Q. Did anybody else have authority to use your Q-stamp?

A. No.

Q. Whoever used your Q-stamp did it without authority, is that correct?

A. Nobody else used my Q-stamp.

Newton said that he had not tried to find out for himself what might have happened, beyond saying that he had talked to someone in Badge Control about what happened to stamps when they were turned in. However, he had clearly discussed the issue with the McDonnell Douglas security department when they conducted an internal investigation after the Paris crash. At one moment, Newton seemed about to reconstruct for the deposition interrogators what he had said to his own firm's security men, but at this point the McDonnell Douglas attorneys intervened and said that the line of questioning was out of order. The McDonnell Douglas position throughout was that the company's own investigation of the mystery of the records had been conducted by qualified lawyers; therefore, all conversations that took place during the investigation were "privileged" under the ancient rule that nothing a person (or a company) says to their (or its) lawyer need be revealed in litigation.

This claim would have been contested by the plaintiff attorneys had the matter gone to trial. However, even if the results of the company's inquiry had been made public, it is doubtful whether they would have been any more revealing. The Douglas divisional president, John Brizendine, referred to it during his own deposition testimony, and the impression he gave was that the results were too indeterminate to be worth remembering. Brizendine gave evidence at some length on the question of functions and responsibilities in the Douglas plant in the light of Ship 29's career. After the crisp, assured, if deeply puzzling, testimony of Shelby Newton, the lawyers dealing with Brizendine found themselves once again in a world of hazy and uncertain recollection.

NOT THAT JOHN Calvin Brizendine's record would suggest any deficiencies in his mental equipment, anymore than his demeanor on the stand suggested inarticulateness. Born in Independence, Missouri, in August 1925, he served as a navy pilot during World War II and then obtained a master's degree in aeronautical engineering from Kansas University, a famous center of such studies. At an early age he developed, he says, an affection for airplanes, particularly Douglas airplanes because "Douglas did things . . . one notch above the rest of the industry." One of his close friends from Kansas got a job as a Douglas test pilot and, in 1950, told Brizendine about a vacancy in the company's flight-testing division. Within ten years Brizendine had worked his way up through Douglas to his first important post as director of the DC-8 program and by the time he joined the DC-10 program as manager in 1968 he had amassed much experience in organizing the production of jet transports. After his period in charge of the DC-10, he was a natural successor to Jackson McGowen as President of Douglas (he was appointed in 1973) and as the Douglas man on the main McDonnell Douglas board. (Dedicating an entire career to one company is something of a tradition in the Brizendine family: John Brizendine's father began work with an oil company as an eighteen cents-an-hour office boy and stayed with the firm for forty-two years.)

According to an official company biography, Brizendine was "responsible for the engineering, flight development and production of the big jetliner" (the DC-10) from May 1968 until after certification. A dictionary definition of "responsible" is: "Answerable, accountable (to another *for* something); liable to be called to account . . . morally accountable for one's action." * The lawyers who questioned Brizendine did not find many things to do with the history of Ship 29 on which Brizendine felt that he could properly be called to account.

In dealing with the general design of the DC-10, he was asked whether the company considered redesigning the floor after Ship One's floor collapsed when the rear cargo door blew off in May 1970. † Brizendine said that he had "no personal knowledge" of whether the design team had thought of doing so, although as they were good engineers he did not doubt they would have considered all the options.

He described his own role as DC-10 program director in exceptionally modest terms:

> If I might explain the relationship of my position and that of the engineers, for example, whose responsibility it was to conceive and execute these designs. As the general manager I was concerned with the end result of their work. And my concern for safety was brought out being satisfied by them that the solution that they had reached was indeed a safe, satisfactory solution. When they explained the bottom line to me, it made sense, the same type of thing had been used successfully on another airplane already flying at the time, and I accepted it as a satisfactory decision. . . .

The last sentence of this passage apparently suggested that Brizendine approved the vent-door idea which was incorporated into the DC-10 locking system because it resembled that on the 747 ("another airplane already flying"). Possibly had he, with his large engineering experience, looked a little above the "bottom line," he would have seen that what was being proposed for the DC-10 was really quite different from the successful Boeing design, and so far from offering extra safety, was actually potentially dangerous. But he was insistent that his role as director had nothing to do with insuring that design work actually was adequately done, although, surely, there is no respected text on management responsibility that would free him of the obligation to do so. Had he personally taken any steps to insure that adequate study had been made of the question of floor strength?

A. I don't recall doing that at all, no.

Q. Do you know of any steps taken by any officer of the Douglas company at that time, to insure that an adequate explanation was made of the possibility of redesigning the floor?

* *Shorter Oxford English Dictionary,* 1968.
† See Chapter 9.

A. I wouldn't think that would be a function of an officer of the company, to instruct the technical staff to study the various options available to it in design. It's the responsibility of the design people to perform those functions, and they are competent, they are delegated that responsibility, and they report the results of their work to the senior management.

It was not easy on Brizendine's account to see what was the actual work of a director of a major Douglas project like the DC-10. Asked what he thought about the crucial decision to change (at a customer's insistence) from hydraulic to electrical actuation—a step undertaken for the sake of economy in moving parts—he said:

A. I don't know what the comparative numbers are.

Had he been involved in any discussions about whether there might be problems in going over from Douglas's and Convair's traditional hydraulic operation to electrical methods?

A. I don't recall.

How about the location and routing of the airplane's control cables. Was the man responsible for directing the program involved in those discussions?

A. Not specifically. I didn't make any design decisions. I was aware how the controls were put in the aircraft. It's traditional.

Could he justify the assertion that it was "traditional" by showing that the same method had been used in previous Douglas aircraft?

A. I can't testify from personal knowledge to all of them.

If Brizendine could remember little about these design decisions, he apparently recalled even less about their outcome. He could not remember whether anyone had ever told him that the locking-pin rigging adjustments had not been performed on Ship 29.

A senior McDonnell Douglas engineer, Marchal Caldwell, had been sent to Paris immediately upon the news of the Turkish Airlines crash—and we know from other sources that it was Caldwell who first discovered that the locking system was misrigged. Brizendine could not recall whether he had ever discussed the matter with Caldwell. He admitted that he had heard Caldwell report on his trip, but could not "specifically recall" what might have been said about the rigging question.

Yet the evidence of the locking-pin adjustment was such as to bear heavily upon Brizendine's own confidently articulated theories about the operation of cargo doors. He had not had any hesitation in telling the lawyers—speaking, of course, with the authority of long experience in airplane engineering—that both the blowout on Ship One and the incident over Windsor simply resulted from human failure ("abuse of the system"). Certainly the baggage handler working on Captain McCormick's plane had used a lot of force on the door handle. But surely the interesting difference at Paris was that Ship 29's cargo door could have been closed with only some thirteen pounds pressure, because the lock-pin travel was so short. Would Brizendine classify that as an "abuse" by the cargo handler?

A. All the facts of that situation are not known to me and I therefore do not have enough information on them . . .

Q. Do you know of any parallels between the Windsor occurrence and the Paris occurrence, in the act of closing the door?

A. There is no way I would have any knowledge of those details.

Q. Did Mr. Caldwell tell you that there were indications of different pin positions?

A. I seem to recall that.

Q. Did you ever see the photographs [of the door]?

A. I don't believe so.

Q. Did you ever read a report about it?

A. More in discussion.

Q. Was the discussion with Mr. Caldwell?

A. He was present, as I recall.

Q. In the two indications of pin protrusion there was one indication of a setting of .031 wasn't there?

A. I would have to look at the rigging instructions to be able to answer that.

Q. You don't know.

A. I don't know.

Q. You never made a study of that.

A. No.

However, his ignorance about what investigations had been made in Paris did not shake his basic assertion that the crash in France really did not have all that much connection with the Windsor incident. Indeed, he seemed to think that the Windsor incident was of no very great significance, other than showing what an excellent airplane the DC-10 was. He did not think that Captain McCormick had done anything very remarkable in bringing the plane down safely, although he did allow that McCormick had "kept his cool and used the plane as it was designed to be used."

He was reminded that McDonnell Douglas had sent a special message to Captain McCormick congratulating him on his performance, but Brizendine thought nothing of the point.

A. I would also tell the guy that waxes the floor in my lobby that he has done a great job, too.

Even though Brizendine's knowledge of the locking-pin adjustments was so slight, he felt confident that the error on Ship 29 was not his company's fault. Indeed, he felt sure that the adjustment *had* been done before Ship 29 left Long Beach, even though the support-plate fitting had not been done, and the records dealing with the work were patently false. What was the basis of his opinion that the pins had been properly rigged at Long Beach, and must have been altered while the plane was in service with THY?

A. The fact that the airplane was manufactured properly.

Q. Is that based on the fact that all the parts required to be in the airplane were in the airplane?

A. I understand what you are driving at, and that is a difficult problem for us. That one part missing doesn't necessarily indict the whole airplane.

Q. There was more than one part missing, wasn't there? Wasn't there a support plate and some bolts and some holes that were supposed to be drilled . . . at the same time those bolts were supposed to have been put on, the rigging was supposed to have been done, isn't that right?

But the lawyers made no impression upon the smooth circularity of Brizendine's evidence. Obviously, any confidence of Brizendine's in the manufacturing process at the Long Beach plant logically had to include confidence in the inspection system. Yet, he showed as little interest in the statements of Evans, Noriega, and Newton (all of whom had completed their evidence before he began) as he apparently did in Caldwell's reports from Paris.

He seemed to have some difficulty remembering which of the three men was which. He had certainly not questioned any of them directly, and had only met Evans at Evans's own request when Evans asked for his stamp to be returned. Brizendine recalled even less about this interview than did Evans (who had at least remembered that after the meeting he—Evans—no longer had any doubt that he must have stamped the false records himself). Brizendine did not know what work Evans was doing at the time of that interview, though perhaps "he was in some administrative job where he was surveying paper or something of that sort."

Brizendine did not admit to any responsibility for Evans's move from inspection to the new job "surveying paper," though he recalled having supported the decision made by some other executive to move Evans after the Paris crash. He did not know that before July 18, 1972, Evans had never inspected any work on the aft cargo door of any DC-10.

He had some vague recollection about Noriega as "basically an electrical inspector," but found it hard to believe that in his twenty-four years at Douglas, Noriega had not done other work.

Both Evans and Noriega, according to other evidence extracted from McDonnell Douglas, had been reprimanded for having misused their stamps, or allowed them to be misused. Brizendine was asked whether Newton had also been reprimanded, despite his refusal to admit to any laxity whatever. Brizendine could not remember.

A. We would have to check his personnel file.

Newton in fact was not reprimanded, which suggests that the company accepted his story. But this apparently happened outside John Brizendine's area of responsibility and control. Brizendine did not seem to have been much concerned with finding out about the falsification of Ship 29's records:

Q. Your investigation didn't reveal [that Noriega had only done electrical inspection work] then?

A. I don't know that. He was classified, I believe, as an electrical inspector.

Q. You don't know whether he did or did not ever have any experience in mechanical work, such as the work set out in service change 812 [fitting of the support plate]?

A. I couldn't tell you whether he did or not.

Q. In your safety system that was there that day—and I believe you said before that safety is number one at Douglas and was at the time—did you have any cross check of any sort that would raise a red flag if a man who did nothing but electrical all of a sudden signs off for a mechanical installation?

A. I am not aware of any cross check that checks stamps to see whether it's electrical or a mechanical stamp.

The only suggestion Brizendine could produce, when confronted with the evidence already in existence, was to say that Noriega might have been given a special assignment. This clashed totally with Noriega's evidence that throughout the whole of the relevant time he was doing electrical work on Ship One. What light could Brizendine throw upon Noriega's suggestion that a superior might have borrowed his stamp?

Q. Did you ever determine what person or persons Mr. Noriega had loaned his stamp to?

A. I don't recall that was ever determined precisely.

Then there was the question of when Newton's stamp had been issued. On the evidence of the documents produced, it appeared possible that Newton's stamp had not even been issued to him at the time when it was supposedly applied to the Ship 29 records. There was an ambiguity in the requisition papers for this particular stamp (never resolved) which suggested that it might not have been issued until August 2. Brizendine was unable to help at all with the interpretation of the requisition document and was asked directly if he knew whether Newton did actually have the stamp in his possession on July 18, 1972. The President answered: "I really don't recall what our investigation showed on that. If it showed anything."

Brizendine's public statement in the immediate aftermath of the Paris crash dwelt upon the shock and unhappiness that it had produced in the Douglas plant, where insuring aerial safety was regarded as a way of life. In a press conference which he gave at Long Beach, on March 25, 1974, Brizendine conceded that the support plate had never been fitted to Ship 29's rear cargo door. Yet, he said, the company's records showed that all the modifications to the door had been done. "This is a circumstance for which we do not yet have an explanation," Brizendine said. "We are investigating this matter vigorously." He backed up that promise with a declaration:

> We know that we have great responsibilities, and we take those responsibilities very seriously. You might remember—as we always do—that a Douglas aircraft is always a Douglas aircraft no matter who owns it, how long it remains in service, or what is done with it long after it's beyond our control. Our planes are part of our lives and part of our identity; that's one reason why we develop and build them with such care.

Giving secret testimony a year later about the missing support plate, Brizendine's attitude was not the same at all. He certainly didn't indicate that his
investigation was conducted vigorously, nor were those "responsibilities" apparently taken very seriously.

THE PLAINTIFF LAWYERS' investigation of the DC-10 and the history
of Ship 29 was a very considerable feat, so far as it went. Very rarely has so
much detail been brought out about the circumstances leading up to a catastrophic accident. At the end of the day, though, the investigation raised more
questions than it answered: it was as though the passions—and human
needs—engendered by the crash were enough to make some temporary rent
in the bland corporate veil. But once McDonnell Douglas and their two
corporate codefendants, General Dynamics and Turkish Airlines, agreed among
them to produce a suitable sum of compensation, then the veil smoothly
replaced itself.

The reality of the situation—discussed more fully in the final chapter of
our story—is that it was not even the three corporations which were the real
defendants in the lawsuit. Behind all of them stood complex alliances of
insurers, tracing back, as most big insurance deals do, to Lloyd's of London.

Most large corporations in the Western world insure themselves at Lloyd's
against legal liability for any accidents that their products may cause. The
essence of the arrangement is that the Lloyd's insurance syndicates agree to
pay, provided that they can direct the conduct of the lawsuit. James Fitzsimons of the New York firm of Mendes and Mount, who appeared in the
matter "for" McDonnell Douglas, is a distinguished member of the group of
American attorneys whose chief business consists in going in on behalf of
Lloyd's to direct the defenses of insured corporations.

The professional skill of such a lawyer is in defending a position just long
enough to discover the real strength of the attack being mounted upon it. At
that point, the interest of the insurance syndicates is to pay up, as expeditiously
as possible, the exact sum necessary to dispose of the claim. They do not
defend lost causes, for in the world of the insurers, there is no such thing.
Even a compensation bill of several hundred million dollars is not going to
break them. A single corporation, facing an overwhelming claim, may have
to fight to the very end, just as an accused man in a criminal case must fight
to the end for his liberty. But the business interest of the insurers is to pay
up what they must and get on with collecting new premiums.

Opposed to the insurance lawyers are a tribe of plaintiff lawyers whose
task is not to discover the truth, but to obtain compensation for those injured
and remuneration for themselves and to do this in the most economic manner
possible. Once they have elicited sufficient damaging information to convince
the Lloyd's attorneys that the maximum obtainable compensation must be
paid, there is no benefit in going further. That point seemed to have been
reached in the DC-10 case in May 1975, when the defendants ceased to con-

test liability for the Paris crash. It is possible, for reasons we will explain later, that there may yet be a full trial of the Paris Air Crash Case. In that event, the plaintiff lawyers will undoubtedly inquire more deeply into how some element of dishonesty—some essential lie, to use Kipling's language—must have been involved in the process that put Ship 29 into the "stream of commerce." For now, the only possible deduction from the evidence of Evans, Noriega, Newton, and Brizendine is that during 1972 the system of "individual responsibility" in the Douglas division of McDonnell Douglas was seriously compromised. "Mister Mac's" not ignoble ideals of teamwork and personal care had simply not proved effective enough.

McDonnell Douglas's attitude toward the falsification of manufacturing records seems to be that the event was an isolated, but totally mysterious failure in an otherwise excellent system. The attitude is understandable, but absurd, because it is impossible to know whether the failure was an isolated one without its causes being understood. One obvious possibility is that control of stamps within the Douglas plant was actually far more lax than the Douglas Quality Standards insisted upon: that people anxious to get work finished on time could find an appropriate stamp to complete the paperwork. Ship 29, when it eventually did find a buyer, found an impatient one, and when the plane was being readied in the late autumn of 1972, there was not very much time to spare. As there appears to have been no general knowledge among workers in the Douglas plant—the gentleman's agreement notwithstanding—that the door modifications were of great importance, it is possible that someone may have decided in late 1972 to "clean up" the paperwork retroactively. Although dishonest, the falsification of the airplane's records might have been seen as an act of little consequence. Indeed, it would probably have never been discovered but for the crash that destroyed Ship 29 and 346 lives.

14

Massacre in the Forest

I have seen many aircraft impacts. I have seen some aircraft torn up as much as this one, but I must say I have never seen any airplane torn up as much as this over such a large area. The pieces were extremely small, very fragmented, and it was scattered over an area half a mile long by one hundred and twenty yards or so wide.
    —CHARLES MILLER,
        former Director of the Aviation Safety Bureau,
        National Transportation Safety Board, Washington, D.C.

IN BRITAIN RUGBY Union Football, a distant cousin to both soccer and American football, is a totally amateur sport. Even at the highest level it is played by enthusiasts who can afford, or negotiate, the necessary time off from work to represent their country. International matches are therefore something of a rarity, and when they do occur, they attract ardent support. On the afternoon of Saturday, March 2, 1974, the English and French "rugger" teams, fifteen men a side, lined up for their annual encounter in Paris. Probably one-third of the ninety thousand spectators who filled the terraces of the Parc des Princes stadium were English and they made enough noise for twice their number. The match ended in a draw, twelve points all, but it was the kind of fast and furious game where the result doesn't seem to matter too much: each side gave as good as they got.

Once the game was over, most of those supporters who had traveled from England to see it would normally have headed for home. The logistics of a journey from Paris to, say London, are always more formidable than the distance between the two cities (215 miles) might suggest, involving passport and customs formalities and, of course, the crossing of the English channel. That particular weekend, the journey promised to be more arduous than usual because engineers employed at London's Heathrow Airport by British European Airways had decided to go on strike. In a month's time BEA was due to merge with BOAC to form British Airways. The strike, by members of the Amalgamated Union of Engineering Workers, was a brief, tactical affair aimed at getting a few more pounds into the men's pockets before the Government declared a wages freeze, which was considered inevitable that year (but in fact never came). All of BEA's flights were grounded and,

236

at Orly Airport in Paris, the other airlines that fly to London were over-whelmed by the sudden demand for seats.

Many of those supporters who tried to get a flight home on Saturday evening were turned away, although in the confusion some passengers got separated from their luggage, which was flown to England without them. Many others didn't even join the battle for seats: things might be a little easier on Sunday and, anyway, Paris is not the worst city in the world to be stranded in for a night.

The twenty-five players and officials from the Bury St. Edmunds Rugby Club in Suffolk had always intended to make a weekend of it. After watching Saturday's international, they had planned to play a match themselves, on Sunday morning, against a team from the Racing Club of Paris. In view of the airline strike and the uncertainty about seats, it seemed prudent to cancel Sunday's kickabout. But there was no question of canceling the plans for Saturday night: Bury St. Edmunds is a pleasant enough country town, but its attractions do not include the sort of varied night life that Paris has to offer.

On the other hand, Wendy Wheal, a twenty-four-year-old photographic model, could not wait to get home to London. Her modeling assignment, in Malaga, Spain, had ended two days early but the BEA strike grounded all planes to England. Her husband suggested that she should stay on in Spain and enjoy some sun until the strike was over. Instead, she headed for Paris in the hope of finding a seat on a London plane from there.

A lot of other travelers were also stranded in Spain by the strike. Margaret Davies, fifty-eight, and her friend Hannah Richards had been on a BEA all-inclusive holiday in the Spanish seaside resort of Torremolinos. There was no direct flight to take them home, but BEA said that if they would travel to either Madrid or Paris, then other airlines would be found to get them back to England: Margaret Davies and Hannah Richards chose to go via Paris.

And so it went. Because of the strike, through chance and circumstance, thousands of people found themselves at Orly Airport on Sunday, March 3, 1974, clamoring to get to London.

Paul Raymond, a senior lecturer for International Computers Ltd. (ICL), had been in Moscow running a training course for the Soviet automobile concern Moskvich to which ICL supplies computers. He had been scheduled to fly back to London on March 1, but the course ran into a few delays and Raymond decided to stay on to complete the syllabus. By the time he was ready to leave on Sunday, all BEA's flights had been canceled and the only practical alternative was to fly to Paris in the hope of finding a connection. The connection he and 213 other people found that Sunday at Orly was Turkish Airlines Flight 981 to London.

HOWEVER AMBITIOUS ITS pretensions, a small international airline like THY cannot possibly afford to support a network of overseas bases. Yet the arrival of every THY flight at a foreign airport inevitably generates a thousand and one tasks that must be accomplished before the airplane can take off again. Passengers, baggage, and cargo will have to be unloaded and, hopefully, replaced with new passengers and their baggage. The airplane will probably need fuel, possibly a little maintenance, and certainly a new supply of precooked meals and coffee. The crew will need to get the latest weather briefing and file a flight plan with air-traffic control. And when all that has been done, the airplane should get a final once-over from expert eyes before it is towed away from the parking ramp and pointed in the general direction of the runway.

Like most other airlines of its size, THY delegates these tasks, on a contractual basis, to more richly endowed airlines and to subcontracting firms which exist at most major airports to supply manpower and machinery to those who need it. At Orly Airport, Paris, much of the logistical support for THY is supplied by the *Société Anonyme* SAMOR which, despite its grand nomenclature, is a private concern, operating under franchise from the French airport authorities.

SAMOR's primary task is to supply muscle—men and machines to unload and load the cargo compartments of arriving aircraft. Although the work is obviously physical, it does require some expertise on the part of the ground crews. SAMOR serves a number of small airlines who between them employ a variety of aircraft and the techniques required, for example, simply to open and close the doors differ from airplane to airplane.

When, in early 1973, THY began to introduce DC-10s onto its scheduled routes, SAMOR's baggage handlers were given brief instructions by the French company on how to cope with this new Goliath. The DC-10 was radically different from the airplanes that they had been used to dealing with: for one thing, the cargo doors were a great deal further from the ground and required steps to reach them and conveyor belts to unload the holds.

And the cargo-door locking mechanisms on the DC-10 were also something very special. It was emphasized to the baggage handlers that closing and locking the cargo doors demanded adherence to a precise routine.

Mohammed Mahmoudi, an Algerian expatriate aged thirty-nine, who has worked for SAMOR since 1968 remembers very clearly the instructions he was given on how to close and lock a DC-10 cargo door. First, he was to activate the electric power circuits and then press a button, situated on a control panel recessed into the fuselage, which would bring the door swinging down onto its rubber seal. After the door had apparently closed he was to hold the button for a further ten seconds to give the actuators time to drive the latches over the spools on the door sill. Of course, it would be impossible for him to see if the latches had actually done their work: that was to be verified by pulling down the locking handle, located in the cargo door itself, alongside the small vent door. If the latches were truly home then the locking handle would move, fairly easily, through its arc into the closed position.

And, as final confirmation that all was well, the vent door would simultaneously close.*

If for any reason the latches had not gone home, the locking handle would encounter resistance and Mahmoudi was told the vent door would refuse to close. He was warned against trying to force the handle down. Mahmoudi did not understand the mechanics and neither could he read the decals on the door which warned against using force on the locking handle because they were printed in English. It didn't seem to matter: locking handle down, vent door closed meant safe. Anything else meant unsafe, and he was to summon his superior.

The final check on a DC-10 cargo door—the visual examination through the peephole of the position of the lock pins—was never made part of Mahmoudi's job. SAMOR's contract with THY stipulated that the airline was responsible for insuring that every plane was safe before it took off. Therefore, reason demanded, somebody from THY should check the peephole.

In Mahmoudi's experience, during 1973 and the first two months of 1974, Osman Zeytin, THY's resident mechanic at Orly, usually did the job. (THY maintains a small staff at Orly Airport to liaise with the contractors.) Sometimes, though, THY flight engineers climbed down from the cockpit to peer through the peepholes.

Nobody told Mahmoudi what to do if Zeytin should be away on vacation —and if the flight engineer should fail to appear.

FLIGHT 981 LEFT Istanbul just after nine o'clock on Sunday morning, March 3, to fly the 1,400 miles to Paris. On board were fifty or so passengers who wanted to travel only as far as the French capital and another 120 who were headed for London.

Most of the London-bound passengers were returning from holidays in Turkey, like the Hart family, originally from Kansas but temporarily living in England. Besides their own three teenage children, Loren Hart, a forty-one-year-old geologist and his wife Alice were looking after the twelve-year-old son of a London friend.

Graham Wood, thirty-seven, and his wife Margaret had left their three children in the care of relatives while they enjoyed a four-day break in Istanbul: the short holiday, flight included, was an incentive prize that Wood had won by selling insurance.

Gene Barrow and Glenda White had also gone to Turkey on a free vacation. The two women, both married and both mothers, earned spare-time cash by selling lingerie at parties which they promoted for friends and neighbors. They were so successful at it the delighted lingerie manufacturers packed them off to the Istanbul Hilton for the weekend as a reward.

Robert and Pauline Milne had spent a more leisurely ten days, walking

* See Appendix B for door diagrams.

around the Turkish countryside. Peter Morris, an up-and-coming young lawyer from Liverpool and his friend Gaynor Osborne had been relaxing, briefly, before Morris committed himself to building up his new law firm. Group-Captain Kenneth Doran and his wife Phyllis had been on a sentimental journey to revisit Istanbul, where Doran had once been the British air attaché.

The Uziyel family—Mose, twenty-four, his wife Rasel, twenty-four, and their nine-month-old daughter Liat—had also been in Istanbul on holiday. Mose now worked as a chemical engineer in London and the Uziyels had been home to show their relatives the new baby.

Naturally, not all of the passengers were holiday-makers: Clara Lux, a remarkable seventy-nine-year-old businesswoman from New York was on a European tour promoting her new book about learning to type; Dr. Charles Bowley had been visiting blood transfusion centers in Turkey; Jack Burtonwood, Joe Freer, and Tadeusz Okuniewski had been in Malatya, in eastern Turkey, for two months, installing machinery in a new textile plant.

And Esther Collen had been in Istanbul planning her marriage to a young Turkish engineer. Now she was on her way home to London to buy a wedding dress.

According to the passengers who left Flight 981 at Paris, the journey from Istanbul that Sunday morning was pleasant and uneventful. Of course, none of those on board could have known that their airplane, McDonnell Douglas's Ship 29, was an unexploded bomb looking for somewhere to go off.

BY MARCH OF 1974, THY had owned and operated Ship 29 for something like sixty-three weeks. For all that time the locking mechanism of the airplane's rear cargo door had been missing the support plate which was supposed to have been fitted to all DC-10s after the Windsor incident. That meant that Ship 29 was, in theory, in essentially the same condition as the American Airlines DC-10 had been—therefore no less, but also no more vulnerable to losing a cargo door in flight.

But by March 3, 1974, Ship 29's rear cargo door contained a far more deadly malignancy than the threat posed by the absence of the support plate.

As explained in the previous chapter, one of the other "fixes" that McDonnell Douglas was supposed to have applied to all THY's DC-10s before delivery was an adjustment to the position of the locking pins. Remember that the main job of the locking pins was to confirm that the latches, which actually held the cargo door closed, had been driven into the "over-center" position with the talons around the spools in the door sill. If the latches were "over-center" then the locking pins, driven by the manual locking handle, would slide over the top of them: if the latches were not "over-center" the locking pins (one pin for each of the four latches) would jam up against metal lugs fitted to the top of the latches. In those circumstances the locking handle would, in theory, become immovable and the baggage handler would be unable to close the vent door.

Of course, Windsor had proved the theory to be nonsense. If the latches did not go "over-center" and the locking pins jammed, a baggage handler could overcome the system by the simple expedient of applying between 80 and 120 pounds of force against the locking handle.

The adjustment to the way in which the locking-pin mechanism was rigged was intended to beat that hazard. The concept was straightforward: the further the locking pins had to travel over the latches when the locking handle was pulled down, the more resistance would be encountered if their journey was interrupted—that is, if the latches had not, in fact, gone "over-center," and were still in the way.

According to McDonnell Douglas, the adjustment needed was a very small one. If the travel of the locking pins was increased by less than a quarter of an inch, 215 pounds of physical energy would be needed to overcome the system. If in addition the support plate was fitted—to prevent the linkage to the vent door from buckling—then the locking handle would be able to resist 430 pounds of force: far more than any human could possibly apply.

But conversely, the opposite was true. If the travel of the locking pins was *reduced,* again by only a fraction of an inch, the force needed to overcome the safety system would be correspondingly *less.*

McDonnell Douglas claims that before delivering Ship 29 in 1972 it increased the travel of the locking pins in the rear cargo door. If that is true then someone must subsequently have tampered with the rigging and reduced the travel of the locking pins way beyond their original (pre-Windsor) position to a point where they barely covered the latches at all. In affidavits that it has filed as part of the Los Angeles litigation, THY vehemently denied that it altered the rigging.

The truth may never be fully established. But what is beyond doubt is that at some time before March 3, 1974, the locking pins on Ship 29's rear cargo door were misrigged. When French investigators examined what was left of Ship 29's rear cargo door, they found two score marks on each of the four locking pins. The score marks, caused by vibration, showed where the pins had sat in relation to the latches during the plane's existence. One set of marks proved that the pins had once been at the pre-Windsor setting when they would have overlapped the latches by .031 of an inch. Of course, according to Service Bulletin 52-37, the travel of the pins should have been extended to increase the overlap to .25 of an inch, but there were no score marks to indicate that the adjustment had ever been made. Instead, the second set of score marks—not as deep as the first set—proved that the travel of the pins had been reduced by three-sixteenths of an inch, to a position where they did not overlap the latches at all.

It is possible, as McDonnell Douglas claims, that the pins were adjusted to the correct (.25) position at Long Beach and then rerigged before vibration had caused them to become scored. It is also possible, as Turkish Airlines claims, that McDonnell Douglas mechanics were responsible for the incorrect rigging, although if that is true, the work must have been done by members of the Douglas support team in Istanbul: according to the experts who examined the pins, the second set of marks were too insubstantial to have been the result of more than a few months of vibration.

Whatever the explanation, the misrigging would have been easy enough to do. To extend the travel of the locking pins a mechanic would only have to uncouple one end of the vertical rod, inside the cargo door, which connects the locking handle with the horizontal tube to which the locking pins themselves are attached. If the mechanic then rotated the vertical rod in one direction, he would *extend* the travel of the locking pins. If he turned it in the opposite direction, he would *reduce* the travel.

Sometime, somewhere, someone turned the rod on Ship 29's rear door the wrong way. From then on a sickly child could have beaten the safety system. Even with the latches only partially closed and with the locking pins jammed up against them, it needed just *thirteen pounds* of physical energy to operate the locking handle successfully and close the vent door. It was only a matter of time until that happened.

PAUL RAYMOND, THE ICL senior lecturer, had spent the previous Christmas at his parents' home in Liverpool. Like most people who work in the computer industry he had some mathematical aptitude and was mildly interested in the arithmetic of chance. During the vacation he and his brother Julian fell into a long conversation about the subject. They concluded that given a sufficient run of time there must be an occasion when a nuclear power station would go wrong and cause enormous damage. Equally, they decided, there would eventually be a crash of a fully loaded jumbo jet. A little over two months later, Raymond became the subject of his own prediction. Our contention is, of course, that given the condition of the rear cargo door there was little that was surprising or unpredictable about the crash of Ship 29.

It was, however, simple bad luck that the DC-10 should have been almost fully loaded for its last flight. Normally THY would have been doing well to have half a plane load on the final leg of the journey from Istanbul. But on March 3 there was considerable competition at Orly for any seat to London. Michael Hannah, a young male model who had been working with Wendy Wheal in Malaga, was probably the last passenger accepted aboard Flight 981. Hannah was anxious to get back to London to see his fiancée: at the barrier he met, by chance, an old friend from his home town of East Molesey, who was competing for the same last seat. Hannah was allowed through but his friend was stopped. "I'm terribly sorry," Hannah called back, hastening toward the plane. Hannah's friend was luckier than he knew. Had he been allowed through the barrier he would have got a seat on board the DC-10. Ironically, in the confusion somebody miscalculated and the plane took off with eleven empty seats.

Flight 981 had been due to depart from Paris at noon but, understandably, the onrush of new passengers caused some delay. Mohammed Mahmoudi had been assigned by his supervisor to work on the DC-10's rear, and smallest, cargo compartment. Having unloaded everything that was destined for Paris, he waited to see whether his compartment would be needed for the new passengers' luggage. It wasn't and he was given the order to close the door.

He followed the procedure he had been taught, and everything seemed to work exactly as it had on the other dozen or so occasions that Mahmoudi had previously closed a DC-10 rear cargo door. Had he been a trained mechanic familiar with the vagaries of the DC-10 door, Mahmoudi might have noticed that the locking handle went home perhaps a little too easily. But all the warnings to him had been against the use of *too much force*.

It would have been impossible for Mahmoudi to see that, inside the door, the latch actuator had stalled and the latches themselves had failed to go beyond the center point of their arc; however the talons gripped the spools in the door sill enough to give the treacherous impression that all was well. Had he looked through the peephole he would have seen that the locking pins were in the "unsafe" position. But that was, specifically, not his job. Osman Zeytin, the THY station mechanic at Orly, was away on vacation in Istanbul. And although there was a Turkish Airlines ground mechanic, Hasan Engin Uzok, on board the DC-10, none of the airport workers saw him or the flight engineer check the peephole.

Through that net of malign circumstances, the airplane, the 335 passengers and the eleven crew on board stood condemned. The only thing that could have saved them at the last moment was the annunciator light on the flight engineer's panel which should have warned that the cargo door was not closed.

The light was designed to remain on until the cargo door had been fully closed, the latches had gone over-center and the locking-pin tube had moved across to trigger a mechanical switch. But on Ship 29 the mechanism had been tampered with.* When Mahmoudi pulled the locking handle down the light in the cockpit went off misleading the crew into believing that all was well.

The Turkish DC-10 lifted off from runway 08 at just after 12:30 P.M.

---

* According to McDonnell Douglas, the locking-pin tube in Ship 29's door had been improperly modified by THY. McDonnell Douglas's supposition is that because of the misrigging of the linkages, the locking-pin tube would have consistently failed to push the microswitch into the off (safe) position even when the door was properly closed. This would have caused the warning light in the cockpit to remain on or, more likely, flicker on and off during flight, creating false alarms. McDonnell Douglas alleges that at some time before the accident, THY "cured" this problem by soldering metal shimms—washers—onto one end of the tube to make its contact with the microswitch more positive. However, because so many shimms were added, the tube would have triggered the microswitch even if it failed to complete its journey; that is, if the locking pins jammed up against the latches. THY has denied that its mechanics added the shimms.

It headed due east to avoid— as the air traffic regulations demand—over-flying Paris and then over the village of Coulommiers, turned north for the flight to London.

The captain, Mejak Berkoz, and his co-pilot Oral Ulusman were, by any standards, experienced flyers, and they provided a flawless takeoff and a smooth climbout. Nothing they said in the routine radio transmissions during the first nine minutes of the flight created any sense of alarm at the air traffic control center at Orly. One of the controllers, Doude Diddier watched the plane's progress on his radar screen. Suddenly without explanation, the dot that represented Flight 981 disappeared.

It was not until the next day that Diddier replayed the video-tape recording of the radar screen and noticed that a minute or so before the dot had disappeared, a tiny sliver had broken from it.

That sliver represented the rear cargo door forty-eight inches tall and forty-four inches wide. At 11,500 feet, the pressure differential inside the DC-10 would have been 4.5 pounds per square inch. Stated more dramatically, there would have been almost five tons of air pressing against the inside of the door, competing directly against the ability of the partially closed latches to hold it shut. Because the latches were not over-center all of that force was transmitted to two bolts, a quarter of an inch in diameter, which held the latch actuator to the inside of the door. The bolts gave way, the latch talons were pulled from the spools and the door blew open, to be ripped from its hinges by the slip stream.

In a precise repeat of the Windsor incident, the DC-10 decompressed rapidly. Air escaped from the rear baggage compartment leaving the cabin floor to cope unaided with the pressure of the air in the passenger cabin. It was too much to bear and sections of the floor caved in. The last two rows of seats in the lefthand aisle—and the six passengers they contained—were plunged into the cavity and ejected through the gaping hole in the fuselage, to fall the two-and-a-half miles to earth. The seats and the bodies landed in a turnip field near the village of St. Pathus. Josef Roxas, who was working in the field, thought that a war had started and that he was caught up in the first air raid.

The remaining events on board Ship 29 during the last one minute and twelve seconds of its existence must be a matter of conjecture, though we can obviously be guided by the experience that was gained over Windsor.

When the portions of the floor collapsed most, if not all, of the control cables running underneath it to the tail would have been jammed or severed. Captain McCormick over Windsor lost everything but for limited command of the ailerons: It is likely that Captain Berkoz over St. Pathus would have been denied even that limited assistance because the cabin floor of his plane was supporting a full load of passengers.

It is probable that the autopilot would have been disengaged and that the throttle levers, controlling the three engines, would have snapped back into the near-idle position. Certainly, as the flight-data recorder witnessed,

Mohammed Mahmoudi, accused by McDonnell Douglas of in effect causing 346 deaths

The crash site in the Forest of Ermenonville, March 3, 1974.

Mejat Berkoz   Oral Ulusman   Huseyin Ozer

The crew of Ship 29 on the fatal flight. As the DC-10 hurled toward the forest Captain Berkoz shouted, "Bring it up, pull her nose up." "I can't bring it up," said the co-pilot Ulusman, "she doesn't respond."

When Captain Jacques Lannier (far right) arrived at the crash site, he was confronted by a nightmare. "On my left, over a distance of four hundred or five hundred meters, the trees were hacked and mangled. . . . In front of me, in the valley, the trees were even more severely hacked and the wreckage even greater. There were fragments of bodies and pieces of flesh which were hardly recognizable. In front of me, not far from where I stood, there were two hands clasping each other, a man's hand tightly holding a woman's hand, two hands which withstood disintegration. . . ." The forest was silent. The spring birdsong had been utterly extinguished.

John Hanessian   Ann Powell   Michael Hannah   Phyllis Gertrude Doran

William Hayes   Harry and Florrie Abrahamson   Lynne Barbara Roberts   Helen and Ann Kelsey

The cold statistics of the world's worst air crash: 346 victims of 21 nationalities; 18 children and one baby died; there were 30 college students; almost two-thirds of the victims were under the age of forty; fifteen entire families were wiped out.

Identification of most of the victims was impossible. Relatives were asked to sift through the debris of personal possessions and bundles of blood-soaked clothing in the hope of finding clues. Most of the human remains were buried together at a mass funeral service near Orly Airport.

Mary Hope (right), widow of a British journalist, began the complex litigation against McDonnell Douglas to obtain damages and also to discover the truth about the crash. Relatives of other victims joined her, and General Dynamics and Turkish Airlines were also sued as co-defendants. The 340 cases are being heard by Judge Pierson Hall (below) in the U.S. District Court in Los Angeles, California. After pretrial hearings the three defendants offered not to contest liability for the crash, but agreeing on the amount of damages to be paid in each case has not been so easily resolved. Total liability for the disaster could run into hundreds of millions of dollars: whatever it costs, most of the money will be paid by insurance syndicates of Lloyd's of London.

the huge airplane slowed, banked to the left and entered into its final dive.*
We shall never know if Berkoz might, in those few seconds, have saved
the DC-10 if like McCormick he had summoned full power from the wing
engines before the plane dived. As it was, once the nose of the crippled airliner had dropped below the horizon, and once the momentum of the dive
had built up, nothing could have saved Flight 981.

It is unpleasant to dwell too much on what it must have been like to be
trapped inside that huge machine as it hurtled toward the ground. There
could have been little doubt in the passengers' minds that their lives were
in awful jeopardy.

We know a little of the events on the flight deck through the cockpit
voice recorder that immortalized the last thoughts of Berkoz, Ulusman, and
the flight engineer, Huseyin Ozer. The recording begins with the sound of
violent decompression followed by the noise of the pressurization system
building up rapidly as it tries, vainly, to replace the escaping air.†

Voices say: *"Oops. Aw, aw."*

A klaxon goes off. Plus nine seconds.

Berkoz: *"What happened?"*

Ulusman: *"The cabin blew out."*

Eleven seconds.

Berkoz: *"Are you sure?"*

Sixteen seconds.

Berkoz: *"Bring it up, pull her nose up."*

Ulusman: *"I can't bring it up—she doesn't respond."*

Twenty-three seconds.

An unidentified voice (probably Ozer): *"Nothing is left."*

Another voice (probably Ulusman): *"Seven thousand feet."*

Klaxon sounds, warning that the plane has gone over the "never-exceed"
speed.

Thirty-two seconds.

Berkoz: *"Hydraulics?"*

Another voice (probably Ulusman): *"We have lost it . . . oops, oops."*

Fifty-four seconds.

Berkoz: *"It looks like we are going to hit the ground."*

Fifty-six seconds.

Berkoz: *"Speed."*

Sixty-one seconds.

Berkoz: *"Oops."* There is no further speech.

Seventy-seven seconds. Sound of initial impact.

The flight-data recorder suggests that during the last few seconds Berkoz

---

* A flight-data recorder is a machine popularly known as the "black box" that records
simultaneously and continuously information about the plane's mechanical activity.
† The official transcript of the conversations has not been released by the French
authorities. This version was obtained from the Turkish pilots' union which had assisted
in translating the dialogue.

or Ulusman juggled with the throttles in an attempt to raise the airplane's nose and then shut off all power. The DC-10 did level out of its dive and was almost horizontal when it reached the roof of the forest but that may have been due to the plane's flying characteristics, rather than the pilots' use of engine power.

Ship 29 hit the ground at 497 miles an hour.

THE COLD STATISTICS of the world's worst air crash deserve to be recorded. The 346 victims were of 21 nationalities. The majority, 177 were British, 56 were Turkish, 48 Japanese, and 25 American. Almost two-thirds of the victims were under the age of forty. One nine-month-old baby and 18 children died. There were 30 college students. Fifteen entire families were wiped out. There were 21 engineers, 3 architects, a priest, 4 doctors, 2 nurses, 8 teachers, 3 accountants, and 3 lawyers.

As a result of the crash, 80 women were widowed and 29 men lost their wives. Fifty-two children lost both their parents and another 207 children lost one parent.

Some of these deaths cost countries, as well as families, dearly. Japan lost 38 young graduates who had been selected by a bank and a trading company as future senior executives: in Japanese business tradition they had been sent on a world tour to prepare them for their new careers. Hubert "Jake" Davies and Geoffrey Brigstocke were Britain's two chief experts in law-of-the-sea. Jim Conway was the secretary of Britain's second largest union, the AUEW.* Dr. Patrick Hutton, who died with his wife Anne, was principal scientific officer at the British Atomic Weapons Research Establishment. And Dr. Charles Bowley of Sheffield was an internationally recognized authority on blood transfusion.

There were people who had spent their lives struggling against great odds, and sometimes danger, only to die just as they were enjoying the rewards. Peter Warnett had overcome a nervous disorder and a painful speech impediment to travel the world as a successful engineering salesman. Tadeuz Okuniewski, born in Poland in 1927, was removed to Siberia by the Soviet Army along with thousands of other Polish men, women, and children, and was finally released by the Amnesty and decanted into Egypt. He had worked himself up to being a service engineer with a British textile engineering company. Eric Kelsey had been born, as Erich Katz, to Austrian Jewish parents in 1922. Aged sixteen, he escaped through Switzerland after the *Anschluss* joined Austria to the Third Reich. He fought in the Royal Artillery, survived wounds, and married an English teacher. After years of work as a translator and then running a fried fish bar, he had in 1973, been able to open his own restaurant in the country town of Malvern, Worcestershire.

---

* Ironically, it was some of Conway's members who called the BEA strike, although they did not receive official backing from the union.

He was returning from his first holiday in years with his wife and their two daughters, Helen, aged nine and Ann, aged six, when they were killed. And then there was the couple, both refugees from unhappy marriages, who had just found new lives together.

It has to be said that the cost—in monetary terms—of many of those lives was cheap. The total compensation paid out by the insurers of McDonnell Douglas, General Dynamics, the Turkish Airlines, and by the FAA will undoubtedly set some kind of record but that is only because of the number of people involved. At the time of writing it seems likely that the damages will reflect the financial and family status of each of the dead, rather than the degree of blame involved in the accident or the particular horror of their deaths. Under those ground rules the old and the young, the single, and the nonrich do not merit very much compensation.

The men whose profession consists of settling the claims that result from airline accidents and other disasters find great favor with such a philosophy. And they are right, at least in saying that no amount of money can compensate for the deaths of 346 people—even if most of them were not rich or particularly eminent, but came in the same category as Jane Austen's Lady Elliot in *Persuasion,* who, "though not the happiest being in the world herself had found enough in her duties, her friends and her children to attach her to life and make it no matter of indifference to her when she was called upon to quit them."

Takehiro Higuchi, a twenty-four-year-old architect from Tokyo, and his wife Atsuko were certainly not indifferent to life when they climbed aboard Flight 981 for its last journey. They had been married just three weeks.

THE VIOLENCE OF SHIP 29's impact tore apart into unrecognizable shreds most of the people who were on board. And by Sunday night it was clear to the French authorities that the THY passenger list would be of little help in identifying the victims: in the confusion at Orly, there had been no time to record the names and addresses of all the passengers. Obviously, identification of the dead was going to require formidable amounts of detective work.

Jean Pierre Lahary is one of the world's greatest experts in the art of reassembling human remains into bodily form. He has no formal qualification in forensic medicine, but, through practice, he has developed some radical techniques which can provide positive identification of the dead where little evidence is apparently available. Two days after the crash Lahary, who is curator of the museum of the Institute of Forensic Medicine in New York, was asked to fly to Paris and employ his grisly expertise on what was left of Ship 29's victims.

Some of us who saw the result of the carnage in the forest believed it would have been less distressing, less of an affront to human dignity, if the remains had been buried where they'd fallen. But cleaning up after death always involves certain formalities—the enactment of wills for example—demanding death certificates which, in turn, require positive identification. Captain Jacques Lannier's policemen, the firemen, and the Red Cross volunteers, therefore diligently collected together the human debris from the crash site.

Forty bodies, including the six ejected from the airplane when the door came off, were found more or less intact and were easily identifiable. The rest had been shattered into approximately 18,000 fragments of which only about 10 percent were of any use to Lahary and Professor Réné Michon, a lecturer in forensic medicine in Paris, who was in charge of the identification process.

Every usable fragment was tagged and numbered and stored in a plastic bag. The details were then fed into a computer and matched against the physical descriptions of all the *supposed* victims, which Interpol, through local police forces, had collected from relatives. In some cases the link that the computer established—for example, between a nicotine-stained finger and a heavy smoker—was a little precarious but it was, at least, a start.

While the computer did its work, Lahary employed a speciality that is all his own—the reconstruction of heads. He would clean what was left of a head, applying cream to the skin to prevent it from hardening. He injected glycerine to retrieve sunken eyes and rebuilt shattered skulls. After some cosmetic touches had been added, the result, often startlingly lifelike, would be photographed. Grotesque though the whole business sounds, the photographs did produce some positive identifications. In all, after six weeks work, Lahary, Michon, and two colleagues were able to persuade a French magistrate to issue 188 death certificates.

Meanwhile, at Jacques Lannier's police station in Senlis, keys, jewelry, rings, watches, and wallets, collected from the site, were laid out on long tables, covered with white paper, for relatives to inspect and hopefully identify. For those who could stomach it, there was also a tour of St. Peter's, the former Catholic church in Senlis, where the victims' clothes lay jumbled on three long trestle tables.

Through the identifications that resulted, and through passports that were found on the site, the magistrate was persuaded to issue another 149 death certificates, bringing the final total to 337.

That means that at least nine people were literally smashed into oblivion. The caveat has to be added because the forensic expert Lahary is convinced to this day that fourteen or fifteen people remain unidentified: he says that he examined pieces from 350 or 351 bodies.

It was—in the circumstances, inevitably—a distressing and imprecise business. One man was listed as a victim by the police on the reasonable grounds that he had made a reservation for Flight 981, his name was on the passenger manifest, and both his passport and his ticket were found at the crash site. Two weeks later he turned up, very much alive at Senlis police station: he had given up his seat on the plane to his fiancée who was carrying his passport in her purse.

And, on Sunday afternoon three weeks after the disaster, Barbara Collen got a telephone call from Paris saying that forensic experts had been able to identify half of her daughter's head. A little while later there was another call saying it wasn't *her* daughter's head after all.

ON TUESDAY, MARCH 5, Charles Miller, then still head of the Aviation Safety Bureau of the NTSB, heard that Ship 29's rear cargo door and six bodies had been found nine miles from the main crash site. The news was ominous because the DC-10 cargo door problem was supposed to have been fixed after the Windsor incident. Yet here was another DC-10 losing a cargo door and this time with catastrophic consequences. Miller left his Washington office and headed for Paris. He was obliged to travel via Boston, and while waiting that night for his connection, he took the opportunity to inspect a DC-10 that just happened to be at the airport. He was disturbed to discover that the peephole in the rear cargo door of this particular plane was covered with grime and difficult, if not impossible, to see through. He was even more disturbed to discover that the ground crews at Boston didn't know what purpose the peephole served.

The following day, March 6, Miller and officials from the FAA and from Douglas who had also traveled to Paris, were allowed to examine Ship 29's door. It was immediately apparent that the support plate was missing and, indeed, had never been fitted.

On March 7 the FAA issued an Airworthiness Directive against the DC-10. It at last made mandatory the modifications to the door which were supposed to have been carried out under Jackson McGowen and John Shaffer's gentlemen's agreement twenty months before. Three weeks later the AD was strengthened by FAA demands for further modifications to the door. Two months after that yet another AD was issued. None of the modifications now made mandatory were new: long before the crash they had been recommended in McDonnell Douglas service bulletins and Ship 29 could and should have benefited from all of them.

On March 26 a Senate subcommittee began hearings into the Paris disaster, and a day later a committee of the House of Representatives followed suit. John Brizendine, the president of the Douglas division of McDonnell Douglas, gave evidence to both committees. He said he welcomed the opportunity "to present our company position and to state our complete and total confidence in the safety and airworthiness of our family of Douglas commercial aircraft including the DC-10 aircraft now operated by 23 airlines throughout the world."

On April 17 at McDonnell Douglas's annual meeting in St. Louis, Missouri, Brizendine's boss Sanford McDonnell made the company's position even clearer. Questioned by newsmen about the Paris crash, he said it had been caused by "human failure"—specifically, the failure of an "illiterate" baggage handler (Mohammed Mahmoudi) to close the cargo door properly.*

Meanwhile, the special committee which Alexander Butterfield, the FAA administrator, had set up to "review thoroughly and carefully the pertinent certification history of the DC-10 as well as investigative reports of all DC-10 accidents and/or incidents which bear even slight similarity to yesterday's crash," was hard at work. The FAA committee's report was written on April 19, 1974. By that time the records of every American airline which owned DC-10s had been examined to determine the number and nature of cargo-door problems that had shown up. The survey covered the six-month period between October 1973 and March 1974 and this is what it revealed:

Problem Area	Number of Occurrences
Inability to open or close or lock and unlock doors	136
Door seals found torn, leaking, missing, or out of place	450
Warning placards found worn, torn, missing, or illegible	129
Electrical malfunctions involving door lock/unlock switches, door warning lights, loose connectors, or circuit breakers	254
Doors frozen open or closed, latch actuators frozen, electrical switches frozen	31
TOTAL	1,000

These problems had been encountered in less than one hundred DC-10s. In other words, in the six months before the crash there had been an average of more than ten cargo-door problems for every DC-10 operating in America.

In its report the FAA committee said that nearly all of the airlines it surveyed "have had repeated door closing latching problems . . . and have had to manually close the doors [with a hand crank] many times."

There is no doubt that McDonnell Douglas would have been aware of these consistent problems because airlines routinely submit maintenance reliability reports to the manufacturers. And, by the time Sanford McDonnell made his statement the corporation was certainly aware that the support plate had never even been fitted to Ship 29's door. In those circumstances it might have been more circumspect for Sanford McDonnell to hedge his bets a little on the cause of the disaster.

Mahmoudi's reaction was one of anger and the union he belongs to threatened to boycott DC-10s at all French airports. To placate the membership an emissary was dispatched from the American Embassy in Paris to

---

* To his regret, Charles Miller had told the Senate committee on March 26 that the baggage handler was illiterate. Mahmoudi could not read the cargo-door warning decals because they were printed in English. However, there is no doubt that some people interpreted "illiterate" to mean persons unable to read and write even their own language.

union headquarters where he apologized, on behalf of the United States Government, for Chuck Miller's reference to illiteracy.

McDonnell Douglas has never publicly withdrawn the slur against Mahmoudi's name, even though in due course it transferred the blame for the crash from his shoulders to General Dynamics and Turkish Airlines, both of which it accuses of negligence. In June 1975, we asked Sanford McDonnell why he hadn't offered Mahmoudi an apology. He said: "Because no one has asked me to."

15

Recovery for Wrongful Death

> *Christian:* "I know what I would obtain; it is ease for my heavy burden. . . ."
>
> *Mr. Worldly Wiseman:* "Why in yonder village (the village is named Morality) there dwells a gentleman whose name is Legality, a very judicious man (and a man of a very good name) that has skill to help men off with such burdens as thine are. . . ."
>
> —JOHN BUNYAN,
> *The Pilgrims Progress*

RECOVERY FOR WRONGFUL death is the name for the legal process by which those who through negligence cause the death of others are made to pay compensation. Because of the number of people it killed, the crash of Ship 29 created one of the largest wrongful deathsuits in legal history. Because Ship 29 was an American-built airplane, operated by a Turkish airline which crashed in France killing people of 21 nationalities, the suit was also one of the most complex ever mounted.

It may seem on the face of it a simple enough proposition that those who are bereaved by wrongful death should be entitled to compensation for their loss. But under English law—of which American law is a branch, although now more vigorous and diverse than the parent trunk—recovery has never been a simple affair. Indeed, for many years the curious position in both Britain and America was that while a breadwinner might win compensation for injuries which *reduced* the family income, the family could not claim one penny in damages if the breadwinner was killed and the income extinguished. This brutal paradox entered into the common law of England and America in 1808 through the decision of an English judge, Lord Ellenborough in the case of Baker versus Bolton.* Mrs. Baker was mortally injured in a stage-coach crash and her widower sued the stagecoach proprietors claiming that he had "wholly lost and been deprived of the comfort, fellowship and assistance of his wife," suffering "great grief, vexation and anguish of mind."

---

* Cited in Harry Street, *The Law of Torts,* Ch. 26. This section draws heavily on Harry Street and Stuart M. Speiser, *Recovery for Wrongful Death.*

Lord Ellenborough told the jury that they must consider only the grief and loss of society that Mr. Baker had suffered during the month between his wife's injury and her death. The account which comes down to us is not very ample but his lordship is reported as saying that "in a civil court, the death of a human being cannot be complained of as an injury." In saying that, he created a precedent which was taken as binding upon the courts of almost every state which derived its legal system from English common law—America, Ireland, Canada, Australia, New Zealand, and (for most purposes) India, and Britain's former African colonies, where common law operates alongside other systems.

Ellenborough was a judge of fierce and reactionary reputation but, in fairness, he was presented with a rather complex problem over Baker versus Bolton. In the misty past when Anglo-Saxon law first coalesced, no distinction was made between intentional and accidental killing. The aim of public policy was simply to prevent individual deaths leading to vengeful chain reactions. To do that the principle of wergild (or man-money—by the same derivation as werewolf) was established:

> If a man killed another, the slayer was to compensate his death by the payment of a certain sum, greater or less according to the circumstances of the case . . . in this manner was every offense considered in the light of civil injury, and the object of the law was to repair the fault, rather than to punish the offender. There was, therefore, no distinction made between things done with deliberate malice and those done in the heat of passion or by inadvertence. . . . *

It was obviously a primitive legal system that might allow a man coolly and with foresight, to calculate the cost of killing another, find the money, and then perform the homicide with impunity. And indeed medieval society, influenced by Christian morality and by a developing conception of the state, gradually imposed upon Anglo-Saxon law a new view of homicide—as an offense against society. By 1400 this change was firmly established in England. A man who caused the death of another was regarded as a felon, an offender against the crown. If the killing was deliberate (and lacked the sanction of war or self-defense), he was liable to execution and the forfeit of all his property to the Crown. Even if the killing was accidental, but involved negligence, it was still treated as an offense, called *homicide per infortunium*. The perpetrator was spared his life but very little else:

> . . . He is not to be discharged out of prison, but bailed till the next term of sessions to sue out his pardon . . . for though it is not his crime, but his misfortune, yet, because the king hath lost a subject, and that man should be more careful, he forfeits his goods, and is not presently absolutely discharged out of prison. . . . †

* Crabb, *English Law*, 1829.
† Lord Hale, *Pleas of the Crown*.

Inevitably, most of those who wronged the Crown at the same time wronged a private interest, and when the king lost a subject, a wife usually lost a husband. But clearly no civil action could usefully be launched against someone who had already lost his goods and, in some cases, his life. Civil actions in England were further inhibited by the so-called merger doctrine which evolved in the 1600s and which said that when the same act amounted to both a felony and a civil wrong, the proceedings should be merged with the criminal prosecution, extinguishing any civil action. (The last vestige of this doctrine lives on in the rule that criminal actions should be heard before any civil actions arising out of the same events. However, it is no longer the case that a man who, say, kills someone with his car can escape paying compensation, if he happens to be convicted of the criminal offense of drunken driving.)

In the meantime, in seventeenth-century American settlements, the concept of wergild lingered on, though it was mixed up with the more modern concepts of homicide. In Massachusetts in 1675:

> James Fford being bound over to this Court to Ansr for his driving a cart over Avigaile King that the child died. After the Court had duely considered the case sentenct him to pay the fine of five pounds to the Country and five pounds mony to its father Samuel King.

With stout common sense the Massachusetts Court of Assistants in 1677 concurred saying that £20 was due to the widow of John Dexter, even though his death (by shooting) resulted from the crime of manslaughter. And as late as 1794 the Supreme Court of Connecticut ruled that damages could be claimed by a man whose wife died at the hands of an incompetent surgeon.

But all this American good sense was swept away by Ellenborough's decision in Baker versus Bolton. It was a briefly argued and sketchily reported case in a local court involving little money and Ellenborough gave no reasoning and cited no authorities. Almost certainly he relied on a careless application of the ancient merger doctrine, although it had no relevance to Baker versus Bolton. Civil suits were only extinguished under the doctrine because there would have been little point in suing a felon whose property had been confiscated by the Crown: nobody suggested in the Baker case that the stagecoach crash was the result of a felony.

Nevertheless, the Ellenborough precedent became entrenched and even today, 168 years later, there is still no clear remedy for wrongful death under the common law of either England or most American states.

However, with the hectic development of the railways in the 1830s and the 1840s and the increasing frequency of accidents, it did become clear that Ellenborough had created an intolerable situation. Obviously the rule led to awful injustice—and to the absurdity that in accidents caused by negligence it was financially more prudent to kill than to injure. (In America, the legend grew that the fire axes carried in early railroad cars were really there to enable brakemen to finish off any wounded passengers who might linger after a crash.)

In those days legal precedent was virtually unbreakable and so there was no remedy but to pass a statute overriding the common law rule. In 1846 the British Parliament passed the first of its Fatal Accidents Acts (known as Lord Campbell's Act) and during the next couple of decades most American states adopted statutes which embraced Campbell's essential principle:

> Whensoever the death of a person shall be caused by wrongful act, neglect or default . . . and the act . . . is such as would (if death had not ensued) have entitled the party injured to maintain an action and recover damages in respect thereof, then and in every such case the person who would have been liable if death had not ensued shall be liable to an action for damages. . . .

The act removed gross injustice but it did not provide a logical solution. The problem with statutory law is that it is difficult to amend and its definitions are vulnerable to political lobbying. As one American authority wrote: "The statutes tend to immobilize the doctrines, and political opposition in legislatures to increasing the burdens upon corporate enterprise has in many states made it difficult to liberalize the statutes." *

Fatal-accident statutes adopted in England and America produced three main effects: they limited the right to sue to immediate family members; they limited the kind of damages that could be awarded; and, in some cases they limited the amount. (In contrast, injury claims under common law are not usually inhibited by arbitrary limits. An employer, for example, can sue if one of his workers is injured, providing he can show loss—but he has no claim if his employee is killed.)

There has been some liberalization of fatal-accident statutes in the United States, but the changes have only been won with painful slowness. When the first major American text on wrongful death was published in 1893 (Tiffany, *Death by Wrongful Act*) there were twenty-two U.S. states which had limits, ranging between $5,000 and $20,000, on the amounts that could be extracted from defendants. In 1966 when Stuart Speiser followed in Tiffany's track, there were still thirteen states, by no means all of them backward or unpopulous, which applied limits. Since then, Colorado, Indiana, Kansas, Maine, Massachusetts, Missouri, Minnesota, New Hampshire, Oregon, South Dakota, Virginia, and Wisconsin have abolished limitation. But West Virginia ($110,000) still retains its limits. It is therefore probably cheaper to kill someone in Huntington than it would be to kill the same person in, say, Detroit or Atlanta.

However, the most remarkable consequence of the fact that wrongful death actions are usually mounted under statutory law is that damages usually are purely *compensatory*. Common law has long possessed a distinct retributive quality, and if in an injury suit the wrongful act is also heavily stained

* McCormick, *Damages*.

with carelessness or greed, then *punitive* damages may be added to those judged necessary to compensate the victim.

In English courts, and those American courts (the majority) which do not allow punitive damages in wrongful death cases, the results can be absurd. To take an example, if a drug company employs an incompetent biochemist and its drugs subsequently injure people, then the company will be liable for compensatory damages under common law. If, however, the company refuses to employ a biochemist at all (noting in its board minutes that the money saved will increase the profits) and subsequently produces defective drugs which injure people, then it may well be liable to punitive damages many times as large as the compensatory damages. But if the drug is so dangerous that it *kills* its users, most jurisdictions in the Anglo-Saxon world will protect the company from punitive damages. The result, in terms of practical litigation, is to diminish the leverage which the plaintiffs' side can bring to bear in an action for wrongful death. Negligence, if provable, usually contains some element of turpitude, and the fear of investigation will thus lie heavily on many defendants in personal-injury suits under the common law. Such fears may be largely ignored when defending a wrongful death action.

One American state which still employs this curious double standard is California. Unlike Texas, Colorado, Kentucky, Montana, Nevada, New Mexico, and South Carolina which all provide, by various means, for punitive damages in wrongful death cases, California's Supreme Court has consistently refused to allow them.* Reckless acts in the Golden State can therefore be less costly if the results are fatal.

Yet, paradoxically, California was the only jurisdiction where most of the families of the 346 victims of Ship 29 could hope to win substantial compensation for what has turned out, on investigation, to be one of the most unnecessary—as well as the most sanguinary—of all aviation disasters.

REGARDLESS OF WHERE air crashes occur, most of their financial cost is usually borne by Lloyd's of London and the crash of Ship 29 is no exception. American insurance companies carried the first $15 million of McDonnell Douglas's product-liability insurance, and 40 percent of the liability for General Dynamics. The Lloyd's aviation-insurance syndicates carried the rest, as well as the total liability for Turkish Airlines and the airplane itself. Within a few days of the disaster there was prima facie evidence that it had been caused by the negligence of at least one and possibly all three corporations, and, therefore, little doubt that they were possibly open to suits for compensation.

One of the prime conditions of insurance contracts is that the insurers shall have control over the conduct of any legal actions, choosing the lawyers and deciding the strategy during negotiation and litigation. In practice this

* South Carolina provides that "where death results from lynching . . . the county in which the lynching occurred . . . shall be liable in exemplary damages of not less than $2,000." In Alabama damages in a death case must be entirely punitive.

means that the consequences of a great many air crashes are moderated by the London law firm of Beaumont & Sons: in particular by Peter Martin, the partner who specializes in aviation and whom we mentioned in Chapter 11.

In accordance with the somewhat Dickensian traditions of English solicitors' firms, Beaumont & Sons occupies a dusty warren of a building in the City of London. There Martin sits under a skylight, behind a huge but ancient desk, with a picture of some long-forgotten biplane behind him. The office does not look like one of the nerve centers of the aviation industry. But it is just that and it has been ever since Major K. M. Beaumont went to Paris in 1919 to help write the Convention of Paris, the first codification of international air law. Today Martin is one of the editors of *Shawcross and Beaumont on Air Law,* the most prestigious compendium on the subject.

In formal terms, Martin usually acts for the airlines that Lloyd's insures, and in the case of Ship 29, he was nominated to act for Turkish Airlines. At the same time, another London law firm, Barlow, Lyde and Gilbert was nominated to act for McDonnell Douglas. Both firms were instructed by Lloyd's to offer compensation to the dependents of those killed in the crash. It is fair to say that both firms were representing closely allied interests.

Lloyd's is an association of wealthy private citizens, arranged in syndicates. Although one syndicate took the lead position in insuring McDonnell Douglas, it did not by any means bear the whole risk. The syndicates trade risks among themselves—much like bookmakers laying off bets—to spread the impact of disaster. Therefore, all three sources of liability—McDonnell Douglas, General Dynamics, and Turkish Airlines—were closely interlinked in the London market, making it natural for Peter Martin of Beaumont's to be very much in sympathy with Denis Marshall, the senior partner at Barlow, Lyde and Gilbert. Right from the start there was complete agreement between them on one central issue. Litigation in America should, if at all possible, be avoided.

Outside the United States the amount of compensation payable for aircraft accidents is limited by the Warsaw Convention of 1929. The convention was constructed for a variety of reasons, including the fear that the nascent airline enterprises of the 1920s might be overwhelmed by suits for damages. But there was also the justification that the causes of air crashes might be so complex (and the cases tried in so many different jurisdictions) as to place almost insuperable barriers in the way of individuals seeking compensation. In effect, the Warsaw Convention struck a bargain. The airlines agreed to accept "strict liability" for all accidents, meaning that claimants would not have to prove negligence, but only damage. On this understanding, the maximum compensation payable for each death was limited to 125,000 (old) French francs—approximately $8,300.

The Convention has since been ratified by most of the countries (including Communist ones) in which airlines operate, and if a person is killed in an air crash in almost any part of the world today, his or her family will be able to claim up to $8,300 (now, because of the change in the value of world currencies, about $10,000) as a matter of right. But, of course, when that sum represents little more than a year's income, in even so dilapidated an economy as Britain's, the benefits of the Warsaw bargain fall rather obviously on the side of the airlines and their insurers. The convention's existence is a demonstration of the inherent tactical weakness of the plaintiffs' side when litigation results from the activities of a highly complex international industry.

Since 1955 some countries have ratified the Hague Protocol, which roughly doubled the maximum compensation. And there are other special arrangements (such as the Montreal Agreement, to which British Airways is a signatory) which increase the possible liability of some airlines even further. But the overall position remains that outside the United States the amount of money that can be claimed from airlines is subject to arbitrary limits, and passengers, whether knowingly or not, agree to those limits when they buy an airline ticket. Inside the United States there are no arbitrary limits.* Although the United States ratified the Warsaw Convention, American courts have consistently found ways to avoid its limitations, chiefly by finding fault with the contract which is said to come into force at the moment of ticket purchase.

Therefore, it can be much more rewarding to press a fatal-accident claim against a U.S. airline, so long as negligence on the airline's part contributed to the cause. The same rules apply when a foreign airliner crashes in American territory or kills U.S. citizens. It is, of course, a very different proposition when a non-U.S. airline crashes outside the United States, killing non-Americans, and only twenty-three of Ship 29's passengers were U.S. citizens. In such circumstances, the litigation can be mounted in America only if the airplane was U.S.-built, and if there is evidence to suggest that the manufacturer—or its subcontractors—was at least partially to blame for the crash. (The liability of manufacturers is not limited by any convention.) However, as better than 80 percent of the Western world's airliners are produced by American firms, there is a very good chance of the first condition being met. And in the United States investigating the conduct of a corporation—in an effort to prove negligence—is made immeasurably easier than in, for example, England or France because of the thoroughness of the pretrial proceedings. Before going to trial the plaintiffs are entitled to examine ("discover") the corporation's records and interrogate witnesses under oath, under the compulsion of the court. The system is designed to allow both sides to assess the strength of the case and, where possible, reach a settlement before trial, saving money and time. The attractions—from the plaintiff's point of view—of suing an American plane maker in the United States are further increased by the probability that by some means or other the airline involved will be citable as a codefendant, robbing it of the protection of the Warsaw Convention even though it is a foreign airline, operating outside the United States.

As may be readily understood, Lloyd's and its representatives, attorneys Peter Martin and Denis Marshall, were therefore eager to offer immediate settlement of any compensation claims arising from the crash of Ship 29—

* Except as we said earlier in West Virginia.

*under whatever system normally applied in the home country of each poten-
tial plaintiff.* This they urged in the name of humanity, saying that settlements
could be made speedily and without rancorous argument. Martin and
Marshall both are fluent in the language of reconciliation, and speak of the
sterility of vengeful inquiry into the causes of disaster. This eloquence was
deployed at some length for the benefit of the present authors, saying that
the tactic was especially relevant to the DC-10 case, because of its great
complexity. The thought was no doubt sincere, though it is difficult to forget
that the financial interests of Lloyd's may be directly benefited when reconciliation triumphs over vengeance. And, of course, what the Lloyd's offer
really amounted to was this: If you will take what we offer you in your home
country, you can have it right away. If you go to America and try to get
more, we will fight, and it will be a long, long time before you will get one
penny.

LORD ELLENBOROUGH'S RULE, the Warsaw Convention, and the sheer
complexity of aviation technology are by no means the only factors that tend
to protect aviation corporations accused of causing wrongful death. Short of
financial collapse or government intervention, a corporation (as its articles of
incorporation usually specify) is perpetual. Time matters to it, but not so
much as time matters to a human family. The dynamics of the situation immediately after an air crash—and, indeed, most human disasters—is that
those bereaved are often in urgent need of money, while at the same time
they have to cope with a unique loss and attempt to rebuild an existence. To
a corporation, or to the interlocked syndicates of Lloyd's, any particular
disaster is only a small part of an ongoing experience. There is much to be
said for clearing up the matter with swift and modest offers. But if that
approach fails, time will invariably be on the corporate side.

It is a curiosity of Anglo-Saxon convention that while it is ethical for
lawyers acting for potential defendants in cases such as that provoked by the
crash of Ship 29 to approach the relatives of the victims with suggestions
about compensation, it is *unethical* for lawyers interested in taking up the
plaintiff side to do so. Convention—strictly enforced by Bar Associations
which can penalize and even disqualify offenders against the code—says that
would-be plaintiff lawyers must wait until they are approached by claimants
and asked for advice. Obviously then, defendant lawyers in civil actions have
something of an advantage.

However, publicity for the crash of Ship 29 was so widespread and so
much of it contained comment about the disadvantages of seeking compensation in "Warsaw countries," that any hopes that Lloyd's had of disposing
of the matter swiftly, modestly, and outside America met with no serious
success. The first relative to file a claim for compensation was Mary Hope,
widow of the journalist Francis Hope, who had been the Paris correspondent
for the *Observer* and the *New Statesman*. Francis Hope came of a well-to-do

family and he was a successful journalist. From the moment her claim was initiated, Mary Hope made it clear, publicly, that her aim was not simply to win compensation but to discover the truth about the crash. For the relatives of other victims, that was another potent argument for not accepting settlement offers from Lloyd's. Mary Hope's claim was filed in the Central District of California on March 28—less than four weeks after the crash—and eventually 339 other claims followed from plaintiffs in Britain, France, Japan, Turkey, and America.*

Mary Hope was able to mount her case with such alacrity because the ineffectiveness of the Warsaw Convention in the United States has meant that a small but highly specialized group of law firms have grown up that make it their business to pursue claims against airlines and airplane manufacturers. They are formidable operators not least because their profession, though difficult, is a highly rewarding one. Invariably they work on the contingent fee system, uniquely American, which allows them to take a share of the damages if they win—and nothing if they lose.† (The best aviation lawyers rarely lose.)

Inevitably in such a large case, most of the leading aviation law firms became involved: Speiser, Krause and Madole of New York and Washington, D.C., Kreindler and Kreindler of New York, and Walkup, Downing and Sterns of San Francisco. But even this distinguished phalanx of attorneys had some difficulty in lodging the case firmly in California. The mere filing of a claim does not guarantee the place of hearing and in April 1974 the Multi-District Litigation Panel—a committee of five Federal judges sitting in Washington—met to decide where such a complex action should be consolidated and tried. McDonnell Douglas, the main defendant, was represented by one of its staff lawyers, by a Los Angeles attorney named Robert Packard and—holding the whip hand—by James Fitzsimons of the New York firm of Mendes and Mount, nominated by Lloyd's.

Mr. Fitzsimons argued at length before the panel that the "law of the domicile" should be applied to the claims for compensation. This was an appalling prospect for the plaintiffs because the 346 victims had, among them, been domiciled in a total of thirty-six jurisdictions, most of which have differing views as to what constitutes liability in a wrongful death case. In 90 percent of those jurisdictions, the plaintiffs would have been required to prove (in various degrees) that the defendants had been actively negligent: under California law it would only be necessary to prove that Ship 29 had been a "defective product." However, Fitzsimons was prepared to contest even the right of aliens to appear in American courts, claiming damages for an accident which had occurred outside the United States. These were sound legal tactics, no doubt, but questionable policy in a world which flies, essentially, in

---

* All of the claims were consolidated under the title of Hope versus McDonnell Douglas.
† Under English law the contingent fee system is considered abhorrent. English lawyers get paid regardless of the result of their efforts. Any lawyer who accepted a share of the winnings would be disqualified from practice and could face criminal prosecution.

American-built aircraft. Most of all, Fitzsimons was anxious to argue that California was *forum non conveniens* for trial, even though most of the documents and most of the witnesses who could testify to the history of the DC-10 were located in or near Los Angeles. It seemed a curious argument because California was at least as convenient as Paris or Istanbul and, arguably, a good deal more so. At a later point in the proceedings, however, Fitzsimons stated quite frankly what it was that his clients found so inconvenient about California. The plaintiffs only wished to go there, he said, because high compensation awards were frequently made in California. Without hesitation, and unanimously, the Multi-District Panel ruled that the trial should take place in Los Angeles.

PIERSON MITCHELL HALL is a senior judge of the U.S. District Court in central California. One of the perquisites of senior (measured by age) Federal judges is the right to choose which cases they will try, and Hall employs this privilege to specialize, almost exclusively, in air-crash litigation. It was therefore inevitable when Los Angeles was fixed as the forum for the Paris Air Crash Case that Judge Hall would elect to preside.

By any standard "Pete" Hall is a remarkable member of the judiciary. Senior Federal judges impose their own schedules and, short of impeachment by Congress, cannot be made to resign or retire. At the age of eighty-two, Hall could justifiably live in virtual retirement on full salary—not least because he has undergone major heart surgery and now depends for life on a plastic aorta and a pacemaker machine (his second). Instead, he chooses to work a grossly overloaded five-day week and at the beginning of 1976 he carried the heaviest case load of any of the 480 District Court Judges in the United States.

What makes Hall even more exceptional is his rare understanding of tragedy, based on personal experience. His father died shortly after he was born and he spent most of his childhood in institutions. His first marriage ended in a bitter and protracted divorce case which left him, in middle age, penniless. And three other wives have died, one of them, Mari Honnold, of leukemia only a few months after they were married in March 1972. (He has since married again.) Not surprisingly then, Hall has an inherent sympathy for people—and, because of his mother's experience, especially wives whose lives are overtaken by disaster.

In 1958 he began hearing his first major air-crash case. It took eight years to settle and the experience left Hall with a profound determination to, as he put it, "do a lot of innovating."

By 1974 he had tried eight other major air-crash cases and settled each of them, on average, in two years. He did it by insisting on rigorous schedules for "discovery" and by setting, at an early stage in each case, a firm trial date. He also pioneered the idea of forming plaintiff committees. It is not uncommon in a major case for there to be a dozen or more lawyers representing different plaintiffs. To avoid an enormous duplication of effort and expense Hall insists that plaintiff lawyers should work together and share tasks.

But, most of all, what Hall did to speed up air-crash cases was to get plain mean when he suspected lawyers, on either side, of stalling. In a 1972 case, for example, Hall believed that a government lawyer—representing the FAA—was wasting time and preventing settlement. He telephoned the lawyer and pointed out that there was a little-used Federal law under which the attorney could find himself personally responsible for the costs caused by "frivolous delay." The case was settled with alacrity.

"I'll tell you why I push these things," says Hall. "Many times the victims are family men who leave a widow and small children. . . . Now if she has to wait eight years for settlement, the time has long gone when she *really* needed the money—when little Johnny was in school or Mary was going off to college. I know how grievous it is for a woman to be in that kind of situation. She takes her husband to the airport and twenty minutes later she is a widow." *

From the onset of the Paris Air Crash Case, it was clear that Hall would need to be very mean indeed to the lawyers on both sides if the victims' families were to get compensation with any speed. (In Hall's opinion, plaintiff lawyers are often the greatest enemies of rapid settlement in air-crash cases because "they're always yelling for more money.") There was, of course, not very much doubt that Ship 29 had been a "defective product," but whose fault was that? The plaintiffs claimed that McDonnell Douglas, General Dynamics, and Turkish Airlines were *all* to blame. Some of the plaintiffs named the U.S. Government as a fourth defendant, alleging that the FAA's negligence had contributed to the cause of the crash. The three main co-defendants each denied liability: each one said that the other two defendants were responsible. So long as those postures were maintained, it was necessary for the plaintiffs to establish the precise degree of liability of each corporation, and that promised to be a long and arduous battle.

There was another, equally formidable, barrier in the way of early settlement. While the Multi-District Panel had fixed California as the forum for the liability issue, that ruling did not necessarily dictate what law should be applied when it came to determining the amount of compensation for each victim. The position of McDonnell Douglas and General Dynamics was—and, at the time of writing, still is—that domiciliary law should apply; in other words, the compensation paid for, say, a British victim should be no more than a British court would allow. On this issue the plaintiffs were deeply divided among themselves. For the English claimants, who represented about half of the total, the application of domiciliary law would have been disastrous. English law limits the amount of compensation payable in a wrongful

---

* Joseph C. Goulden, *The Benchwarmers: The Private World of the Powerful Federal Judges.* Goulden's book was extremely critical of many Federal judges but he had nothing except praise for Hall.

death case to the exact amount of *financial* loss which a plaintiff can show that he or she has suffered. Many of the English plaintiffs were parents whose adult children had been killed: because they had not been dependent on their children for financial support their claims, under English law, were worth virtually nothing. On the other hand, the 120 or so plaintiffs from France, Japan, and Turkey were in favor of domiciliary law because all three countries allow "material damages" (for loss of support) *and* "moral damages"—which is monetary compensation for the grief bereaved relatives have suffered.

For its part Turkish Airlines claimed that its liability—if any—was limited by the Warsaw Convention. Although, as we have said, U.S. courts have consistently found ways to avoid the Convention's limitations, the issue has never been definitely ruled on. Obviously, if the Turks were to end up with the major share of the blame for the crash of Ship 29, the airline (or, more accurately, the Lloyd's syndicate which insured it) would have little to lose by taking the Warsaw issue to the U.S. Supreme Court. To make matters even more complicated Turkish Airlines added another dimension to the lawsuit by filing a claim against McDonnell Douglas—for the loss of Ship 29 itself.

With the litigation enveloped by these horribly complex issues it looked, in the summer of 1974, as though Peter Martin and the other Lloyd's lawyers could well be right: by going to America to sue, the victims' relatives might have to wait for years and years before receiving one penny in compensation. But the discovery process which U.S. courts employ can be a very powerful weapon in the hands of a judge like Hall who intends to get compensation paid while—to paraphrase him—little Johnny is still in school. The questions of law were, of course, largely academic until responsibility for the crash had been determined. To appropriate blame it was necessary to inquire deeply into the history of the DC-10, and of Ship 29 in particular, and under a rigorous schedule imposed by Hall, that is what the plaintiff lawyers proceeded to do.

McDonnell Douglas, General Dynamics, and the FAA all produced from their records thousands of documents for scrutiny. (Turkish Airlines was under the same compulsion but took several months to produce its records and even then some documents were censored "in the interests of security of the Turkish state.") Among the General Dynamics paperwork the lawyers found the Applegate memorandum with its chilling prediction of a DC-10 crash, and J. B. Hurt's impervious reply about the "interesting legal and moral problem." McDonnell Douglas's papers contained the Hazell and McCabe reports on the deficiencies of Turkish Airlines, and the records of Ship 29 with their false claims that the rear cargo door had been modified.

In September 1974, the plaintiff lawyers began the interrogation of witnesses. Compared to a trial, the deposition process—as it is called—is conducted with a degree of informality, usually in the offices of one of the law firms involved. The witness and lawyers from both sides sit around a table and there are no court officials present. The sessions are limited to about five hours a day and there are regular breaks for refreshment and other needs. But this apparently leisurely atmosphere is misleading. The witnesses give evidence under oath and every word is taken down by a court reporter with the sanction that the testimony may, one day, be used in evidence at a trial. The intent of the plaintiff lawyers, of course, is to extract from witnesses statements or admissions that will be damaging to the defense and from which there can be no retreat.

THE DC-10 PRETRIAL hearings were held in the offices of Robert Packard on Wilshire Boulevard, Los Angeles. Packard formally acted for McDonnell Douglas, but like James Fitzsimons of Mendes and Mount, New York, he was actually hired in the case by Lloyd's. The first to be deposed in the Paris case were McDonnell Douglas executives and employees, including John Brizendine, Bill Gross, who is now manager of the DC-10 program, and Ray Bates, Douglas's Director of Engineering. The questioning was extensive: Brizendine, for example, gave evidence for eleven days and the typewritten record of his deposition ran to over fifteen hundred pages. It was also, at times, fairly hostile. Edward Evans, the Douglas inspector who stamped off the support plate modification on Ship 29, was at one point reduced to tears.

The men from McDonnell Douglas were followed by witnesses from the FAA. Then the plaintiff attorneys turned their attention to General Dynamics and began questioning Convair executives about the subcontractual relationship with Douglas and, of course, about the Applegate memorandum. It was at that point that the four defendants capitulated.

J. B. Hurt of Convair was being questioned and, naturally enough, the interrogation by the plaintiff lawyers was focusing on his memorandum about the "interesting legal and moral problem" which Applegate had raised. He made no apology for writing his memorandum. At the time Applegate raised the issue, he insisted, McDonnell Douglas was already aware of the potentially catastrophic threat to the DC-10 posed by the weakness of the floor.

The day that evidence was given, a plaintiff attorney received a telephone call from a lawyer representing one of the American insurance syndicates involved in the litigation. If the deposition proceedings could be immediately adjourned, it might be possible for the defendants to come to agreement among themselves about settlement.

A couple of days later, on May 25, 1975, James Fitzsimons, formally representing McDonnell Douglas, went before Judge Hall. He said he was speaking on behalf of all four defendants who were now prepared not to contest liability for the crash of Ship 29. He pointed out that not contesting liability was not the same as admitting liability—although the distinction can hardly be called a large one. In any event, the defendants were prepared to begin discussing the amount of compensation to be paid for each victim on exactly the same basis as they would have done if the plaintiffs had gone to trial on liability and won.

During the settlement negotiations the three main defendants, McDonnell Douglas, General Dynamics, and Turkish Airlines, would act as one, said Fitzsimons. The insurers of the three corporations had agreed among them as to how the total compensation bill would be shared, but the details of that agreement would remain secret. (The essentials of the agreement are as follows: the U.S. Government contributed a fixed sum which we believe to be $3 million; Turkish Airlines agreed to pay an average of $30,000 for every passenger killed, a total of $10,050,000; General Dynamics agreed to pay a fixed percentage—between 15 and 20 percent, we believe—of the total compensation bill and McDonnell Douglas agreed to pay the balance.)

The irresistible question, of course, is why did the defendants, with one accord, suddenly end their resistance to the charges of negligence which they had fought solidly for more than a year?

One reason could be that in April 1975 one plaintiff lawyer, Jerry Sterns, thought of a way of unraveling what he saw as the hopeless complexity of the litigation. So long as Turkish Airlines was a co-defendant with the two California manufacturers any trial would be bedeviled by the Warsaw Convention issue. Although, as we have said, U.S. judges have consistently found ways to get round the convention's limitations, the U.S. Supreme Court has yet to make a definite ruling: in a case as big as this one there was little doubt that if Hall made a ruling on the issue it would be challenged, by one side or the other, all the way to the Supreme Court. In those circumstances the case against *all* the defendants would go unresolved for years. Sterns therefore proposed, and Judge Hall agreed, that there should be *separate* trials. The case against Turkish Airlines would be put on the back burner, as it were, while the plaintiffs proceeded against McDonnell Douglas and General Dynamics.

Part of the answer may lie in a document of more than two hundred pages which was submitted to Judge Hall on March 10, 1975, by Donald Madole, chairman of the Plaintiffs' Discovery Committee. The document summed up the work done over the previous months by Madole and his colleagues, and it invited Hall to rule that the case against McDonnell Douglas, General Dynamics, and Turkish Airlines was so overwhelming as to require him to summarily pronounce them liable for the crash without going to trial on the matter.

It has to be said that there was very little chance, if any, of the motion being granted. Obviously, Summary Judgment denies the defendant a right to trial and that is something that judges in most jurisdictions are loathe to do. And, in due course, Madole's motion was dismissed. But obtaining judgment was not really the purpose of the motion. Madole's aim was to apply pressure on the three main defendants by revealing, for the first time in public, the scope and the power of the charges against them. Toward the end of 1974, Judge Hall had issued a so-called gag order, making secret the depositions and documents that had been gathered during discovery, to prevent pretrial publicity. Madole's motion, which quoted extracts from the confidential material, automatically became available to the public and press as soon as it was filed. The extent of the indictment may be gauged if the Table of Contents alone is recited:

I. AFFIDAVIT OF DONALD W. MADOLE IN SUPPORT OF PARTIAL MOTION FOR SUMMARY JUDGMENT ON LIABILITY.

1.0 *The Direct and Immediate Cause of the Paris Air Crash Was the Inadvertent Opening of the Aft Bulk Cargo Door at Altitude*
   1.1   The Accident
   1.2   Why the Cargo Door Opened
   1.21 Operation of the Aft Cargo Door
   1.22 Previous Accidents:
         a. Fuselage No. 1 Incident (May 29, 1970);
         b. The Orbit of Reasonable Foreseeability: Windsor Accident (June 12, 1972)

2.0 *Gross Negligence of Defendant McDonnell Douglas*
   2.1   Service Bulletins—Documented Duty
   2.1a Service Bulletin A52-35 (Exhibit 2016)
   2.1b Service Bulletin 52-37
   2.1c Service Bulletin 52-49 (MD Exhibit 10007)
   2.2   MD Failed to Install Crucial Service Bulletins on the Accident Aircraft and Such Failure Was a Contributing and Proximate Cause of the Crash in Paris on March 3, 1974.
   2.21 Crucial Service Bulletins Were Not Installed on Hull 29
   2.22 Reports Were Received from the Field Concerning the Malfunction of the Cargo Door Closing Mechanism or Malfunction Indications.
   2.23 The Evidence Revealed the Incontrovertible Physical Fact That the Latch Actuator Was Not Fully Extended, and That the Latches Were Not Safe.
   2.24 It Was Determined from an Examination of the Lock Pins on the Aft Cargo Door Involved in the Paris Wreckage That Three of the Four Lock Pins on the Aft Cargo Door Latching Mechanism Had Been Adjusted to an Unsafe Position.
   2.25 Tests Done by MD Show That the Ship 29 Cargo Door was Improperly Adjusted or "Misrigged."

3.0 *MD Failed to Comply with Certain Federal Aviation Regulations (FARs) Which Were Intended to Insure That Jet Transport Aircraft Were Constructed in a Manner for Safe Public Transportation, and That Such Violations Were a Contributing and Proximate Cause of the Paris Air Crash on March 3, 1974.*
   3.1   Violation of FAR 25.1309
   3.2   The Testimony Show That MD Violated FAR 25.365(e) Which Violation Was a Contributing and Proximate Cause of the Paris Air Crash.
   3.3   The Testimony Show That MD Violated FAR 25.783(d) Which Violation Was a Contributing and Proximate Cause of the Crash

4.0 *Design Defects*
   4.1   The DC-10 Was Defectively Designed, Manufactured and Developed by MD in Conjunction with GD in a Defective Condition.
   4.2   GD Jointly Manufactured, Developed and Produced the DC-10 with Design Defects

5.0 *Testimony, Evidence and Documents Show That the Crash of Hull 29 Was Due to and Proximately Caused by the Wilful Misconduct of THY.*

   5.1   The Testimony That THY Dispatched Hull 29 in Paris on March 3, 1974 in a Defective Condition by Wilfully Refusing and/or Neglecting To Make Use of the Viewing Port Which Would Have Provided It With an Indication That the Door Was not Safely Locked and That Such Refusal Was a Contributing and Proximate Cause of the Crash.

   5.2   The Testimony, Evidence and Documents Show That THY Wilfully and Carelessly Adjusted the Locking Pins on the Aft Cargo Door on Hull 29 in Such a Condition as To Provide No Warning Indication of an Unsafe Door.

   5.3   The Testimony, Evidence and Documents Show That THY Wilfully Installed the Aft Cargo Door Actuator Without Making Certain Required Adjustments and That Such Failure Was a Contributing and Proximate Cause of the Crash.

II. MEMORANDUM OF POINTS AND AUTHORITIES IN SUPPORT OF MOTION FOR PARTIAL SUMMARY JUDGMENT

1.0 *Preliminary Statement*

   1.1   The Defendants are Jointly and Severally Liable

   1.21 Summary Judgment is Appropriate in the Instant Litigation

   1.22 Once the Party Moving for Summary Judgment Has Demonstrated That There Is No Genuine Issue as to Any Material Fact and That He Is Entitled to a Judgment as a Matter of Law, the Opposing Party Has the Burden of Demonstrating the Existence of a Genuine Issue as to a Material Fact.

2.0. *The Acts and Omissions of Defendants MD and GD Contributed to and Proximately Caused the Crash.*

   2.1   Admissions from Agents, Servants and Employees of Defendants MD and GD are Competent Evidence Against the Said Defendants.

   2.2   Post Accident Tests by MD Are Admissible in California To Show That the Failure of MD to Install Certain Service Bulletins Was a Contributing and Proximate Cause of the Crash.

   2.3   Standard of Care

   2.31 The FARs

   2.32 Manufacturers Strict Liability

   2.33 The Omissions and Commissions of Defendants MD and GD Were a Contributing and Proximate Cause of the Crash

3.0 *The Testimony, Evidence and Documents Show That Turkish Airlines (THY) Wilfully Dispatched Hull 29 on the Morning of March 3, 1974 in a Defective Condition; Wilfully Failed to Comply with Approved Maintenance Procedures; and That Such Wilful Misconduct Was a Contributing and Proximate Cause of the Crash.*

   3.1   Wilful Conduct

   3.2   THY Was a Supplier of a Defective Product Is Subject To Strict Tort Liability—Warsaw Is Not Relevant.

III. PROPOSED FINDINGS OF FACT AND CONCLUSIONS OF LAW

IV. TABLES OF CASES AND AUTHORITIES

V. CERTIFICATE OF SERVICE

Publicity for the plaintiffs' allegations was not the only thing that Madole's "unsuccessful" motion achieved. In order to resist summary judgment each of the three main defendants had to place on the public record some answer to the accusations. The charges that they made against each other in doing so were damning.

In its response McDonnell Douglas set out to show that General Dynamics had "from the beginning . . . played a major role in the design effort" of the DC-10.

> The contractual obligations eventually worked out between McDonnell Douglas and General Dynamics required General Dynamics to take exception to design philosophy that it felt was incorrect or improper. As a result of these arrangements General Dynamics played a major role in the design of the original cargo door and cargo door latching system: . . . Indeed, the "cargo door group" was almost 100% General Dynamics personnel. . . .

Plainly, that was not how General Dynamics saw it.

> . . . As an independent contractor, General Dynamics is not responsible for design defects created by Douglas. . . . In the present litigation, the product General Dynamics supplied was fuselage sections which it manufactured to specifications and loads provided to it by Douglas. Douglas was responsible for the assembly of the aircraft, for the installation of the vent door in the aft cargo door, for the design and installation of the subsequent changes to the vent door and for assisting THY (Turkish Airlines) in the maintenance and repair of hull number 29. . . . Under these circumstances, Douglas is clearly responsible for the aircraft as it reached THY. . . .

General Dynamics went on to argue that when it had delivered the fuselage of Ship 29 to Douglas in the summer of 1971—the year before the Windsor incident—"it was not clear that the vent door, which had just been designed by Douglas, was obviously bad." McDonnell Douglas had subsequently developed the cargo-door modifications and ". . . expressly undertook to itself install these modifications on hull Number 29."

The General Dynamics argument continued:

> The evidence further shows that apparently the crucial alterations were not made by Douglas upon hull No. 29, and instead three different Douglas employees deliberately did not do the tasks assigned to them and instead counterfeited the paperwork so that it was fraudulently made to appear that the changes had been made.

One of the most serious allegations made by the plaintiffs was, of course, that McDonnell Douglas and General Dynamics had both known, long before Paris, that if a DC-10 took off with an improperly locked cargo door the results could be lethal. In view of the emergence of the Applegate and Hurt correspondence General Dynamics could hardly deny knowledge. But, because the Applegate memorandum was never sent to McDonnell Douglas, the

vital question was did Douglas know of the danger? In its response to Madole's motion General Dynamics was categorical: "It [GD] did warn Douglas
in writing of the potential floor collapse and resulting catastrophe in the event
of an opening of the cargo door in flight. . . ." (Emphasis added.)

For its part, McDonnell Douglas did not claim to be unaware of the general
danger:

> In 1967, the FAA and the manufacturers knew that in the event a cargo
> door blew off or inadvertently opened in flight, deflection of the cabin floor
> could occur. Designed safety focused not on strengthening or further venting
> the floor, but on making the door opening in flight extremely remote.

The corporation attempted to justify that approach to design safety by
arguing that the alternative defense against the dangers of decompression had
only become a serious consideration *after* Paris:

> It is only in 1975, through a rule-making proceeding, that the FAA is
> even considering requiring that the floors of all wide-bodied jets made by all
> manufacturers be strengthened and/or further vented to avoid collapse fol-
> lowing rapid depressurization.

In view of the FAA's attempts before the Paris crash (1973 and February
13, 1974) to get McDonnell Douglas as well as Boeing and Lockheed to
strengthen wide-bodied floors, this seemed a dubious claim.* But if, for the
sake of argument, the statement were taken at face value it obviously begged
a question: With Ship 29's safety reliant on the integrity of the cargo-door
locking system, how "remote" was the chance of that door coming open in
flight?

McDonnell Douglas conceded, in its response, that the support plate was
missing from Ship 29's rear door, "although the McDonnell Douglas manufacturing records indicate to the contrary." However McDonnell Douglas was
adamant that the most important modification—the rerigging of the locking
pins had been carried out correctly on Ship 29's door before the plane was
delivered to Turkish Airlines. Ship 29's manufacturing records said that the
locking pins had been correctly adjusted and there was "no evidence to suggest that the McDonnell Douglas paperwork was similarly erroneous with
respect to the . . . pin setting. Douglas contends that the lockpins left the
factory with [the correct] setting accomplished. McDonnell Douglas further
contends that the lockpins were readjusted to the [incorrect] setting by THY."
As a result, the force needed to move Ship 29's locking handle was only
thirteen pounds. Even if the support plate had been fitted, the force required
to defeat the safety system would have been only thirty-three pounds, and so
"the missing support plate was of no consequence at all."

To add to Turkish Airlines' apparent culpability, McDonnell Douglas accused the airline of "habitually" tampering with the cockpit warning system

* See Chapter 9.

by adding shims to the locking tube (see Chapter 14, page 243) and of failing to check the peephole in Ship 29's cargo door before the fatal flight. All things considered:

> The conduct of THY . . . introduced an unforeseeable element that amounted to a legally superseding cause of the Paris crash. That conduct was so highly extraordinary as to cut off liability for any earlier negligence allegedly committed by Douglas or any original defects allegedly incorporated in the aircraft. Douglas had no reason to suspect that a knowledgeable airline like THY would alter the aircraft to defeat the safety features incorporated in it.

Turkish Airlines denied the allegations. The cockpit warning system had not been tampered with and the airline did not "rig, adjust, re-rig or 'mis-rig' " the locking-pins. THY also said that it had never been warned by McDonnell Douglas, or anybody else, of the consequences of in-flight decompression. Finally, the airline asked a significant question: How could McDonnell Douglas have properly adjusted the locking-pin when on the day the work was supposed to have been done, "the engineering drawings specifying that rigging had not yet been issued"?

It is difficult to imagine how the plaintiffs' cause could have been better served than it was by those three responses to the Motion for Summary Judgment, taken together. Each corporation denied its own liability, of course. But inherent to all of the defense arguments was the admission that Ship 29 crashed and 346 people were killed because the rear cargo door was unsafe. Therefore, Ship 29 was, undoubtedly, a "defective product."

With that supreme fact established there could no longer be any doubt that the victims' relatives were entitled to compensation for wrongful death. What remained at issue was which corporation should pay the compensation or, if more than one corporation was to blame, how the cost should be shared. And here the important point was, whatever the outcome, Lloyd's of London would have to pay the majority of the bill because Lloyd's carried the majority of the risk for all three corporations.

Negligence cases that go for trial before a jury—and certainly an American jury—tend to be vastly more expensive to the defense than out-of-court settlements, if liability is proved. In the circumstances, it was hardly surprising that the insurance men should have decided to sue for peace.

A few weeks later, in August 1975, Judge Hall removed another of the barriers that had threatened to delay settlement. First—and to the satisfaction of the English plaintiffs—he ruled against domiciliary law: the plaintiffs were entitled to California-style damages, he said. But Hall also went a very long way toward satisfying those Japanese, French, and Turkish plaintiffs who had wanted domiciliary law. Their objection to California law had been, of course, that it did not allow for "moral damages." California law does recognize that compensation should be paid for "loss of society" but traditionally the state's courts have interpreted that provision somewhat narrowly.

Hall announced that he intended to "liberalize" the interpretation, by applying California law *and* federal law. Under federal law, "loss of society" includes the loss of love, affection, care, attention, companionship, comfort, and protection.

Hall's ruling is likely to become a landmark in air-crash litigation. He has long argued that air-crash cases, usually complex enough in themselves, should be tried under federal law to achieve a measure of consistency. Air-crash cases usually are tried in federal courts in the U.S. because they invariably involve "diversity of citizenship"—meaning that all the parties to the action do not live in the same state. However, a federal court is normally bound by the laws of the state in which it sits. As a consequence, it is possible for federal courts sitting in, say, Oregon and Alabama to try the same issue and yet reach totally different verdicts.

In his ruling in the Paris air-crash case, Hall said that because the federal government wrote the laws which govern aviation in the United States, it had an "interest" in all air-crash litigation. That interest was just as great as the interest of a state which, in wrongful-death law, was primarily concerned with deterring the manufacture and sale of defective products within its borders.

AS THIS MANUSCRIPT is completed—March 1976—it seems that it may, after all, be too early to talk in terms of a plaintiffs' victory. Nine months have passed since liability for the crash ceased to be an issue, yet only eight of the victims' families have so far received compensation. The Paris Air Crash Case as a whole still appears to be a fair way yet from its conclusion because two serious difficulties have emerged.

First, not all the plaintiff families are willing in principle to accept compensatory damages alone. Two of the plaintiff attorneys, Lee Kreindler and Jerry Sterns, believe that the failures which led to the crash of Ship 29 were sufficient to win punitive damages from two of the defendants, McDonnell Douglas and General Dynamics. As we have said, California law has not to date allowed punitive damages in wrongful-death cases, but Kreindler and Sterns believe the law can be successfully challenged. One avenue open to them, they say, is to contest the California wrongful-death statute on the grounds that it is unconstitutional because the Fourteenth Amendment of the U.S. Constitution says, in part, that no state shall "deny to any person within its jurisdiction the equal protection of the laws." California law is certainly unequal as it stands at the moment because in negligence cases the injured can claim punitive damages whereas the relatives of the dead cannot.

It is possible that Judge Hall would be in sympathy with that argument. In ruling that both California and federal law would be applied he did declare that the federal interests were as much at stake as state interests. However, whether this opinion would lead him to overrule state law—as opposed to "liberalizing" it—remains to be seen. Even if he did take that radical step, it is by no means certain that the decision would be upheld by the federal appeals court.

It might also be possible for Kreindler and Sterns to mount a challenge against the California wrongful-death statute under a 1970 Supreme Court decision in the case of Moragne versus States Marine Lines Inc. In that case the Supreme Court ruled that Lord Ellenborough got his common law wrong —and that his rule should be abolished—with an eloquence that was almost worth waiting 162 years to hear:

> Where existing laws impose a primary duty, violations of which are compensable if they cause injury, nothing in ordinary notions of justice suggests that a violation should be non-actionable simply because it was serious enough to cause death. On the contrary, that rule has been criticized ever since its inception, and described such terms as "barbarous." . . .

The Supreme Court said that there was no intellectual justification for the Ellenborough rule whatever, and that it had arisen through a misapplication of the felony-merger doctrine. That doctrine had made sense in England when all those who caused death automatically lost their property to the Crown, but that doctrine had

> long since been thrown into discard even in England, and . . . never existed in this country at all. . . . The American courts never made the inquiry whether this particular English rule, bitterly criticized in England, "was applicable to their situation," and it is difficult to imagine on what basis they might have concluded that it was.

However, the Moragne case did not necessarily establish a right to sue for wrongful death under common law—on the same basis as injury actions— because it dealt with a maritime case. Again, it remains to be seen whether all American courts including those in California will take the opportunity to bring wrongful death into line with the rest of twentieth-century law.

There is no doubt whatever that McDonnell Douglas and General Dynamics will resist punitive damages and this means as things stand now that there will, after all, have to be a full trial before a jury to decide liability, and it is presently set for October 1976. Of course, that fact alone may bring about a resolution. In theory punitive damages and compensation are quite separate, but in the practical business of legal negotiation they are closely related. If most plaintiffs find the level of compensation sufficient, the urge to travel a longer and more arduous route fades considerably and vice versa. It is, therefore, a nice calculation for the insurers: to offer the least possible sum that will satisfy this large galaxy of claimants, without provoking among them such resentment as to cause a large proportion to split off and set about establishing a legal precedent which must, in the end, be highly damaging to all insurance interests.

The prospects of that happening are not, at the moment, very bright because—and this is the second difficulty that has emerged—even those families which agreed to forego any claim they might have to punitive damages are far from satisfied with the offers of compensation that have been made.

In the summer of 1975, when the defendants agreed not to contest liability, Speiser, Krause and Madole of New York * and Washington, D. C. who are representing nearly half of the plaintiffs, advised their clients that it was in their interest to begin negotiating sums on the basis of uncontested liability. In the six months that followed half a dozen cases were settled but it became increasingly clear that there was a vast disparity between the two sides on what "California-style" compensation should amount to. The only escape from this impasse was to put the matter to a California jury so that both the plaintiffs and the defendants could "test the temperature of the water."

The first trial—on the subject of compensation alone—began in Los Angeles before Judge Hall on January 27, 1976. At issue was the amount of compensation that should be paid for a young London couple, David and Phyllis Kween, who were returning from a holiday in Turkey when they were killed in the crash of Ship 29. They left behind two daughters Melissa, now aged four and Lauren, three. During pretrial negotiations the defendants— McDonnell Douglas and General Dynamics—had offered $300,000 to settle the case.† The plaintiff attorneys had asked for $700,000.

When the trial began the jury, five women and one man, was told that it would last for four or five days. In fact it took the most part of five weeks, and it only ended when it did because Judge Hall curtailed what he called "all the huffing and puffing."

The most rancorous argument was over what future the two little Kween girls were entitled to. Their mother was an American by birth and since the tragedy, Melissa and Lauren had lived with their maternal grandparents in Seattle, Washington. Jerry Sterns, for the plaintiffs, argued that the compensation should be sufficient to pay for a full-time nanny and for the girls' support and education in America until they reach the age of twenty-one.‡ James Fitzsimons, for McDonnell Douglas, saw the girls' future very differently. They were, he said, only entitled to the kind of upbringing that they would have received had their parents survived.

Lloyd's had obtained expert advice, from a Queens Counsel in London, on what those prospects might have been. It was not an optimistic document. The QC, Edward Machin wrote:

---

* Speiser's clients are also represented by Butler, Jefferson, and Fry of Los Angeles. James Butler and his partners are not aviation specialists but they are experienced general practitioners in the personal-injury field, and were especially necessary to the plaintiff team because of their expertise and standing in Southern California. Butler's offices —which, coincidentally, are in the same building as the offices of Robert Packard, the McDonnell Douglas/Lloyd's attorney—served as the headquarters for the Plaintiffs' Committee.

† Turkish Airlines was not a defendant because it had already settled the Kween case by negotiation for $60,000.

‡ The Kween family was formally represented by a Seattle law firm, but Jerry Sterns agreed to conduct the plaintiffs' case on their behalf. As we have said, Sterns's own clients are still, at the time of writing, pressing for punitive damages.

> I have some considerable personal experience of this kind of family, having myself lived in North West London for many years. They are hardworking and closely knit, and the evidence which I have read suggests that this family [the Kweens] were no exception. The children of such a family are usually anxious to earn for themselves at the earliest possible opportunity, and it is most usual, especially in the case of girls, for them to enter a family business, if there is one, as soon as they leave school. There would certainly be no question, in this family, of the daughters enjoying a university career. . . . I think that the most one could envisage, had the parents lived, would have been that the children might have taken a short secretarial course after leaving school and before becoming wage-earners.

Mr. Machin's confidence about the Kweens's lack of educational ambition is remarkable in that Mrs. Kween was a college graduate and a teacher. On the proposition that the Kween children would not long have been dependent upon their parents, Machin estimated that the maximum compensation payable under English law, the Fatal Accidents Act, was a total of $80,000. From that figure, Machin wrote, a portion of what the children had inherited from their parents could be deducted: There was the proceeds from the sale of the family home, he said, $2,000 worth of silverware and a couple of cars which "I cannot believe" were worth less than $2,000.

Machin was also asked by Lloyd's to determine how much compensation David Kween's parents were entitled to. At the time of the disaster David was in partnership with his father Cecil, running three tobacconists' shops. They hoped to build up a chain of shops but, since David's death, Cecil had felt unable to go through with the plans. The plaintiffs claimed that Cecil was entitled to compensation: Machin "unhesitatingly concluded" that, under English law, Kween could not claim one penny.

During the trial the defense said that it was willing to pay the Kween children far more than English law dictated but nothing like the figures the plaintiffs were talking about. The jury was singularly unimpressed. "They [the defense] lost me when they said it would cost seven dollars a week to feed a child," said one juror afterward. "I couldn't get hogswill for that."

The verdict was $879,450 for the death of David Kween and $630,500 for Phyllis. The total, $1,509,950, was more than five times the amount that Lloyd's had offered and more than double what the plaintiffs would have been willing to accept in settlement before trial. The jury said that their award included an unspecified amount for the children's loss of "love and affection" and compensation for David's father. (It was left to Judge Hall to decide, in private, how the total compensation should be divided between the children and their grandfather.)

The plaintiff lawyers were understandably jubilant. "We have been telling Lloyd's all along what a California jury would do to them but they wouldn't believe it," said one of the plaintiff team. "Now, if they will only sit around a table and talk realistically we can get all these cases settled in a couple of weeks."

There was no immediate sign that Lloyd's was prepared to become "more realistic." After the verdict had been announced Len Kirby, Chief Press Officer at Lloyd's, telephoned London newspapers to give his views. Kirby intended his remarks to be "off the record"—meaning nonattributable—and "for guidance" but *The Sunday Times* (London) refused to accept that condition and printed what he said:

> Early and generous settlement was offered in these cases well in excess of British legal precedents. But a lot of people are looking on it as a bonanza. Certain people are disgusted by the morbid, money-grubbing exercise.
> Underwriters have tried to come to amicable settlements, but people are being be-dazzled. They think they are going to hit the jackpot. It's distressing that people are being greedy . . . it's human nature.

Lloyd's, or more formally, McDonnell Douglas and General Dynamics, could, of course, appeal the Kween verdict or try to persuade the plaintiffs to settle for somewhat less than the $1.5 million. At the time of writing the verdict stands. (It is certain that the plaintiffs will not now accept $700,000 in settlement because they had to bear their share of the costs of a five-week trial.) And, Kirby's remarks notwithstanding, there was some hint in March 1976 that the attitude of Lloyd's was shifting: the day before the second test case was due to go to trial—again, on the issue of compensation only—the defendants settled for "substantially more" than they had previously offered.*

It is not very profitable to put estimates of litigation results into a manuscript that must spend several months going through the press. Legal negotiation often proceeds for a long time with glacial slowness: then suddenly the imminence of trial (the punitive damages trial is currently set for October) causes a rapid thaw and abrupt agreement. Equally, plaintiffs can tire of the struggle or become so desperate for money that they are forced to settle. It is, therefore, possible that by the time these words are read there will have been financial settlements in most, and even all, of the cases.

But, if compromise is unreachable, the Paris Air Crash Case could drag on for years, not least because there is a limit to how much even a judge like Hall can do to force a conclusion. The Kween trial left him exhausted and his doctors ordered a complete six-week rest.

There is also the point that some of the plaintiffs are seeking the truth about the cause of the crash as much as money. That may only be discoverable through a full liability trial.

After the Kween verdict, and after Kirby of Lloyd's had made accusations about "money-grubbing," Mary Hope, who initiated the lawsuit in California, said:

> I'd like to emphasize that our suit is not about money, but about justice. Compensation must of course be paid, and be based on some realistic ap-

---

* The case concerned the amount of compensation payable for the death of Milton Safran, forty-nine, an antique dealer from South Carolina who left a widow and four children. Mrs. Safran asked for the amount of the settlement to be kept secret.

proximation to the loss of financial support suffered. But, more importantly, we wanted to get the facts about the crash, and punish those responsible. McDonnell Douglas built a lethal aircraft, and Lloyd's insured it—so don't let them talk to me about greed.

AT THE END of the story, it seems natural to return to the parallel of the *Titanic*—a great venture, launched amid strenuous competition, and brought to disaster by the pride and confidence of its builders. Both cases will probably stand for a long time in engineering history as classic proofs of the fact that machines, however ingenious, are only devices for overcoming the physical universe, and that universe is coldly unforgiving of mistakes. It is not any accident that we have quoted Rudyard Kipling more than once in this book, for Kipling is pre-eminently the poet who understood machines as symbols of humanity's stubborn attempt to create order from confusion.

But as he knew, the machines remain cold, to the end of the day. ("Remember please, the Law by which we live./We are not built to comprehend a lie./We can neither love, nor pity, nor forgive./If you make a slip in handling us, you die.") The faults and flaws of machines are caused by the human agencies which bring them into existence—and these, as he knew, and we should know, are not uniform in their strength. They are not, of course, weaker or stronger than their mechanical offspring—but different and not predictable.

Kipling's *Hymn of Breaking Strain,* cited as the epigraph of Chapter 10, reminds us that physical materials can always be justly matched to the demands made on them, but that

> . . . the gods have no such feeling
> Of justice toward mankind.
> To no set gauge they make us,
> For no laid course prepare,
> And presently o'er take us
> With loads we cannot bear
> —*Too merciless to bear.*

It is a grim poem. But it recognizes, at the same time, the courage and splendor of the builders' enterprise, and in terms which are readily applicable to the history of commercial aviation. Parts of this we have touched on, from Cayley's original vision down to the present day, and it would be an astonishing cynicism that could represent it, overall, as other than a splendid enterprise. By aiming high, by attempting great feats of physical control, aviators have always placed themselves in the way of great dangers, just as did the people before them who took Joshua Slocum's view that "the sea was made to be sailed over," and set out to prove it.

> We hold all Earth to plunder
> All Time and Space as well.

Too wonder-stale to wonder
At each new miracle.
Then, in the mid-illusion
Of God-head in our hand
Falls multiple confusion
On all we thought or planned
—*The mighty works we planned.*

It is, however, important to remember that they were, and remain, mighty works, and not petty or unimportant enterprises. In some form or another, they will continue.

Epilogue

THE PLANE THAT Douglas built has yet to repay the $1.5 billion that was invested in its development, but in many ways the DC-10 has already achieved remarkable success. Airlines in America use more DC-10s than any other wide-bodied jet and around the world there are now some 210 in service. On an average day the DC-10 fleet carries ninety thousand people on six hundred scheduled flights to 136 cities. And in less than five years, eighty-two million people have flown on a DC-10 in safety and considerable comfort. In that same period, four DC-10s have been destroyed in accidents, but only in the Paris crash did the passengers lose their lives. Only the Boeing 747 has surpassed that record and there can be little doubt that the DC-10 will stand in aviation history as irrefutable testimony to the supremacy of American technology.

Of course, this book has concentrated on the other, less inspiring side of the DC-10 story: the failure of its designers to provide a proper defense against in-flight decompression; the failure of McDonnell Douglas's management to respond to repeated warnings—and two actual demonstrations—of the hazard; the chilling predictions of the catastrophe that eventually occurred; the falsification of Ship 29's records; the hasty and premature promotion to the jumbo-jet league of Turkish Airlines; and the consistent failure of the FAA to protect public safety. Through a combination of some or all of those events, 346 people were killed and thousands of other lives were placed in unnecessary jeopardy.

How could it have happened? Throughout the process of conception, design, and manufacture, the DC-10 benefited from the unrivaled plane making experience of Douglas and the financial strength of McDonnell, perhaps the best-run aerospace company in America. It also benefited from the vast resources of the American subcontracting industry and, in particular, the expertise of corporations such as Convair, Ling Temco Vougt, and General Electric. Its design was sanctioned by the FAA and the plane apparently met all of the Federal Airworthiness Regulations which, by and large, dictate the standard of aviation safety in the Western world. And yet, to use the lawyers' language, a defective product entered the stream of commerce. In the circumstances we have described it is, surely, insufficient to simply ascribe the blame to human failure, and leave it at that.

What failed was the system, the most expert, the most sophisticated, and the most rigorous system of its kind in existence. And what makes the failure doubly disturbing is that its causes are still imperfectly understood.

Mainly through the efforts of the plaintiff lawyers in the Paris Air Crash Case, we do know a great deal about the history of Ship 29 and its brief careeer. But, as we have said before, the job of plaintiff lawyers is to extract monetary compensation and that process does not necessarily demand that the whole truth be established. As things stand, the litigation in Los Angeles has left many important questions unanswered.

We have tried to plug a few gaps in the story and provide some essential background but journalists, lacking the power to subpoena witnesses and documents, cannot dig as deeply into the DC-10 affair as the circumstances demand.

The only body with the necessary power and authority is the Congress of the United States. It is true that both the Senate and the House of Representatives have already partially inquired into the Paris crash but both sets of hearings were held in the immediate aftermath when the state of knowledge was primitive. And although the House of Representatives provided trenchant criticisms of the FAA—as well as some useful corrective recommendations—Congress as a whole has shown little persistence. It is sobering to reflect on how much more zeal has been expended on Capitol Hill in investigating Lockheed's penchant for bribery.

There are several reasons why Congress should reexamine the DC-10 affair.

First, the creation of Ship 29's false records remains a deeply mysterious event. After the Paris crash the FAA investigated the manufacturing inspection system at Long Beach and *apparently* found it to be sound. (The caveat must be used because neither McDonnell Douglas nor the FAA will reveal details of the investigation or say how the conclusion was reached.) But what the FAA did not do was establish how it was possible for essentially the same inspection system to break down in the summer of 1972. The implication of John Brizendine's testimony in Los Angeles is that McDonnell Douglas doesn't know the answer either. In those circumstances, is it really possible to be confident about the integrity of the Long Beach system? And if two planes—Ships 29 and 47—could acquire false records, how many other Douglas aircraft of the same vintage might have been affected?

On a more general front, it is clear from an examination of the DC-10 affair that there are serious weaknesses in the American regulatory system. The FAA has gone some way toward putting its own house in order— Airworthiness Regulations have been tightened up, gentleman's agreements have been banned, the DC-10 cargo door is now as fail-safe as it should have been in the first place, and jumbo-jet floors are being strengthened— but it cannot do the entire job unaided. For example, the FAA must still pursue its dual role as part promoter and part policeman of the aviation industry. A House of Representatives subcommittee agreed as long ago as January 1975 that this dual mission sometimes created conflict * :

* The subcommittee was set up by the House Committee on Interstate and Foreign Commerce to review FAA performance.

> There have been instances when appropriate FAA actions in furtherance of air safety have been unreasonably delayed, or omitted entirely, because of an oversolicitous attitude on the part of some within the agency concerning the economic well-being of the aircraft industry or the air carriers.

Yet Congress has neither passed nor even prepared the legislation that is necessary to change the FAA's charter and remove that possible conflict.

More crucially, the same House subcommittee urged a reassessment of the Designated Engineering Representative system which allows aircraft manufacturers to monitor and approve the quality of their own work. As we explained in Chapter 10, designated representatives—McDonnell Douglas employees, temporarily wearing an FAA hat—played an enormous role in the process that led to certification of the DC-10: of the 42,950 "conformity inspections" that were made on behalf of the FAA, 31,895 were carried out by Douglas employees.

There has been no suggestion—and we do not mean to imply—that the designated representatives at Long Beach abused their trust or were willfully negligent. However, the fact remains that the system removes a vital check: a man who approves, say, a design test on behalf of the company that employs him is unlikely to revise his opinion when that test is presented to him, in his honorary FAA capacity, as evidence of airworthiness.

Yet, according to the FAA, designated representatives have been used more and more as aircraft have increased in complexity. The agency simply does not have the workforce to do the job because it doesn't have the budget. Only Congress can provide the remedy.

Finally, a congressional inquiry should consider the threat posed to aviation safety by the fierce competitive pressures which followed upon the creation of three rival American jumbos. There has been no serious inquiry into the effects of unlimited competition in the aviation industry since the committee of wise men, on which James McDonnell sat, reported in 1973 that the present policy was unworkable.

IN REVIEWING WHAT has happened to the main characters in the DC-10 story it seems appropriate to turn once more to the words of Rudyard Kipling: "Your glazing is new and your plumbing's strange/But otherwise I perceive no change." * As this is written the DC-10 is again surrounded by controversy, this time concerning the safety of its engine, the CF-6 manufactured by General Electric. The NTSB believes that the engine is susceptible to "catastrophic" failure; General Electric says that the charge is "untrue" and that the NTSB's language is "inflammatory"; McDonnell Douglas says, in effect, that there is no danger and the FAA seems ambivalent.

The conflict follows an accident at John F. Kennedy Airport, New York, in November 1975, when a DC-10 owned by Overseas National Airways

---

* Rudyard Kipling, "A Truthful Song."

was destroyed by fire.* The plane had been accelerating along a runway when the starboard engine ingested a flock of seagulls and disintegrated. The pilot aborted takeoff and brought the DC-10 to a shuddering halt. Fortunately, all of the 139 people on board were airline employees, familiar with evacuation procedures, and they escaped before a series of explosions ripped through the plane.

Bird-strike is a common enough hazard at airports and modern airliners, including the DC-10, are designed to fly safely even if an engine does fail during takeoff. What concerned the NTSB about the Kennedy accident was the *disintegration* of the CF-6 engine and the catastrophic fire that followed.

At the beginning of April 1976, it published its report on the Kennedy accident which urged the FAA to take a minatory line against the CF-6. The report recommended that airlines around the world should be warned of the danger, that the engine should be immediately retested, that whatever modifications were necessary should be made to CF-6s already in service as well as to new engines, and that until the problem is cured there should be patrols amounted at every airport to chase away birds before any plane fitted with CF-6 engines takes off. (All DC-10s except those operated by Northwest Airlines have CF-6 engines. At present the only other aircraft which have it are thirteen models of the European A-300B Airbus, two 747s belonging to KLM, and three 747s owned by the U.S. Air Force.)

As a routine courtesy General Electric was warned five days in advance of what the NTSB's report would say. The company's first reaction was an attempt to get publication delayed. When that failed GE issued its angry rebuttal of the charges. The company argues that CF-6 engines have operated for a total of four million hours and that there have only been three incidents, including the one at Kennedy, where bird-strike caused more than superficial damage.

In May the FAA agreed with the NTSB that modifications were necessary and said that it intended to make them mandatory. However, it has given General Electric until June 1, 1977 to modify these CF-6 engines already in service: the chance of there being another "catastrophic" failure before then is apparently thought by the FAA and General Electric to be "statistically remote."

Meanwhile the airbus race goes on, although it might now be more accurately described as the jumbo race. Lockheed has a long-range TriStar, and Boeing a smaller version of the 747 with the result that all three American manufacturers now compete for any and every wide-bodied order. None of them is doing very well.

In 1975 Lockheed failed to sell a single TriStar, and the production rate has been cut back from thirty-six to nine planes a year. Although the company made money overall in 1975, the TriStar program lost more than

---

* ONA is an American charter airline. Six weeks after the Kennedy accident, ONA's remaining DC-10 crashed on landing at Istanbul. Nobody was hurt but the plane was destroyed.

$93 million. In total 154 TriStars have been sold since 1969. Lockheed needs another 150 orders to recoup the original development and tooling costs, and with the ramifications of the bribery scandal still reverberating around the world, the prospects look decidedly grim.

McDonnell Douglas has, to date, collected 230 firm orders for the DC-10 but needs to sell another 170 to retrieve its original investment. It won fourteen new orders in 1975 but most of those were conditional and depend on airline business picking up. The production rate at Long Beach has also been cut back substantially.

In terms of numbers, Boeing is still ahead in the race with 301 firm orders for 747s. Production is running at the rate of thirty planes a year.

The Convair division of General Dynamics still holds the main sub-contract for the DC-10. Toward the end of 1974, the terms were renegotiated, in General Dynamics' favor, and the company is now beyond its break-even point in the program.

There have, on the face of it, been some important changes at the FAA but whether they are significant, or largely cosmetic, remains to be seen. Perhaps the most hopeful development was the appointment by President Ford of an impressively qualified Administrator. John McLucas, who took over the job in November 1975, is an engineer with a master's degree in science. He was formerly Secretary of the Air Force and, before that, chief executive officer of the Mitre Corporation, M.I.T.'s think-tank.

The NTSB was finally made an entirely independent agency by Congress in April 1975, but it continues to be run by a board of five political appointees. John Reed, the chairman, was not renominated when his term in office expired in January 1976 and he was replaced, at President Ford's request, by Webster Todd. Todd has owned several small aeronautical engineering companies. He worked at the Nixon White House for two years and, in 1972, for CREEP.

As for some of the other people in our story:

    —"Mister Mac" remains chairman of McDonnell Douglas but devotes an increasing amount of his time to the affairs of the United Nations;

    —Sanford McDonnell is still president of the corporation;

    —Donald Douglas Sr. lives in retirement near Malibu in Southern California. He is honorary chairman of Douglas Aircraft Company but hasn't attended a board meeting since the merger;

    —Donald Douglas Jr. has also retired and lives about one mile from his father. For a "modest fee" he attends McDonnell Douglas board meetings four times a year;

    —Jackson McGowen lives in Solvang, California, and grows grapes;

    —John Brizendine remains president of the Douglas division and is a member of the McDonnell Douglas board;

    —David Lewis continues his spectacular career at General Dynamics;

    —Dan Applegate is still a senior executive with the Convair division;

    —Arvin Basnight and Richard Sliff both retired from the Western Region of the FAA in December 1974 and now act as occasional consultants to airplane manufacturers;

—Richard Spears lost his job as general manager of the NTSB in March 1976 and became, as before, consultant to the chairman;

—Bryce McCormick is still flying DC-10s for American Airlines;

—John Shaffer owns a moving company in Washington, D.C. He is also a director of the Beech Aircraft Corporation and a consultant to the chairman of TRW. Despite the unanimity of judgments by the Congress, the French Commission of Inquiry, the NTSB, and the FAA itself on the role of the FAA under his leadership, Shaffer remains convinced that his handling of the DC-10 affair was "unimpeachable" and that the attacks on his conduct have been grossly unfair. He says we would have sued for libel but for some advice he was given by a former attorney general: "When you step in the shit just stand still. Don't track it around."

Authors' Note

WORK ON THIS book began a few hours after the crash of Ship 29 on March 3, 1974, and continued more or less nonstop for two years. With colleagues from *The Sunday Times* (London) we spent days at the site of the disaster witnessing the salvage of the debris and we went to the mass funeral of many of those who died. In the following weeks we spoke to those who witnessed the crash, those who were bereaved by it, those who cleared up after it, and those who were charged with finding its cause. We interviewed at length Mohammed Mahmoudi, who closed Ship 29's rear cargo door at Orly, and one of us traveled on the flight deck of a surviving Turkish Airlines DC-10 to better understand the complexities of operating this huge aircraft. We also made a point of examining the locking mechanism of an actual DC-10 cargo door.

In the United States we spent months talking to aviation executives, aircraft engineers and designers, pilots and mechanics, regulators, lawyers and politicians, bankers and Wall Street brokers. We also followed, as closely as possible, the progress of the litigation in Los Angeles. (Indeed, in a minor way, we became involved in that litigation. We wanted to attend the pretrial hearings where witnesses from McDonnell Douglas and General Dynamics were being questioned. After the defense lawyers had refused us access, we sought an order from the U.S. District Court in Los Angeles on the grounds that pretrial proceedings are—or, under the U.S. Constitution, should be— a matter of public record. Our application was opposed by the attorneys for McDonnell Douglas and, after listening to the arguments for three days, Judge Peirson Hall rejected our claim. Subsequently, at McDonnell Douglas's request, Judge Hall issued a "gag" order preventing lawyers and witnesses from discussing the case. As things turned out this prohibition did not seriously hamper our inquiries, although we should state that, to our knowledge, Judge Hall's gag order has not been actively breached.)

Throughout our investigation we drew heavily on the expertise and patience of a great many people who work, directly or indirectly, for the aviation industry. We would like to give credit to those who helped but some would find themselves compromised and others were given specific promise of confidence. It therefore seems best to make do with an expression of thanks which is no less sincere because it is generalized.

We can, however, safely acknowledge our debt to Boeing and Lockheed who allowed us great freedom to inquire into the histories of their companies

285

and their aircraft, and we are grateful for similar reasons to Rolls-Royce.

We must also express our gratitude to McDonnell Douglas. While the company consistently refused to talk about the Paris tragedy, the design of the cargo door, the history of Ship 29, or its operation by Turkish Airlines, we were able to spend four days at Long Beach and St. Louis, in August 1975, questioning executives about other aspects of the DC-10's history. For allowing those interviews, and for their courtesy, we are grateful to Sanford McDonnell the corporate president, and to John Brizendine, president of the Douglas division.

Surprisingly, there is a great scarcity of useful literature about the history of the aviation industry, but we did obtain many leads and ideas from the columns of *Aviation Week* and *Business Week* in the United States and *Flight International* in Britain. We also borrowed ideas from *Fortune* magazine articles on the Douglas takeover and the Rolls-Royce–Lockheed bankruptcy crisis. The NTSB, the FAA, Britain's Civil Aviation Authority, ICAO, the U.S. Civil Aviation Bureau, and IATA all provided valuable information.

Of the other people who helped us, Captain Jacques Lannier and the men of the Senlis Gendarmerie deserve special mention. The professional interests of journalists and policemen often conflict, particularly at the site of major disaster, but Lannier showed that they can be reconciled. His men allowed access but not intrusion, and they gave answers when they had them. In return the journalists avoided, by and large, hampering either the salvage work or the investigation. That cooperation made what was an unpleasant enough job a little easier on everyone.

There are also professional colleagues to whom our thanks are due. Peter Williams of Thames Television, London, conducted three important interviews: with John Shaffer, with Mr. and Mrs. Collen whose daughter Winifred died in the Paris crash, and with Mr. and Mrs. Kaminsky who were on board the American Airlines DC-10 during the Windsor emergency. Antony Terry, Paris correspondent for *The Sunday Times* (London) provided an enormous amount of assistance, as did Metin Munir in Ankara. Patrick Forman, the pilot among us, provided expert help and Elaine Handler spent nine months conducting the survey into airline safety. Our London assistant Christine Walker did a thousand jobs and kept us sane.

In the United States we owe thanks to Nicholas Fraser and Rene Schapiro, who made inquiries for us, and especially to the staff of the New York office of *The Times* (London) newspapers: our chief U.S. researchers Laurie Zimmerman, Georgetta Moliterno, and Diane Condon. Joe Petta remains the best teleprinter operator in the world and Robert Ducas the most helpful manager in the newspaper business. Ducas has acted as midwife to five previous *Sunday Times* (London) books: he says that the birth of this one was the most troublesome but he's made that claim four times before. We wish him luck with his next confinement. Also in New York we owe thanks to our editor Roger Jellinek.

Finally, and most of all, we have to thank Harold Evans, editor of *The*

Sunday Times (London), whose generosity and courage made this book possible.

Paul Eddy
Elaine Potter
Bruce Page

June 1976

Aircraft Engines

IN EXPLAINING HOW aircraft engines work it is necessary first to destroy the popular misconception that propulsion results from the thrust of engine exhaust pushing against the atmosphere. In fact, whatever the nature of the plane, powered flight is a practical demonstration of Sir Isaac Newton's third law of motion: "for every force acting on a body there is an opposite and equal reaction." In aircraft propulsion the "body" is air and the job of an aircraft engine is to accelerate columns of it backwards. The inevitable reaction is the creation of equal forces which thrust the engine—and the airframe attached to it—in the opposite direction.

One of the most familiar examples of the reaction principle is the garden water sprinkler. Here the "body" is water which is accelerated by forcing it through nozzles. As a consequence the sprinkler mechanism rotates in the opposite direction. (Fig. 1)

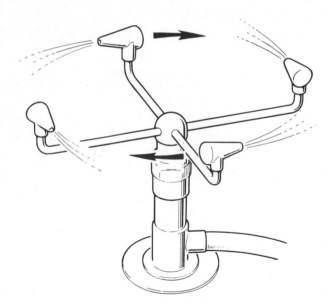

(Fig. 1)

Another simple illustration of the principle is provided by the toy balloon which, when the air or gas escapes through the nozzle, rushes away in the opposite direction. And similarly a rocket motor works by expelling gas at extremely high velocity through an exhaust nozzle. Basically, a rocket is a chamber in which chemical fuels are burned causing them to expand very rapidly. If the chamber was sealed all of the pressures of the expanding gases would cancel each other out and there would be no thrust (short of an explosion). (Fig. 2)

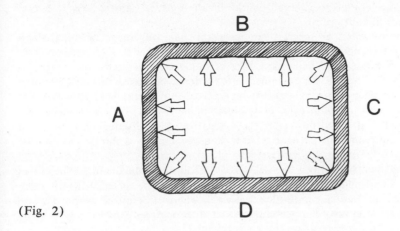

(Fig. 2)

But because one wall of the chamber is, in effect, opened there is no pressure to counter the forces on surface A and the result is net thrust. (Fig. 3)

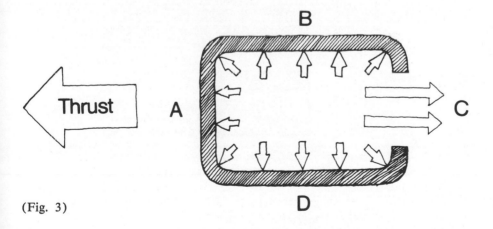

(Fig. 3)

The disadvantage of a rocket is that the "body" which provides the reaction, and therefore the thrust, is gases which are produced exclusively by burning chemical fuels. All of this fuel has to be carried inside the motor which means the rocket is very large in proportion to the payload it can carry. Aircraft engines, on the other hand, utilize as the "body" air which is readily available in the atmosphere. (For this reason, of course, aircraft engines cannot work at all outside the atmosphere, and it is why spaceships are driven by rocket motors.)

Although they appear very different, piston engines with propellers and all jets employ exactly the same basic principles in that both propel their aircraft by thrusting columns of air backwards. The only essential difference is that propellers give a small acceleration to a large quantity of air: most jets give a large acceleration to a small quantity of air.

To turn a propeller, an aircraft engine again takes advantage of the free air supply. Inside the engine, large quantities of air are mixed with hydrocarbon fuels and burned to cause rapid expansion. The resulting energy rotates the propeller-shaft through pistons. As we explained in Chapter 2, at speeds below 350 miles per hour a propeller is a very efficient way of driving an airplane. However, air is not a very dense medium even at sea-level pressure and when a propeller becomes too powerful the airflow over the blades begins to disintegrate. Instead of being thrust backwards in an orderly flow some of the air is flung laterally and thrust is lost.

Jet engines, which produce a rocket-like reaction are much more efficient at high speeds. The simplest example is the ramjet, in which the speed of the aircraft's passage forces air into a combustion chamber where it is mixed with fuel and violently heated. The resulting gases are expelled through an exhaust nozzle. (Fig. 4)

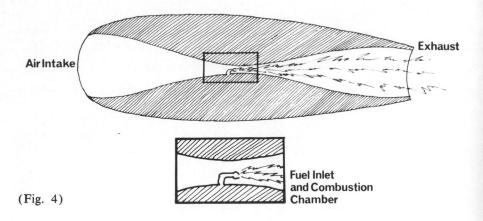

(Fig. 4)

The reaction creates pressures on the backward-facing walls of the combustion chamber which are much greater than the pressures of air on the intake walls. The result is a powerful reaction drive. (Fig. 5)

The obvious problem with the ramjet is that it cannot start an airplane from rest since the intake of air depends exclusively on the plane's forward speed. Some other prime mover (or a separate launch device) is therefore needed to accelerate the airplane until its airflow becomes self-sustaining. Most conventional jet engines use a gas turbine.

The turbine is connected through a shaft to a compressor which is a kind of multi-bladed propeller. To start the engine from rest, auxiliary power is applied to the turbine which rotates the compressor at very high speed. The blades of the

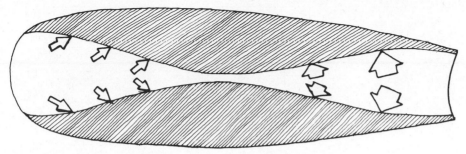

(Fig. 5)

compressor force air into combustion chambers where it is mixed with fuel and ignited. (Fig. 6, p. 292)

On its passage through the engine the air accelerates to a velocity of up to 2,000 feet per second, or about 1,400 miles per hour. Some of this enormous energy is given up to drive the turbine—and thus the compressor, to keep the whole process going—and the rest is expelled through the exhaust nozzle, providing a reaction drive. The amount of energy that is consumed driving the turbine as opposed to providing thrust depends on the forward speed of the aircraft: the faster the plane is traveling, the greater the ram effect which automatically reduces the amount of work the compressor has to do. The problem with pure turbo-jets is that they do not start to become efficient as propulsion units until the aircraft attains a speed of about 450 miles per hour. At speeds lower than that, most of the energy produced by the engine is consumed in driving the turbine—60 percent at 350 miles per hour, 70 percent at 250 miles per hour, and so on. Only at speeds well above those attained by commercial airliners do pure jets become really efficient. For example, it is not until the forward airspeed has reached 800 miles an hour that 75 percent of the energy produced by a pure turbo-jet engine is available to provide thrust.

Obviously then at medium speeds (between 350 and 450 miles per hour) neither piston/propeller engines nor pure jets work very well. One solution which engine designers produced in the late 1950's was the turbo-prop which combines some of the qualities of the other two engines.

The arrangement is exactly the same as in a pure jet except that the turbine drives a propeller as well as a compressor. The propeller provides propulsive energy by thrusting a column of air backwards. The compressor then forces some of the air into combustion chambers where it is mixed with fuel and ignited. After sufficient energy has been extracted for the turbine's needs the surplus is expelled to provide a reaction drive that adds to the thrust of the propeller. This is an elegant and very efficient power unit for speeds of up to 450 miles per hour. But after that the propulsive efficiency of the turbo-prop declines rapidly as the propeller provides less and less thrust. (Fig. 7, p. 293)

The overwhelming advantage of the by-pass engine—which made the present generation of jumbo jets possible—is that it is efficient at all speeds above two hundred miles per hour. The by-pass engine has a propeller—in the form of a multi-bladed fan—but it is encased inside a cowling which ensures that the air flow remains orderly even at very high speeds. Having passed through the blades of the fan, most of the air by-passes the core of the engine which means that none

292

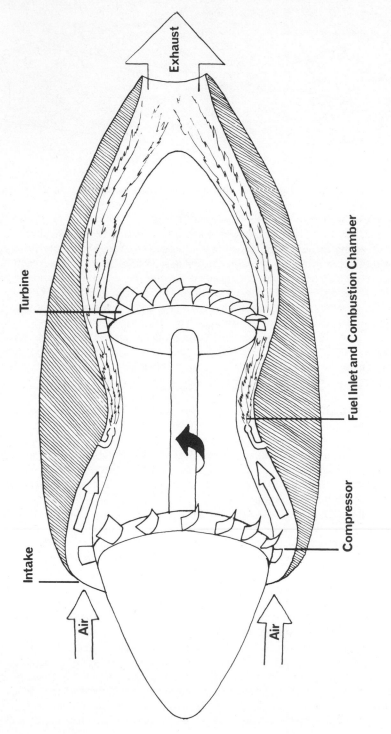

Exhaust

Turbine

Fuel Inlet and Combustion Chamber

Compressor

Intake

Air

Air

(Fig. 6)

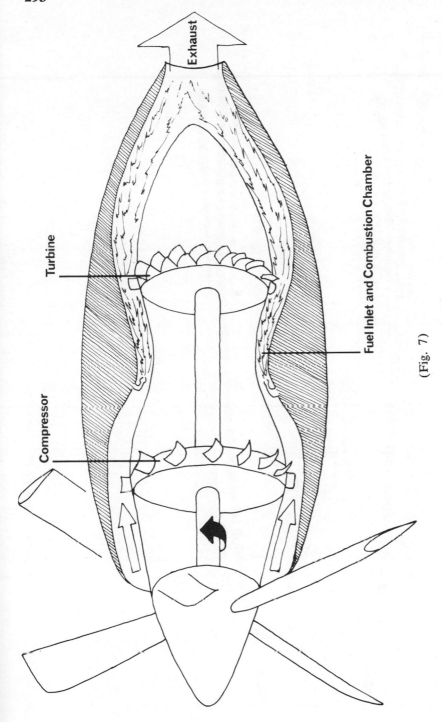

Exhaust

Turbine

Compressor

Fuel Inlet and Combustion Chamber

(Fig. 7)

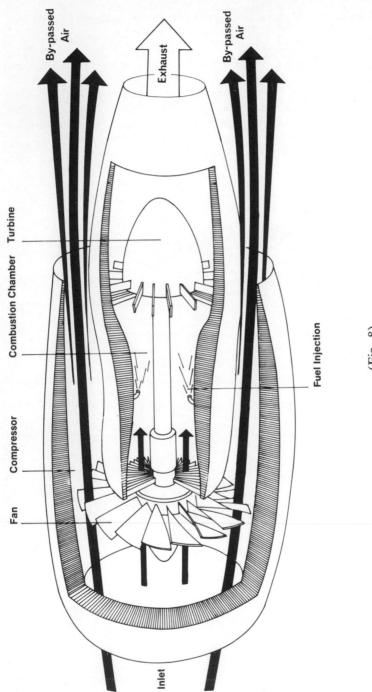

By-passed Air

Exhaust

By-passed Air

Turbine

Combustion Chamber

Compressor

Fan

Inlet

Fuel Injection

(Fig. 8)

of its energy is "wasted." The remainder is passed through compressors, as in a conventional jet, to power the turbine which drives the compressors and the fan. The surplus energy is expelled through the exhaust, providing a reaction drive to add to the thrust of the fan. (Fig. 8)

The ratio of air that by-passes the engine depends on the diameter of the fan. Early by-pass engines like the Pratt and Whitney JT-7D which powers most models of the Boeing 707 and the DC-8 have ratios of little more than 1–1. But jumbo jet engines like the Rolls-Royce RB-211 have fans of over ten feet in diameter and provide a by-pass ratio of around 4.7. That means they can deliver three times the thrust of pure jets.

Cargo-Door Latching Systems

IN A PRESSURIZED airline a cargo door that is not properly locked is as potentially dangerous as an unexploded bomb. The design of the latching system that holds the door shut is therefore crucial: it must be foolproof in operation and as fail-safe in flight as technology and ingenuity will allow. But, because a cargo door is likely to be opened and closed several times a day under all sorts of conditions and by different operators, the latching system must also be simple to operate and easy to maintain.

In Chapter 8 we described the C-latch system which Boeing originally designed for the KC-135 tanker and which, in refined versions, has been used in subsequent Boeing aircraft up to and including the 747. The drawings below show —in purely diagramatic terms—the basic principles of the system.

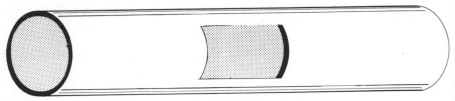

(Fig. 1)

A C-latch is a hard steel tube which for about one inch of its length has had half its barrel cut away. Depending on the size of the door, four or more of these latches are fitted to the bottom edge. (Fig. 1)

When the door has been closed down against its seal, each C-latch fits up against a round steel bar—its latch spool—attached to the door sill in the hull proper. (Fig. 2)

Then an electric actuator revolves all the C-latches together through 180 degrees so that the belly of the C in each case slides around to the other side of the spool (Fig. 3). As an additional safeguard Boeing devised a locking system. Above the C-latches they mounted a rod inside of the door that can slide fore and aft, paralel to the doorsill, latches, and spools. The rod carries a locking pin for each C-latch. If the C-latches are correctly in place the locking pins slide across behind them. The build-up of pressure inside the aircraft after take off forces the latches even more firmly against the spools and the door cannot open in flight.

The locking pins are driven into place manually, by the door operator pulling down on a handle. When, *and only if*, the locking pins complete their journey, the final journey, the final movement of the handle closes a vent-door in front of the

C-LATCH: OPEN, DOOR CLOSING

(Fig. 2)

C-LATCH LOCKED DOOR SHUT

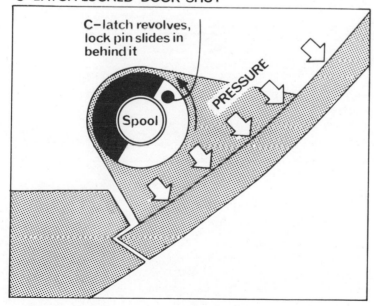

(Fig. 3)

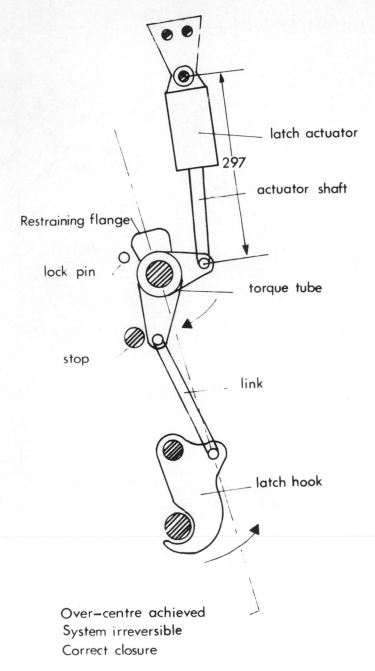

latch actuator

297

actuator shaft

Restraining flange

lock pin

torque tube

stop

link

latch hook

Over–centre achieved
System irreversible
Correct closure

(Fig. 4)

operator's face—confirming that all is well. If, however, the C-latches have failed to revolve through their half circle the pins will jam up against them and the vent door *cannot* close.

The locking system devised by McDonnell Douglas for the DC-10 seems very similar in basic principle, but the original design contained fundamental discrepancies. (Fig. 4)

The DC-10 latch (four or more to a door) shaped at the bottom something like the talon of a bird is attached to a torque tube that is revolved by an electric actuator. Figure 4 (taken from the French Commission of Inquiry report about the Paris crash) shows how the latches work when they have been correctly closed. As the torque revolves, the latch hook is driven down over its spool while the top of the latch swings through an arc of about 90 degrees until it comes to rest against a metal stop. Provided the top part of the latch passes beyond the center point of its arc (shown by the broken line) the force created by pressurization will be transmitted to the door structure.

Experience proved, however, that it was possible for the latch hooks to grip the spools to give the impression that the door was secure—but without the top part of the latches actually going over-center. In that case the door would be held shut only until the pressure inside the plane had built up to dangerous levels. To eliminate the hazard the DC-10 was given a locking-pin system similar to the Boeing design but with one vital—and dangerous difference. (Fig. 5, p. 300)

After the door had been lowered into place the electric actuator (1) inside the door revolved the latch bar (2) and drove the latches over center (3). The operator was then required to pull down a metal handle (4) which closed the vent-door (5) by revolving a torque tube and *simultaneously* drove the locking pins behind metal lugs on each latch. That was achieved by lifting a vertical rod (6) which, through a bell crank (7) moved the locking-pin tube (8) from right to left. In theory if the latches were not fully home the pins would jam up against the metal lugs and the resistance would be transmitted back to the handle: in theory also the vent-door would not close. The Windsor incident, however, proved that in practice if the locking pins jammed, the handle could be forced down and the vent-door could close if the torque tube buckled. McDonnell Douglas's first major modification to the DC-10 door after Windsor was to design a support plate (9) to prevent the torque tube from distorting. This plate was not fitted to Ship 29.

Another modification devised by McDonnell Douglas called for the linkages in the system to be rerigged so that the locking pins would be driven further over the top of the latches. This would have the effect of increasing the amount of resistance should the latches not be fully home.

In the original design, the locking pins were required, in the closed position, to overlap the lugs on top of the latches by ⅟₃₂ of an inch (A). The modification after Windsor called for the amount of travel to be increased so that the locking pins would overlap by ¼ of an inch (B). On Ship 29's rear cargo door somebody misrigged the pins so that even in the fully closed position they barely covered the lugs at all (C). (Fig. 6, p. 301)

With the support plate fitted and the rigging correctly done the amount of force required to overcome the system would have been over 400 pounds—an impossible feat for even an exceptionally strong adult. Because Ship 29 did not have the support plate, and because the locking pins had been misrigged, the amount of

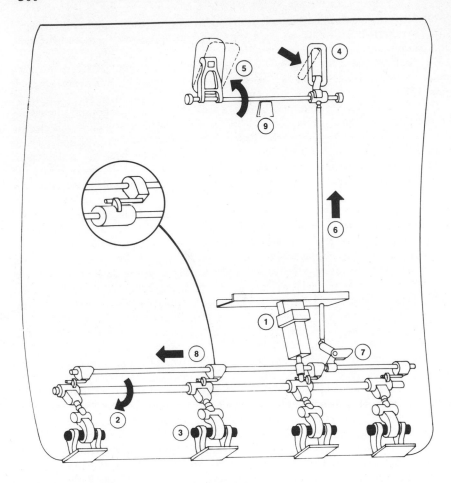

(Fig. 5)

force required to overcome the safety system was only thirteen pounds. Figures 7 and 8, pp. 302-303 (also taken from the official French report) show what caused the Paris disaster. Although the lock pins jammed up against the lugs—or flanges—on top of the latches, Mahmoudi was still able to close the vent door without using any force.

Because the top part of the latches had failed to go over-center all of the forces created by pressurization after take off were transmitted to the electric actuator. At 11,500 feet the two bolts holding the actuator to the door sheared, and the door blew open.

After the Paris disaster the FAA issued an airworthiness directive making mandatory the fitting of a "closed-loop" system to all DC-10 cargo doors. This employs the Boeing principle that the vent-door should only be able to close as a *consequence* of the locking pins completing their journey.

As before, pulling down on the handle (1) lifts the vertical rod (2) which,

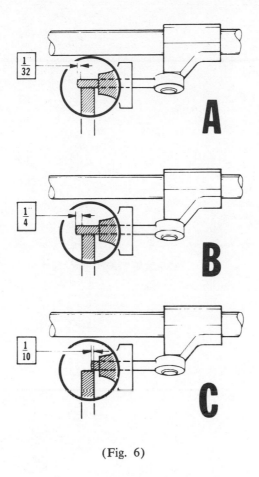

(Fig. 6)

through a bellcrank (3), drives the locking-pin bar (4) from right to left. There is, however, no longer a torque tube connecting the handle to the vent-door. If the latches are home then the pins (5) will move freely across. A second bellcrank (6) transmits the energy to a second vertical rod (7), which revolves the torque tube (8) and closes the vent-door (9). If the pins jam, because the latches are not closed, the vent-door cannot move no matter how much force is applied to the handle. (Fig. 9, p. 304)

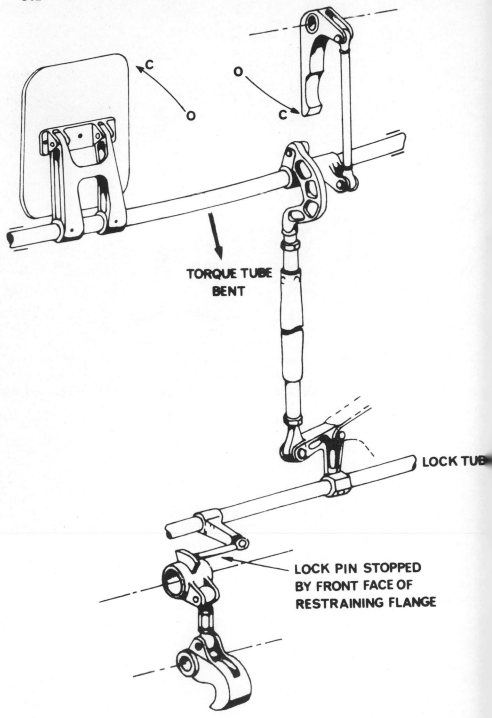

C

O

O

C

TORQUE TUBE
BENT

LOCK TUB

LOCK PIN STOPPED
BY FRONT FACE OF
RESTRAINING FLANGE

(Fig. 7)

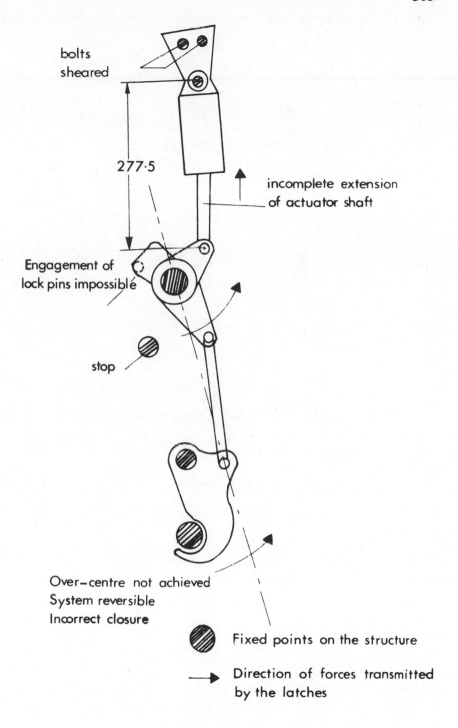

bolts
sheared

277·5

incomplete extension
of actuator shaft

Engagement of
lock pins impossible

stop

Over–centre not achieved
System reversible
Incorrect closure

Fixed points on the structure

Direction of forces transmitted
by the latches

(Fig. 8)

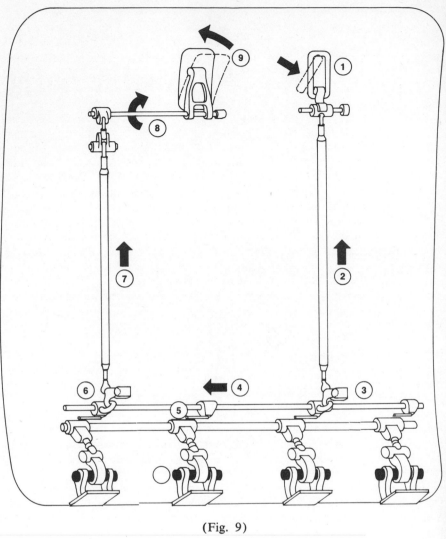

(Fig. 9)

Appendix C
Airline Safety

IN THE LAST twenty-five years over thirty thousand people—passengers, crew, and bystanders—have been killed in aircraft accidents throughout the world. About two-thirds of the victims were traveling on regular passenger services operated by national flag carriers or recognized airlines which can fairly be described as household names.

The conclusions about the comparative merits of leading airlines (see Chapter 11) were based on our examination of each carrier's record during the twenty-five years between 1950 and 1974. We chose that time span because most of the major airlines flew continuously throughout those years and because it is long enough for the freak bad periods to be balanced by the good. The last table of this appendix shows which airlines have, in our opinion, done a consistently good job—and those which haven't. However, first it is necessary to detail how we reached those conclusions. We should also enter some caveats.

First, and most important, we have *not* taken into account the causes of accidents. In our analysis each accident serves to worsen the overall record of the airline involved—even if the airline was in no way to blame for the crash. Obviously, this may mean that some of the airlines in our survey have been unfairly discriminated against, although we believe there have been few cases where it could be argued that the circumstances which led to a crash were *totally* beyond the control of the airline. Bombs, bad weather, inferior airports, mountainous terrain, careless air traffic controllers, and defective airplanes are all constant potential hazards which an airline can and should guard against: some airlines have done so immaculately.

The list that follows is a chronology of the accidents we have been able to trace. Our main source was the British Civil Aviation Authority's *World Airline Accident Summary*. It is undoubtedly the most comprehensive published record of its kind, but it is not without errors and omissions and we have therefore attempted to verify the data it provided.

Where possible we have cross-checked the CAA's records with the investigation report on each individual accident. (Generally speaking, aircraft accidents are investigated by a government body, or a government-appointed team, in the country where the accident occurred and, of course, the resulting reports can differ considerably in the quality and accuracy of the information they provide. Because of that we have only relied on reports produced by expert bodies—such as the NTSB's Bureau of Aviation Safety in the United States—or reports circulated by the International Civil Aviation Organization on the grounds that the information was probably reliable.)

In some cases, where facts were sparse, we consulted newspaper accounts. But it has to be said that because of the immediacy with which such accounts are written, the information has to be treated with caution.

We also consulted reference books, but, again, with caution.

Finally, we wrote to seventy-four airlines inviting them to verify and comment on the information we had obtained. Fifty-six airlines cooperated in compiling the survey, sixteen failed to reply and two—KLM Royal Dutch Airlines and the USSR carrier Aeroflot—replied by refusal to give any information.

In general we have omitted accidents involving air taxi services unless the company was operating a scheduled service. We have also omitted accidents that overtook freight flights, training flights, demonstration flights, and ferry and delivery flights unless we have been able to establish that passengers were on board or that bystanders were killed or that the aircraft collided with another airplane, the occupants of which were killed.

In all of the tables used in this appendix, the following short forms have been used.

AIRLINES:

Aer Lingus	Irish Airline	CAT [A]	Compagnie Aerien de Transport
Aero.	Aerolinea(s) (Spanish for airline)	CAT [B]	Civil Air Transport
Aeroflot	USSR airline	CBF	Corporacion Boliviana de Formento
AIA	Associacion Internacional de Aviacion	CDA	Compania Dominicana de Aviacion
AIDA	Associacion Interamericana de Aviacion	COHATA	Compagnie Haitienne de Transport
ALA	Aerotransporte Litoral Argentino	C P Air	Canadian Pacific Airlines
ALIA	Royal Jordanian Airline Corp.	Cruzeiro	Brazilian Airline
Alitalia	Italian Airline	CSA	Czechoslovak Airline
ANA	Australian National Airline	CTC	Compania de Transportes Caribbean
Ansett	Airlines of Australia	DTA	Departamento de Transporte Aereo
AREA	Aerolineas Ecuatorianas S.A.	EAAC	East African Airways Corp.
ATESA	Aero Taxis Ecuatorianos S.A.	El Al	Israeli Airline
ATSA	Aero Transportes S.A. (Mex.)	EPA	Eastern Provincial Airways
AVENSA	Aerovias Venzolanas S.A.	FEAT	Far Eastern Air Transport Corp.
AVIACO	Aviacion y Comercio S.A.	FKKK	Fujita Koko Kabushiki Kaisa
AVIANCA	Aerovias Nacionales de Colombia S.A.	FOA	Filipinas Orient Airways Inc.
AVIATECA	Empresa Guatemalteca de Aviacion	Garuda	Indonesian Airline
AVISPA	Aerovias del Pacifico	IAC	Indian Airlines
BCAL	British Caledonian Airlines	IAT	International Air Transport
BCPA	British Caribbean Pacific Airways	Iberia	Spanish Airline
BEA	British Airways European Division	JAL	Japan Air Lines
BIAS	Belgian International Air Services	JAT	Yugoslav Airlines
BOAC	British Airways Overseas Division	KLM	Royal Dutch Airlines
BUCI	British United (Channel Islands) Ltd. (now British Island Airways, Ltd.)	LAB	Lloyd Aereo Boliviano S.A.
		LACSA	Lineas Aereas Costarricenses S.A.
BWIA	British West Indian Airlines	LADE	Lineas Aereas del Estado
CAAC	Central African Airways Corp.	LADECO	Linea Aerea del Cobre S.A.
		LAN	Chilean Airline
		LANICA	Lineas Aereas de Nicaragua S.A.

LANSA	Linea Aerea Nacional S.A.
LAP	Linea Aerea Paulista
LAU	Lineas Aereas Unidas
LAV	Linea Aeropostal Venezolana
LOT	Polish Airline
Lufthansa	German Airline
MALEV	Hungarian Airline
MEA	Middle East Airlines Air Liban
NAB	Navegacao Aerea Brasil
NAC	New Zealand National Airways Corp.
PAB	Panair do Brasil
PAL	Philippine Air Lines
Pan Am	Pan American Airways
PIA	Pakistan International Airlines
PRINAIR	Puerto Rico International Airlines Inc.
RAI	Reseau Aerien Interinsulaire
SAA	South African Airways
SABENA	Belgian Airline
SAC	Servicios Aereos Cochabamba
SAESA	Servicios Aereos Especiales S.A.
SAHSA	Servicio Aereo Hondurena S.A.
SAM	Sociedad Aereonautica de Medellin Consolidada S.A.
SAP	Société Alps
SAS	Scandinavian Airlines System
SATCO	Servicio Aereo de Transporte Cooperativo
SATENA	Servicio Aereo Territorios Nacionales
Saudia	Saudi Arabian Airlines
SAVCO	Servicio de Aviacion Consolidada
SAVG	Sociedade Anonima Viacao Gaucha

SEA	Empresa Servicios Especiales Aereos
SFH	Société France Hydro
SIR	Société Indochinese de Ravilaillment
SITA	Société Internationale de Telecommunications Aeronautiques
STA	Société de Travail Aerien
STAAP	Société de Transport
STAEO	Société de Travail Aeriene Europeu
STAT	Société Trans Atlantique Arienne
TAA	Trans-Australia Airline
TABSA	Transportes Aereos Benianos S.A.
TABSO	Bulgarian Airline
TAC	Transportes Aereos Colombianos
TACA	Transportes Aereos Centro Americanos
TAG	TAG Air
TAI	Transport Aeriennes Intercontinenteaux
TAM	Transportes Aereos Militares (same name in Bolivia and Peru)
TAN	Transportes Aereos Nacionales
TAO [A]	Transportes Aereos Orientales
TAO [B]	Aerolineas TAO
TAP	Transportes Aereos Portugueses
TAROM	Rumanian Airline
TASSA	Transporte Aereo del Sur S.A.
THY	Turkish Airline
TOA	Domestic Airlines Co. Ltd.
TWA	Trans World Airlines
UAB	Union of Burma
UAT	Union Aerien de Transports
UTA	Union de Transports Aeriens

VARIG — Viacao Aerea Rio-Grandense S.A.
VASP — Viacao Aerea Sao Paulo S.A.
VIASA — Venezolana Internacional de Aviacion S.A.

COUNTRIES:

Afg.	Afghanistan		GB [1]	Great Britain, colonies and protectorates
Alg.	Algeria		Gre.	Greece
Arg.	Argentina		Guat.	Guatemala
Aus.	Austria		HK	Hong Kong
Austrl.	Australia		Hol.	Holland
Belg.	Belgium		Hond.	Honduras
Bol.	Bolivia		Hung.	Hungary
Braz.	Brazil		Ice.	Iceland
Bul.	Bulgaria		Ind.	India
Cam.	Cambodia		Indon.	Indonesia
Camer.	Cameroun		It.	Italy
Can.	Canada		Jam.	Jamaica
Ch.	Chile		Jap.	Japan
Colmb.	Colombia		Jord.	Jordan
Czech.	Czechoslovakia		Leb.	Lebanon
Den.	Denmark		Mex.	Mexico
DR	Dominican Republic		MR	Malagasy Republic
Ecu.	Ecuador		NG	New Guinea
E. Ger.	East Germany (German Democratic Republic)		Nic.	Nicaragua
Egy.	Egypt		Nig.	Nigeria
Ethio.	Ethiopia		N. Ire., GB	Northern Ireland
Fin.	Finland		Nor.	Norway
Fr.	France		N. VN	North Vietnam (Democratic Republic of Vietnam)
GB	Great Britain		NZ	New Zealand
			Pak.	Pakistan
			Phil.	Philippines
			Pol.	Poland
			Port.	Portugal
			Rhod.	Rhodesia
			Rum.	Rumania
			SA	South Africa

Saud.	Saudi Arabia
S. Ire.	Southern Ireland (Republic of Ireland)
S. Kor.	South Korea (Republic of Korea)
Sp.	Spain
S. VN	South Vietnam (Republic of Vietnam)
Swe.	Sweden
Swit.	Switzerland
Tai.	Taiwan (Nationalist China)
Thai.	Thailand
Turk.	Turkey
US	United States
USSR	Union of Soviet Socialist Republics
Venz.	Venezuela
W. Ger.	West Germany (Federal Republic of Germany)
WI	West Indies
Yug.	Yugoslavia

AIRPLANES:

AC	Aerocommander Division, North American Rockwell Corp. (US)
AN	Antonov (USSR)
Avro	Avro Whitworth Division, Hawker Siddeley Aviation Ltd. (GB)
B	The Boeing Company (US)
BAC	British Aircraft Corporation Ltd. (GB)
BE	Beech Aircraft Corporation (US)
BR	Bristol Aircraft Ltd. (GB)
C	Curtiss-Wright Corporation (US)
CL	Canadair Ltd. (Can.)
Con.	Consolidated Aircraft Co. (US)
CV	Convair Division, General Dynamics Corporation (US)
DC	Douglas Aircraft Co., McDonnell Douglas Corporation (US)
DH	De Havilland Aircraft Co. Ltd. (GB)
DH(A)	De Havilland Aircraft PTY. Ltd. (Austrl.)
DH(C)	De Havilland Aircraft of Canada Ltd. (Can.)
EMB	EMBRAER—Empresa Brasileira de Aeronautica S.A. (Braz.)
F	VFW-Fokker
FH	Fairchild Industries Inc. (US) (F-27 built under license in US)
G	Grumman Corporation (US)
HP	Handley Page Ltd. (GB)
HS	Hawker Siddeley Aviation Ltd. (GB)
IL	Ilyushin (USSR)
JU	Junkers (W. Ger.)
L	Lockheed Aircraft Corp. (US)
Lear	Gates Learjet Corp. (US)
M	Martin Marietta Aerospace (US)
MBB	Messerschmitt-Bolkow-Blohm (W. Ger.)
N	Noorduyn Norseman Aircraft Ltd. (Can.)
Nihon	Nihon Airplane Mfg. Co. Ltd. (Jap.)
Nord	Nord-Aviation (Fr.)
PA	Piper Aircraft Corp. (US)
SA	Sud-Aviation (Fr.)
Saab	Saab-Scandia Aktiebolag (Swe.)
Short	Short Brothers & Harland Ltd. (GB)
TU	Tupolev (USSR)
VC	Vickers Aircraft Co. (GB)
VC(C)	Vickers Aircraft Co. (Can.)
YAK	Yakovlev (USSR)

CHRONOLOGICAL LIST OF AIR DISASTERS FROM 1950 TO 1975

DATE	AIRLINE	AIRCRAFT AND LOCATION	KILLED
1950			
Jan 24	PAL (Phil.)	DC-3, disappeared between Iloilo–Manila, Phil.	4
Jan 24	STAT (Fr.)	DC-3, hit mountain near Tamatave, MR	14
Feb 9	—	DC-3, hit mountain near Cali, Colmb.	4
Feb 27	CSA (Czech.)	DC-3, near Prague, Czech.	5
Mar 7	Northwest (US)	M-202, on landing, Minneapolis, Minnesota, US	13 2 *
Mar 12	Fairflight (GB)	Avro-Tudor, on landing, near Cardiff, Wales, GB	80
Mar 25	THY (Turk.)	DC-3, fire on landing, Ankara, Turk.	16
Apr 16	World Air (GB)	—, hit Mt. Hohgant, Swit.	6
May 2	AVIANCA (Colmb.)	DC-3, near Chimborazo, Ecu.	15
May 24	LANSA (Colmb.)	DC-3, hit volcano, near Pasto, Colmb.	25
May 30	Aero. Brasil (Braz.)	DC-3, near Ilheus, Braz.	13
Jun 5	Westair (US)	C-46, ditched in Atlantic Ocean	28
Jun 9	New Tribes (US)	DC-3, Fonseca, Colmb.	15
Jun 12	Air France (Fr.)	DC-4, on landing, Bahrain, Persian Gulf	46
Jun 14	Air France (Fr.)	DC-4, on landing, Bahrain, Persian Gulf	40
Jun 23	Northwest (US)	DC-4, Benton Harbor, Michigan, US	58
Jun 24	Northeast (GB)	—, on takeoff, West Riding, GB	4
Jun 26	ANA (Austrl.)	DC-4, on takeoff, York, West Austrl.	28
Jun 29	Argonaut (US)	C-46, door opened, passenger fell out, Virginia, US	1
Jul 9	Aigle Azur (Fr.)	DC-3, on takeoff, Casablanca, Morocco	22
Jul 17	Indian National Air (Ind.)	DC-3, southeast Patharkot, Nepal	22
Jul 18	DTA (Angola)	DC-3, near Bocoio, Angola	9

DATE	AIRLINE	AIRCRAFT AND LOCATION	KILLED
Jul 24	Air Liban (Leb.)	DC-3, shot down over northern Israel	3
Jul 28	PAB (Braz.)	L-Constellation, on landing, Porto Alegre, Braz.	50
Jul 29	CAT ᴬ (Fr.)	BR-170, exploded over Sahara Desert	26
Jul 30	SAVG (Braz.)	—, Alegrete, Braz.	10
Aug 18	Gander (Can.)	N-Norseman, on take off, Gander Lake, Newfoundland, Can.	2
Aug 21	American (US)	DC-6, fuselage puncture, Eagle, Colorado, US	1
Aug 30	—	N-Norseman, Timagami, Ontario, Can.	5
Aug 31	TWA (US)	L-Constellation, Cairo, Egy.	55
Sep 4	Robinson (US)	DC-3, on take off, Utica, New York, US	16
Sep 14	Iran Air (Iran)	DC-3, on take off, Teheran, Iran	8
Sep 21	JAT (Yug.)	DC-3, on landing, Zagreb, Yug.	10
Oct 13	Air Atlas (Fr.)	DC-3, on takeoff, Casablanca, Morocco	4
Oct 13	Northwest (US)	M-202, Almelund, Minnesota, US	6
Oct 17	BEA (GB)	DC-3, on takeoff, London, GB	28
Oct 31	BEA (GB)	VC-Viking, on landing, London, GB	28
Nov 3	Air India (Ind.)	L-Constellation, hit Mt. Blanc, Fr.	48
Nov 7	Northwest (US)	M-202, on landing, Butte, Montana, US	21
Nov 13	Curtis Reid (Can.)	DC-4, hit mountain, Grenoble, Fr.	52
Nov 23	Faucett (Peru)	DC-3, Cuzco, Peru	9
Dec 1	Iran Air (Iran)	DC-3, near Qom, Iran	8
Dec 8	IAT (Fr.)	DC-4, after takeoff, Bangui, Central African Rep.	46
Dec 8	Air Atlas (Fr.)	DC-3, hit mountain in Pyrenees, Fr.	5
Dec 13	Indian Airlines (Ind.)	DC-3, Katagiri, Ind.	20
Dec 14	VASP (Braz.)	DC-3, on takeoff, hit house, Ribeirao Preto, Braz.	1 3 *
Dec 15	AVENSA (Venz.)	DC-3, on takeoff, near Valera, Venz.	31
Dec 30	Aero. Argentinas (Arg.)	DC-3, near Cobo, Arg.	17

DATE	AIRLINE	AIRCRAFT AND LOCATION	KILLED
1951			
Jan 12	—	DH-Dove, disintegrated over Ixopo, SA	12
Jan 13	Air Carriers (GB [1])	DC-3, hit mountain, Thai.	10
Jan 14	National (US)	DC-4, on landing, Philadelphia, Pennsylvania, US	7
Jan 16	Northwest (US)	M-202, Reardan, Washington State, US	10
Jan 17	Alitalia (It.)	Marchetti-95B, Civitavecchia, It.	14
Jan 20	TAM (Peru)	DC-3, Huilyo, Peru	16
Jan 28	Transoceanic (GB [1])	Short-Solent, on takeoff, Marsaxlokk Bay, Malta	1
Jan 29	—	DH(C)-2, Lake Tessier, Saskatchewan, Can.	4
Jan 31	Icelandair (Ice.)	DC-3, on landing, near Reykjavik, Ice.	20
Feb 3	Air France (Fr.)	DC-4, Mt. Cameroun, Nig.	29
Feb 6	—	Avro-Anson, on takeoff, Northeast Territories, Can.	2
Mar 2	Mid-Continent (US)	DC-3, on landing, Sioux City, Iowa, US	16
Mar 11	Pacific Overseas (Thai.)	DC-4, on takeoff, near Mt. Butler Is., HK	26
Mar 22	Cruzeiro (Braz.)	DC-3, on landing, Florianopolis, Braz.	3
Mar 22	LANSA (Colmb.)	DC-3, near Hato Nuevo, Colmb.	30
Mar 26	Aero. Argentinas (Arg.)	DC-3, on takeoff, Rio Grande, Arg.	11 2 *
Apr 6	Southwest (US)	DC-3, hit hill, Santa Barbara, California, US	22
Apr 9	Siamese Airways (Thai.)	DC-3, on landing, Cape D'Aguilar, HK	16
Apr 25	Cubana (Cuba)	DC-4, midair collision, Key West, Florida, US	39 4 *
Apr 28	United (US)	DC-3, on landing, Fort Wayne, Indiana, US	11
May 18	VASP (Braz.)	DC-3, Rancharia, Braz.	6

DATE	AIRLINE	AIRCRAFT AND LOCATION	KILLED
Jun 6	NAB (Braz.)	DC-3, on landing, Rio de Janeiro, Braz.	2
Jun 22	Pan Am (US)	DC-4, Monrovia, Liberia	40
Jun 29	JAT (Yug.)	JU-52, near Zagreb, Yug.	14
Jun 30	United (US)	DC-6, Fort Collins, Colorado, US	50
Jul 12	LAP (Braz.)	DC-3, on landing, Rio do Sul, Braz.	33
Jul 14	Airtaco (Swe.)	L-14H, on takeoff, Stockholm, Swe.	4
Jul 16	Qantas (Austrl.)	DH(A)-3 (Drover), NG	7
Jul 21	C P Air (Can.)	DC-4, disappeared between Vancouver, Can.–Tokyo, Jap.	37
Jul 29	LAB (Bol.)	C-46, on takeoff, Cochabamba, Bol.	7
Aug 24	United (US)	DC-6, on landing, near Oakland, California, US	50
Sep 8	VASP (Braz.)	DC-3, on takeoff, hit house, Sao Paulo, Braz.	13 3 *
Sep 12	SAP (Fr.)	DC-3, west of Balearic Islands, Sp.	39
Sep 17	Real (Braz.)	DC-3, Parati, Braz.	10
Oct 8	ATSA (Mex.)	DC-3, Blanco, Mex.	8
Oct 15	SAA (SA)	DC-3, near Kokstad, SA	17
Oct 17	Queen Charlotte (Can.)	Con.–PBY-5A, hit Mt. Benson, Vancouver, Can.	23
Oct 24	JAT (Yug.)	DC-3, on landing, near Skopje, Yug.	12
Oct 28	—	C-46, on takeoff, Ciudad Flores, Guat.	27
Nov 5	Transocean (US)	M-202, Tucumcari, N. Mexico, US	1
Nov 14	LOT (Pol.)	IL-12, on takeoff, near Lodz, Pol.	16
Nov 19	THY (Turk.)	DC-3, on landing, near Cairo, Egy.	5
Nov 21	Deccan (Ind.)	DC-3, on landing, Calcutta, Ind.	16
Nov 27	LAU (Mex.)	C-39, on takeoff, San Luis, Mex.	13
Dec 13	New Guinea (NG)	DH-Dragon, Papua, NG	3
Dec 16	Miami Airlines (US)	C-46, Elizabeth, New Jersey, US	56
Dec 22	Misrair (Egy.)	SA-Languedoc, on landing, near Teheran, Iran	22

DATE	AIRLINE	AIRCRAFT AND LOCATION	KILLED
Dec 29	Continental Chart. (US)	C-46, Little Valley, New York, US	26
Dec 30	Transocean (US)	C-46, on approach, Fairbanks, Alaska, US	4

1952

DATE	AIRLINE	AIRCRAFT AND LOCATION	KILLED
Jan 1	Air France (Fr.)	JU-52, on takeoff, Andapa, MR	6
Jan 10	Aer Lingus (S. Ire.)	DC-3, near Lake Gwyant, S. Ire.	23
Jan 19	Northwest (US)	DC-4, on landing, Sandspit, British Columbia, Can.	36
Jan 22	American (US)	CV-240, on landing, Elizabeth, New Jersey, US	23 7 *
Feb 4	SABENA (Belg.)	DC-3, near Kikwit, Zaire	15
Feb 11	National (US)	DC-6, on takeoff, Elizabeth, New Jersey, US	29 4 *
Feb 16	Hunting Air (GB)	VC-Viking, hit hill, near Burgio, Sicily, It.	31
Feb 19	STAAP (Fr.)	Con.-Liberator, Yaounde, Camer.	9
Feb 19	Deccan (Ind.)	DC-3, on landing, Nagpur, Ind.	3
Feb 28	PAB (Braz.)	DC-3, on landing, Uberlandia, Braz.	8
Mar 3	Air France (Fr.)	SA-Languedoc, on takeoff, Nice-le-Var, Fr.	38
Mar 22	KLM (Hol.)	DC-6, on landing, near Frankfurt, W. Ger.	42
Mar 22	Maritime (Can.)	DC-3, disappeared, between Brunswick–Labrador, Can.	4
Mar 26	Aeroflot (USSR)	—, on landing, hit military aircraft, Moscow, USSR	70
Mar 29	TACA (El Salvador)	DC-3, hit mountain, near San Felipe, Venz.	12
Mar 30	PAL (Phil.)	DC-3, on takeoff, Baguio, Phil.	10
Apr 9	JAL (Jap.)	M-202, hit Mihara volcano, Jap.	37
Apr 11	Pan Am (US)	DC-4, on takeoff, San Juan, Puerto Rico	52

DATE	AIRLINE	AIRCRAFT AND LOCATION	KILLED
Apr 18	North Continent (US)	C-46, on landing, near Whittier, California, US	29
Apr 29	Pan Am (US)	B-Stratocruiser, Carolina, Braz.	50
Apr 30	Deccan (Ind.)	DC-3, near Safdarjung, Delhi, Ind.	9
May 5	Fred Olsens (Nor.)	DC-3, near Telemark, Nor.	10
May 13	VASP (Braz.)	DC-3, on takeoff, Sao Paulo, Braz.	5
Jun 14	Morton Air (GB)	DH-Consul, ditched in English Channel	6
Jul 10	—	DC-3, ditched in Atlantic Ocean, off Salvador, Braz.	5
Jul 27	Pan Am (US)	B-Stratocruiser, door opened, passenger fell out, Rio de Janeiro, Braz.	1
Aug 12	TAN (Braz.)	DC-3, exploded midair, Palmeiras, Braz.	24
Aug 21	Airwork (GB)	HP-Hermes, near Trapani, Sicily, It.	7
Oct 14	Aero. Brasil (Braz.)	DC-3, on landing, Porto Alegre, Braz.	14
Nov 11	UAT (Fr.)	DC-4, near Lake Chad, Chad	5
Dec 6	TABSO (Bul.)	IL-2, between Sofiya-Varna, Bul.	18
Dec 6	Cubana (Cuba)	DC-4, on takeoff, near Findley Field, Bermuda	37
Dec 21	Syrian Air (Syria)	DC-3, near Damascus, Syria	9
Dec 25	AVENSA (Venz.)	DC-3, on takeoff, near Caracas, Venz.	3
Dec 25	Iran Air (Iran)	DC-3, on landing, Teheran, Iran	27

1953

Jan 5	BEA (GB)	VC-Viking, on landing, Belfast, N. Ire., GB	27
Jan 7	Associated (US)	C-46, hit mountain, near Fish Haven, Idaho, US	40
Jan 7	Flying Tiger (US)	DC-4, on landing, Issaquah, Washington State, US	7

DATE	AIRLINE	AIRCRAFT AND LOCATION	KILLED
Jan 26	Alitalia (It.)	DC-3, near Sinnai, Sardinia, It.	19
Feb 2	Skyways (GB)	Avro-York, disappeared over North Atlantic	39
Feb 7	UAT (Fr.)	DC-4, forest near Bordeaux, Fr.	6
Feb 14	National (US)	DC-6, ditched into Gulf of Mexico	46
Mar 3	C P Air (Can.)	DH-Comet-1, near Karachi, Pak.	11
Mar 13	Orient Airways (Pak.)	CV-240, hit mountain, Tripura Territory, Ind.	16
Mar 17	Aigle Azur (Fr.)	DC-3, on landing, near DaNang, S. VN	8
Mar 20	Transocean (US)	DC-4, on approach, Alvarado, California, US	35
Mar 29	CAAC (Rhod.)	VC-Viking, disintegrated, Tanzania	13
Apr 10	Air France (Fr.)	JU-52, on takeoff, Miandrivazo, MR	4
Apr 10	Caribbean Int (GB[1])	L-Lodestar, on takeoff, Kingston, Jamaica, WI	13
Apr 10	Real (Braz.)	DC-3, near Anchieta Is., Braz.	26
Apr 14	Miami Airlines (US)	DC-3, near Selleck, Washington State, US	7
Apr 16	Aigle Azur (Fr.)	DC-3, near Hanoi, N. VN	30
Apr 20	Western (US)	DC-6, on takeoff, San Francisco, California, US	8
May 2	BOAC (GB)	DH-Comet-1, on takeoff, near Calcutta, Ind.	43
May 9	Indian Airlines (Ind.)	DC-3, on takeoff, Palam, Maharashtra, Ind.	18
May 11	C P Air (Can.)	Con.–PBY-5A, on landing, Prince Rupert, British Columbia, Can.	2
May 17	Delta (US)	DC-3, near Marshall, Texas, US	19
May 24	Meteor Air (US)	DC-3, St. Louis, Missouri, US	6
Jun 15	LACSA (Costa Rica)	DC-3, near San Ramon, Costa Rica	11
Jun 15	LAN (Ch.)	L-Lodestar, near Copiapo, Ch.	7
Jun 16	Aigle Azur (Fr.)	DC-3, Pakse, Laos	25
Jun 17	PAB (Braz.)	L-Constellation, on landing, Sao Paulo, Braz.	17

DATE	AIRLINE	AIRCRAFT AND LOCATION	KILLED
Jul 12	Transocean (US)	DC-6, east of Wake Island, Pacific Ocean	58
Aug 2	Orient Airways (Pak.)	DC-3, on takeoff, Sharjah, Trucial States	1
Aug 3	Air France (Fr.)	L-Constellation, near Yazihan, Turk.	4
Sep 1	Air France (Fr.)	L-Constellation, hit Mt. Cemet, Fr.	42
Sep 1	Regina (US)	DC-3, near Vail, Washington State, US	21
Sep 16	American (US)	CV-240, on landing, near Albany, New York, US	28
Sep 25	THY (Turk.)	DC-3, on takeoff, near Ankara, Turk.	6
Sep 28	Resort (US)	C-46, on landing, Louisville, Kentucky, US	25
Oct 13	Air Maroc (Morocco)	—, on takeoff, Tangier, Morocco	1
Oct 14	SABENA (Belg.)	CV-240, on takeoff, near Frankfurt, W. Ger.	44
Oct 19	Eastern (US)	L-Constellation, on takeoff, Idlewild, New York, US	2
Oct 29	BCPA (US)	DC-6, on approach, San Francisco, California, US	19
Nov 3	LAB (Bol.)	DC-3, on landing, near Tarabuco, Bol.	28
Dec 12	AVIACO (Sp.)	BR-170, Somosierra Mountains, Sp.	23
Dec 12	Indian Airlines (Ind.)	DC-3, on takeoff, Nagpur, Maharashtra, Ind.	13
Dec 15	Misrair (Egy.)	VC-Viking, on takeoff, near Cairo, Egy.	6
Dec 19	SABENA (Belg.)	CV-240, on landing, near Kloten, Swit.	1

1954

Jan 5	THY (Turk.)	DC-3, near Lapseki, Turk.	4
Jan 10	BOAC (GB)	DH-Comet-1, near Elba, It.	35
Jan 11	AVIANCA (Colmb).	DC-3, near Manizales, Colmb.	21
Jan 14	CSA (Czech.)	DC-3, on takeoff, near Prague, Czech.	15

DATE	AIRLINE	AIRCRAFT AND LOCATION	KILLED
Jan 14	PAL (Phil.)	DC-6, on landing, Rome, It.	16
Feb 26	Western (US)	CV-240, near Wright, Wyoming, US	9
Mar 13	Aigle Azur (Fr.)	C-46, shot down, Dien Bien Phu, S. VN	?
Mar 13	BOAC (GB)	L-Constellation, on landing, Singapore	33
Mar 26	Aeromexico (Mex.)	DC-3, on approach, Monterrey, Mex.	18
Apr 3	THY (Turk.)	DC-3, after takeoff, near Adana, Turk.	25
Apr 8	Air Canada (Can.)	DC-3, midair collision, Moose Jaw, Saskatchewan, Can.	35 2 *
Apr 8	SAA (SA)	DH-Comet-1, on takeoff, near Naples, It.	21
Apr 13	SIR (Fr.)	L-18, Xieng-Khouang, Laos	16
Apr 23	Aero. Argentinas (Arg.)	DC-3, Sierra del Vilgo, Arg.	25
Apr 30	Darbhanga (Ind.)	DC-3, on takeoff, near Calcutta, Ind.	5
May 22	NAC (NZ)	DC-3, near Paraparaumu, NZ	2
May 31	TAN (Braz.)	DC-3, hit Cipo Mt., Braz.	19
Jun 8	STAEO (Fr.)	DC-3, near Da Nang, S. VN	4
Jun 19	Swissair (Swit.)	CV-240, ditched in English Channel near Folkstone, GB	3
Jul 23	Cathay Pacific (HK)	DC-4, shot down off Hai-nan Tao Is., China	10
Aug 9	AVIANCA (Colmb.)	L-Constellation, on takeoff, near Lajes, Azores	30
Aug 16	Air Vietnam (S. VN)	BR-170, near Pakse, Laos	47
Aug 22	Braniff (US)	DC-3, near Mason City, Iowa, US	12
Aug 23	KLM (Hol.)	DC-6, ditched in North Sea, off IJuuiden, Hol.	21
Sep 5	KLM (Hol.)	L-Super Constellation, on takeoff, ditched in River Shannon, S. Ire.	28
Sep 12	Cruzeiro (Braz.)	DC-3, on landing, ditched in bay, near Rio de Janeiro, Braz.	6
Nov 16	TAM (Peru)	DC-3, disappeared between Pucallpa–Lima, Peru	24

DATE	AIRLINE	AIRCRAFT AND LOCATION	KILLED
Dec 4	Air Laos (Laos)	DC-3, near Luang Prabang, Laos	28
Dec 18	LAI (It.)	DC-6, on landing, ditched in Jamaica Bay, New York City, US	26
Dec 22	Johnson (US)	DC-3, on approach, near Pittsburgh, Pennsylvania, US	10
Dec 25	BOAC (GB)	B-Stratocruiser, on landing, Prestwick, Scotland, GB	28
Dec 29	Aeroflot (USSR)	—, near Moscow, USSR	45
Dec 31	Aeroflot (USSR)	—, on takeoff, Irkutsk, USSR	17

1955

DATE	AIRLINE	AIRCRAFT AND LOCATION	KILLED
Jan 12	TWA (US)	M-202, collision on takeoff, Covington, Kentucky, US	13 2 *
Jan 31	AIDA (Colmb.)	Con.–PBY-5A, on landing, Caqueta River, Colmb.	5
Feb 2	Indian Airlines (Ind.)	DC-3, on takeoff, near Nagpur, Maharashtra, Ind.	10
Feb 5	Nigeria Airways (Nig.)	BR-170, near Calabar, Nig.	13
Feb 13	SABENA (Belg.)	DC-6, Reatini Mountains, It.	29
Feb 19	TWA (US)	M-404, Albuquerque, Mexico, US	16
Feb 26	—	DH-Dove, on takeoff, Formosa, Arg.	7
Mar 6	Real (Braz.)	DC-3, on landing, Conquista, Braz.	5
Mar 7	—	L-Lodestar, near Ciudad de Valles, Mex.	7
Mar 8	Mexicana (Mex.)	DC-3, Jalisco State, Mex.	26
Mar 9	AVIANCA (Colmb.)	DC-3, near Trujillo, Colmb.	8
Mar 20	American (US)	CV-240, on landing, Springfield, Missouri, US	13
Mar 26	Pan Am (US)	B-Stratocruiser, ditched in Pacific Ocean off Oregon, US	4
Apr 11	Air India (Ind.)	L-Constellation, bomb explosion, off Sarawak, South China Sea	19

DATE	AIRLINE	AIRCRAFT AND LOCATION	KILLED
Apr 18	UAT (Fr.)	DH-Heron, between Yaounde-Doula, Camer.	12
May 18	EAAC (Kenya)	DC-3, hit Mt. Kilimanjaro, Tanzania	20
May 21	LAV (Venz.)	DC-3, ditched in Caribbean Sea, near Barcelona, Venz.	4
Jun 16	PAB (Braz.)	L-Constellation, on approach, Asuncion, Paraguay	16
Jun 18	Tigres (Mex.)	C-46, on takeoff, Leon, Mex.	18
Jul 14	LAU (Mex.)	DC-3, hit mountain near Oaxaca, Mex.	22
Jul 17	Braniff (US)	CV-340, on landing, Chicago, Illinois, US	22
Jul 27	El Al (Israel)	L-Constellation, shot down, Bulgarian–Greek border	58
Aug 4	American (US)	CV-240, near Rolla, Missouri, US	30
Aug 6	Aeroflot (USSR)	—, near Voronezh, Ukraine, USSR	25
Aug 26	Cruzeiro (Braz.)	DC-3, hit Caparao Mountain, Braz.	13
Sep 2	UAB (Burma)	DC-3, on takeoff, near Meiktila, Burma	9
Sep 10	SFH (Fr.)	Latecoere-631, Cameroun Mountains, Camer.	16
Sep 17	Associated (Can.)	BR-170, near Thorhild, Alberta, Can.	2
Sep 21	BOAC (GB)	CL-Argonaut, on landing, near Tripoli, Libya	15
Oct 2	Faucett (Peru)	DC-4, Peru	19
Oct 6	United (US)	DC-4, Medicine Bow Peak, Wyoming, US	66
Oct 10	JAT (Yug.)	CV-340, on landing, near Vienna, Aus.	7
Nov 1	United (US)	DC-6, bomb explosion, near Longmont, Colorado, US	44
Nov 17	Peninsular (US)	DC-4, on takeoff, Seattle, Washington State, US	28
Dec 1	Cruzeiro (Braz.)	DC-3, on takeoff, near Belem, Braz.	6
Dec 19	—	DC-3, Sao Paulo, Braz.	26
Dec 21	Eastern (US)	L-Constellation, on landing, Jacksonville, Florida, US	17

DATE	AIRLINE	AIRCRAFT AND LOCATION	KILLED
1956			
Jan 17	Quebecair (Can.)	DC-3, Labrador, Can.	4
Jan 18	CSA (Czech.)	DC-3, near Torysa, Czech.	22
Feb 18	Scottish Air (GB)	Avro-York, on takeoff, Zurrieq, Malta	50
Feb 20	TAI (Fr.)	DC-6, on approach, near Cairo, Egy.	52
Feb 24	Syrian Airways (Syria)	DC-3, after takeoff, explosion, near Aleppo, Syria	19
Mar 21	Indian Airlines (Ind.)	DC-3, on landing, Tezpur, Assam, Ind.	2
Apr 1	TWA (US)	M-404, on takeoff, Pittsburgh, Pennsylvania, US	22
Apr 2	Northwest (US)	B-Stratocruiser, on takeoff, near Seattle, Washington State, US	5
Apr 18	PAB (Braz.)	Con.–PBY-5A, near Parcentins, Braz.	3
Apr 30	Scottish Air (GB)	Avro-York, on takeoff, Stansted, Essex, GB	2
May 15	Indian Airlines (Ind.)	DC-3, on landing, Katmandu, Nepal	14 1 *
May 24	AVIATECA (Guat.)	DC-3, hit mountain, near Panzos, Guat.	30
Jun 13	Piedmont (US)	DC-3, door opened, passenger fell out, North Carolina, US	1
Jun 20	LAV (Venz.)	L-Super Constellation, ditched in Atlantic Ocean, off New York City	74
Jun 24	BOAC (GB)	CL-Argonaut, on takeoff, Kano, Nig.	32
Jun 30	United (US)	DC-7 midair collision, Grand Canyon, Arizona, US	58
Jun 30	TWA (US)	L-Super Constellation	70
Jul 9	Air Canada (Can.)	VC-Viscount, fuselage puncture, Flat Rock, Michigan, US	1
Jul 16	Aero. Argentinas (Arg.)	DC-3, near Rio Cuarto, Arg.	18
Aug 8	UAB (Burma)	DC-3, near Thazi, S.E. of Mandalay, Burma	12

DATE	AIRLINE	AIRCRAFT AND LOCATION	KILLED
Aug 25	LAB (Bol.)	DC-3, on landing, La Paz, Bol.	2
Aug 29	C P Air (Can.)	DC-6, on landing, Cold Bay, Alaska, US	15
Sep 9	Jordan Int. (Jord.)	C-46, on takeoff, near Amman, Jord.	1
Oct 28	—(Egy.)	C-46, disappeared over Mediterranean, Damascus, Syria–Cairo, Egy.	16
Nov 5	Britavia (GB)	HP-Hermes, on landing, near Basingstoke, GB	7
Nov 7	Braathens (Nor.)	DH-Heron, hit mountain, near Tolga, Nor.	2
Nov 15	Guest (Mex.)	DC-4, Puerto Somoza, Nic.	25
Nov 17	Pacifico (Colmb.)	DC-3, hit El Rucio Mt., Colmb.	36
Nov 23	LAI (It.)	DC-6, on takeoff, hit house, near Orly Airport, Paris, Fr.	34
Nov 24	CSA (Czech.)	IL-12, on takeoff, near Eglisau, Swit.	23
Nov 27	LAV (Venz.)	L-Constellation, on approach, near Caracas, Venz.	25
Dec 8	AIDA (Colmb.)	Con.–PBY-5A, hit mountains, Caqueta, Colmb.	14
Dec 9	Air Canada (Can.)	DC-4, near Hope, British Columbia, Can.	62
Dec 22	JAT (Yug.)	CV-340, on landing, Munich, W. Ger.	3
Dec 22	LAI (It.)	DC-3, hit mountains, Milan–Rome, It.	21

1957

DATE	AIRLINE	AIRCRAFT AND LOCATION	KILLED
Jan 6	American (US)	CV-240, on landing, Tulsa, Oklahoma, US	1
Jan 11	LADE (Arg.)	VC-Viking, on takeoff, near Buenos Aires, Arg.	17 / 1 *
Jan 23	LANICA (Nic.)	DC-3, hit Concepcion Volcano, Nic.	16
Feb 1	Northeast (US)	DC-6, on takeoff, LaGuardia Airport, New York City, US	20
Feb 23	—(Braz.)	JU-2, Sao Paulo state, Braz.	6

DATE	AIRLINE	AIRCRAFT AND LOCATION	KILLED
Mar 2	Alaska (US)	DC-4, hit hill, near Blyn, Washington State, US	5
Mar 9	AVIANCA (Colmb.)	DC-3, Trujillo Mountains, Colmb.	15
Mar 14	BEA (GB)	VC-Viscount, on landing, Manchester, GB	20 2 *
Mar 18	LAB (Bol.)	DC-3, Mt. Sayary, Bol.	19
Apr 7	VARIG (Braz.)	C-46, on takeoff, Bage, Braz.	40
Apr 8	Air France (Fr.)	DC-3, on takeoff, Biskra, Alg.	34
Apr 10	Real (Braz.)	DC-3, hit mountain, near Anchieta Is., Braz.	26
Apr 20	Air France (Fr.)	L-Super Constellation, window broke, between Iran–Turk.	1
May 1	British Eagle (GB)	VC-Viking, on takeoff, near Basingstoke, GB	34
May 9	AVIACO (Sp.)	BR-170, on landing, near Barajas, Sp.	37
May 13	LADE (Arg.)	VC-Viking, near Bariloche, Arg.	16
Jun 14	LOT (Pol.)	IL-14, on approach, Wnukowo, USSR	9
Jun 23	Pacific Western (Can.)	DC-3, near Port Hardy, British Columbia, Can.	14
Jun 24	Adastra (Austrl.)	L-Hudson, on approach, Horn Is., Queensland, Austrl.	6
Jul 1	PIA (Pak.)	DC-3, Charlakhi Is., Bay of Bengal, Ind.	24
Jul 3	Kenting (Can.)	L-Hudson, near Fort Rupert, Quebec, Can.	4
Jul 16	KLM (Hol.)	L-Super Constellation, on takeoff, Biak Is., NG	58
Jul 25	Western (US)	CV-240, bomb punctured fuselage, Daggett, California, US	1
Aug 11	Maritime (Can.)	DC-4, near Issoudun, Quebec, Can.	79
Aug 15	Aeroflot (USSR)	IL-14, on landing, hit factory, Copenhagen, Den.	23
Aug 30	Honduras Air (Hond.)	DC-3, after takeoff, near Juticalpa, Hond.	12
Sep 3	AVENSA (Venz.)	DC-3, Penas Blancas Hills, Venz.	8

DATE	AIRLINE	AIRCRAFT AND LOCATION	KILLED
Sep 15	Northeast (US)	DC-3, on landing, New Bedford, Massachusetts, US	12
Oct 2	Pacific Western (Can.)	VC(C)-Stranraer, on takeoff, British Columbia, Can.	4
Oct 3	Lebanese Int. (Leb.)	C-46, after takeoff, near Beirut, Leb.	27
Oct 23	BEA (GB)	VC-Viking, on landing, Belfast, N. Ire., GB	7
Oct 28	Iberia (Sp.)	DC-3, near Getafe, Sp.	21
Nov 3	Herfurtner (Den.)	DC-4, on takeoff, Dusseldorf, W. Ger.	6 1 *
Nov 4	TAROM (Rum.)	IL-14, on landing, Moscow, USSR	4
Nov 8	Pan Am (US)	B-Stratocruiser, ditched in Pacific Ocean	44
Nov 15	Aquila (GB)	Short-Solent, on takeoff, Isle of Wight, GB	43
Nov 15	AVIACO (Sp.)	DH-Heron, on landing, Majorca, Sp.	4
Nov 21	Straits Air (NZ)	BR-170, on takeoff, Christchurch, NZ	4
Dec 8	Aero. Argentinas (Arg.)	DC-4, near Bolivar, Arg.	61
Dec 10	Northern Wings (Can.)	DC-3, on landing, Jean Lake, Quebec, Can.	3
Dec 11	PAL (Phil.)	DH(C)-3, on takeoff, Ozamiz, Phil.	2
Dec 31	Aero. Argentinas (Arg.)	Short-Sandringham, on takeoff, Buenos Aires, Arg.	8

1958

DATE	AIRLINE	AIRCRAFT AND LOCATION	KILLED
Jan 20	TAM (Bol.)	DC-3, hit mountains, near La Paz, Bol.	11
Feb 2	Loide (Braz.)	DC-4, on takeoff, Rio de Janeiro, Braz.	5
Feb 6	BEA (GB)	DH-Ambassador, on takeoff, Munich, W. Ger.	23
Feb 19	RAI (Fr.)	Con.–PBY-5A, near Raiatea Is., Tahiti	15
Feb 27	Manx (GB)	BR-170, hit mountain, near Winter Hill, Berkshire, GB	35

DATE	AIRLINE	AIRCRAFT AND LOCATION	KILLED
Mar 7	Misrair (Egy.)	VC-Viking, near Port Said, Egy.	8
Mar 24	Indian Airlines (Ind.)	DC-3, near Katmandu, Nepal	20
Mar 25	Braniff (US)	DC-7, on takeoff, Miami, Florida, US	9
Apr 4	AVIACO (Sp.)	DH-Heron, on landing, Barcelona, Sp.	16
Apr 6	Capital (US)	VC-Viscount, on landing, Freeland, Michigan, US	47
Apr 7	AREA (Ecu.)	DC-3, Chugchilan Mountains, Ecu.	32
Apr 8	TAO ᴬ (Ecu.)	JU-52, after takeoff, Quito, Ecu.	3
Apr 21	United (US)	DC-7, midair collision, Las Vegas, Nevada, US	47 2 *
May 15	PIA (Pak.)	CV-240, on takeoff, Palam, Maharashtra, Ind.	21 2 *
May 18	SABENA (Belg.)	DC-7, near Casablanca, Morocco	65
May 20	Capital (US)	VC-Viscount, midair collision, near Brunswick, Maryland, US	11 1 *
May 31	Air France (Fr.)	DC-3, near Moliere, Alg.	15
Jun 2	Aeromexico (Mex.)	L-Constellation, on takeoff, Guadalajara, Mex.	45
Jun 2	Pan Am (US)	B-Stratocruiser, fuselage puncture, Manila, Phil.	1
Jun 16	Cruzeiro (Braz.)	CV-440, on landing, Curitiba, Braz.	21
Jun 25	Indian Airlines (Ind.)	DC-3, near Damroh, Assam, Ind.	5
Aug 9	CAAC (Rhod.)	VC-Viscount, hit mountain, Cyrenaica Region, Libya	36
Aug 11	Loide (Braz.)	DC-4, near Carapo, Braz.	10
Aug 12	All Nippon (Jap.)	DC-3, ditched in sea near Izu Is., Jap.	33
Aug 14	KLM (Hol.)	L-Super Constellation, Atlantic Ocean off S. Ire.	99
Aug 15	Northeast (US)	CV-240, on landing, Nantucket, Massachusetts, US	25
Aug 29	Pacific Western (Can.)	DH(C)-2, Northwest Territories, Can.	4
Sep 5	Loide (Braz.)	C-46, on approach, near Campina Grande, Braz.	13

DATE	AIRLINE	AIRCRAFT AND LOCATION	KILLED
Sep 9	Flying Tiger (US)	L-Super Constellation, near Tokyo, Jap.	8
Oct 14	LAV (Venz.)	L-Super Constellation, hit Mt. Cedro, Venz.	23
Oct 15	TAM (Bol.)	DC-3, near Villa Montes, Bol.	20
Oct 17	Aeroflot (USSR)	TU-104, near Kanash, USSR	65
Oct 22	BEA (GB)	VC-Viscount, midair collision, Nettuno, It.	31
Nov 1	Cubana (Cuba)	VC-Viscount, on landing, Nipe Bay, Cuba	17
Nov 3	Yemenite Air (GB [1])	DC-3, hit mountain, near Perugia, It.	8
Nov 9	Aero. Topografica (Port.)	M-Mariner, missing, North Atlantic Ocean	42
Dec 4	AVIACO (Sp.)	SA-Languedoc, Guadarrama Mountains, Sp.	21
Dec 26	UAT (Fr.)	DC-6, on takeoff, Salisbury, Rhod.	3
Dec 30	VASP (Braz.)	Saab-Scandia, on takeoff, Rio de Janeiro, Braz.	21

1959

Jan 6	S.A. de Honduras (Hond.)	DC-3, hit Pena Blanca Mountain, Guat.	5
Jan 8	Southeast (US)	DC-3, hit mountain near Bristol, Tennessee, US	10
Jan 11	Lufthansa (W. Ger.)	L-Super Constellation, on landing, Rio de Janeiro, Braz.	36
Jan 16	Austral (Arg.)	C-46, on landing, Mar del Plata, Arg.	51
Jan 21	Bolivar (Venz.)	C-46, hit La Culata Peak, Venz.	4
Jan 22	Air Jordan (Jord.)	CV-240, near Wadi es Sir, Jord.	10
Feb 1	General (US)	DC-3, near Kerrville, Texas, US	3
Feb 3	American (US)	L-Electra II, on landing, New York City, US	65

DATE	AIRLINE	AIRCRAFT AND LOCATION	KILLED
Feb 17	THY (Turk.)	VC-Viscount, on landing, Gatwick Airport, London, GB	14
Mar 5	TACA (El Salvador)	VC-Viscount, on takeoff, Managua, Nic.	15
Mar 13	AVIACO (Sp.)	BR-170, on landing, Mahon, Minorca, Sp.	1
Mar 29	UAT (Fr.)	Nord–Noratlas, near Banga, Chad	9
Mar 29	Indian Airlines (Ind.)	DC-3, near Hailakandi, Assam, Ind.	24
Apr 17	Tigres (Mex.)	C-46, exploded near Bahia de Kino, Mex.	26
Apr 29	Iberia (Sp.)	DC-3, near Valdemeca, Sp.	28
May 2	Austria Flug-dienst (Aus.)	DC-3, Alfabia Mountains, Majorca, Sp.	5
May 12	Capital (US)	L-Constellation, on landing, Charleston, West Virginia, US	2
May 12	Capital (US)	VC-Viscount, near Baltimore, Maryland, US	31
May 13	T.A. Peruanos (Peru)	C-46, hit hills, northeast Peru	12
May 15	Aero. Argentinas (Arg.)	DC-3, on takeoff, Mar del Plata, Arg.	10
May 21	Interpol (Ch.)	C-46, hit mountain, near Chimbote, Peru	8
Jun 1	Aero. Nacionales (Costa Rica)	C-46, attacked by fighter aircraft, Nic.	60
Jun 23	AVIANCA (Colmb.)	DC-4, near Mt. Baco, Peru	14
Jun 26	TWA (US)	L-1649, near Milan, It.	68
Aug 3	Kalinga (Ind.)	DC-3, near Sagong, Ind.	6
Aug 19	Transair (GB)	DC-3, hit Mt. Monseny, Sp.	32
Aug 27	Aero. Argentinas (Arg.)	DH-Comet IV, near Asuncion, Paraguay	2
Sep 8	Mexicana (Mex.)	DC-3, bomb explosion, Mex.	2
Sep 23	VASP (Braz.)	Saab-Scandia, on takeoff, Sao Paulo, Braz.	20

DATE	AIRLINE	AIRCRAFT AND LOCATION	KILLED
Sep 24	Reeve Aleutian (US)	DC-4, near Great Sitkin Island, Alaska, US	16
Sep 24	TAI (Fr.)	DC-7, on takeoff, Bordeaux, Fr.	54
Sep 29	Braniff (US)	L-Electra II, near Buffalo, Texas, US	34
Oct 29	Olympic (Gre.)	DC-3, hit mountain, near Avlon, Gre.	18
Oct 30	Piedmont (US)	DC-3, near Charlottesville, Virginia, US	26
Nov 16	National (US)	DC-7, ditched in Gulf of Mexico, near New Orleans, Louisiana, US	42
Nov 21	Ariana Afghan (Afg.)	DC-4, on takeoff, Beirut, Leb.	24
Dec 1	Allegheny (US)	M-202, near Williamsport, Pennsylvania, US	25
Dec 8	SAM (Colmb.)	C-46, near Cartagena, Colmb.	45
Dec 13	Aeroflot (USSR)	IL, hit mountain, between Kabul, Afg.–Tashkent, USSR	29
Dec 22	VASP (Braz.)	VC-Viscount, midair collision, near Rio de Janeiro, Braz.	32 10 *
Dec 31	LAB (Bol.)	DC-3, near San Jose de Chiquitos, Bol.	11

1960

DATE	AIRLINE	AIRCRAFT AND LOCATION	KILLED
Jan 3	Indian Airlines (Ind.)	DC-3, near Taksing, Ind.	9
Jan 6	National (US)	DC-6, bomb explosion, near Bolivia, North Carolina, US	34
Jan 18	Capital (US)	VC-Viscount, near Charles City, Virginia, US	50
Jan 19	SAS (Swe.)	SA-Caravelle, near Munich, W. Ger.	42
Jan 21	AVIANCA (Colmb.)	L-Super Constellation, on landing, Montego Bay, Jam.	37
Jan 26	T.A. Timor (Port.)	DH-Heron, between Austrl.–Timor	9
Feb 5	LAB (Bol.)	DC-4, near Cochabamba, Bol.	59
Feb 25	Real (Braz.)	DC-3, midair collision, near Rio de Janeiro, Braz.	26 35 *

DATE	AIRLINE	AIRCRAFT AND LOCATION	KILLED
Feb 26	Alitalia (It.)	DC-7, near Limerick, S. Ire.	34
Mar 15	AIDA (Colmb.)	Con.–PBY-5A, near El Refugio Airport, Colmb.	6
Mar 16	All Nippon (Jap.)	DC-3, collision on landing, Nagoya, Jap.	3
Mar 17	Northwest (US)	L-Electra II, near Cannelton, Indiana, US	63
Mar 19	SAM (Colmb.)	C-46, Cordoba Province, Colmb.	25
Apr 5	LANICA (Nic.)	C-46, near Guina, Nic.	2
Apr 12	Cruzeiro (Braz.)	DC-3, on takeoff, Pelotas, Braz.	10
Apr 19	LACSA (Colmb.)	C-46, on landing, Bogota, Colmb.	32
Apr 22	Sobelair (Belg.)	DC-4, hit mountain, near Bunia, Zaire	35
Apr 28	LAV (Venz.)	DC-3, bomb explosion, near Calabozo, Venz.	13
May 15	Balair (Swit.)	DC-4, hit Mt. Marra, Sudan	12
May 17	Transamerican (Arg.)	C-46, near Mendoza, Ch.	10
Jun 10	TAA (Austrl.)	F-27, ditched in sea near Mackay, Queensland, Austrl.	29
Jun 14	Pacific Northern (US)	L-Constellation, hit Mt. Gilbert, Alaska, US	14
Jun 24	Real (Braz.)	CV-340, ditched in sea, near Rio de Janeiro, Braz.	54
Jun 25	LACSA (Colmb.)	DC-3, near San Jose, Costa Rica	1
Jul 10	Gulf Air (Bahrain)	DC-3, near Sharjah, United Arab Emirates, Persian Gulf	16
Jul 11	Mission (US)	DC-3, hit Mt. Pichincha, Ecu.	18
Jul 14	Northwest (US)	DC-7, ditched in sea, near Manila, Phil.	1
Jul 31	Deutsche Flug (W. Ger.)	CV-240, near Rimini, It.	1
Aug 1	TAM (Bol.)	DC-3, hit mountain, near La Paz, Bol.	6
Aug 15	Pacific Western (US)	DH(C)-2, near Lorna Lake, British Columbia, Can.	4
Aug 17	Aeroflot (USSR)	IL-18, near Kiev, Ukraine, USSR	35
Aug 29	Air France (Fr.)	L-Super Constellation, on landing, Dakar, Senegal	63

DATE	AIRLINE	AIRCRAFT AND LOCATION	KILLED
Sep 7	Aero. Argentinas (Arg.)	DC-6, near Salto, Uruguay	31
Sep 19	World (US)	DC-6, hit mountain, Guam Island, US	80
Sep 26	Austrian Airlines (Aus.)	VC-Viscount, near Moscow, USSR	31
Sep 28	Mexicana (Mex.)	DC-3, near Juchitepec, Mex.	6
Sep 29	Egyptair (Egy.)	VC-Viscount, near Elba Is., It.	21
Oct 4	Eastern (US)	L-Electra II, on takeoff, Boston, Massachusetts, US	62
Oct 14	Itavia (It.)	DH-Heron, hit Mr. Capanne, Elba Is., It.	11
Oct 28	Northwest (US)	DC-4, on landing, Missoula, Montana, US	12
Oct 29	Arctic Pacific (US)	C-46, on takeoff, Toledo, Ohio, US	22
Nov 7	AREA (Ecu.)	F-27, hit mountain, near Quito, Ecu.	37
Nov 23	PAL (Phil.)	DC-3, near Mt. Baco, Phil.	33
Dec 7	TAN (Braz.)	C-46, near Cachimbo, Braz.	15
Dec 16	TWA (US)	L-Super Constellation midair collision, Staten Is., New York, US	45
Dec 16	United (US)	DC-8	83 6 *
Dec 22	PAL (Phil.)	DC-3, on takeoff, Cebu, Phil.	28

1961

Jan 3	Finnair (Fin.)	DC-3, on landing, Koivulahti, Fin.	25
Jan 24	Garuda (Indon.)	DC-3, near Burangrang, Indon.	21
Feb 3	Garuda (Indon.)	DC-3, disappeared, off Madura Is., Indon.	26
Feb 15	SABENA (Belg.)	B-707, near Brussels, Belg.	72 1 *
Mar 9	LAV (Venz.)	DC-3, near Paramo Curumo, Venz.	12

DATE	AIRLINE	AIRCRAFT AND LOCATION	KILLED
Mar 28	CSA (Czech.)	IL-18, near Nuremberg, W. Ger.	52
Apr 3	LAN (Ch.)	DC-3, near Lastima, Ch.	24
May 10	Air France (Fr.)	L-1649, near Ghadames, Libya	78
May 30	KLM (Hol.)	DC-8, ditched in Atlantic Ocean off Port.	61
Jun 12	KLM (Hol.)	L-Electra II, near Cairo, Egy.	20
Jun 13	Air Cameroun (Camer.)	DC-4, on takeoff, Douala, Camer.	5
Jun 17	CDL (W. Ger.)	DC-4, on landing, Kano, Nig.	1
Jun 30	Transcontinental (Arg.)	C-46, on landing, Buenos Aires, Arg.	24
Jul 11	United (US)	DC-8, on landing, Denver, Colorado, US	17 1 *
Jul 12	CSA (Czech.)	IL-18, hit power line on landing, Casablanca, Morocco	72
Jul 19	Aero. Argentinas (Arg.)	DC-6, near Azul, Arg.	67
Aug 6	MALEV (Hung.)	DC-3, near Budapest, Hung.	27 3 *
Aug 9	Cunard Eagle (GB)	VC-Viking, near Stavanger, Nor.	39
Sep 1	TWA (US)	L-Constellation, on takeoff, Chicago, Illinois, US	78
Sep 5	Ethiopian (Ethio.)	DC-3, near Sendafar, Ethio.	5
Sep 10	President (US)	DC-6, near Limerick, S. Ire.	83
Sep 12	Air France (Fr.)	SA-Caravelle, on landing, Rabat, Morocco	77
Sep 17	Transair (Swe.)	DC-6, on landing, Ndola, Zambia	16
Sep 17	Northwest (US)	L-Electra II, on takeoff, Chicago, Illinois, US	37
Sep 23	THY (Turk.)	F-27, on landing, Ankara, Turk.	28
Oct 7	Derby (GB)	DC-3, hit Mt. Canigou, Pyrenees, Fr.	34
Nov 1	PAB (Braz.)	DC-7, near Recife, Braz.	45

DATE	AIRLINE	AIRCRAFT AND LOCATION	KILLED
Nov 8	Imperial (US)	L-Constellation, near Richmond, Virginia, US	76
Nov 23	Aero. Argentinas (Arg.)	DH-Comet IV, on takeoff, Campinas, Braz.	52
Nov 30	Ansett (Austrl.)	VC-Viscount, on takeoff, Sydney, New South Wales, Austrl.	15
Dec 21	BEA (GB)	DH-Comet IV, on takeoff, Ankara, Turk.	27
Dec 24	Kodiak (US)	G-21A, on takeoff, Kodiak Is., Alaska, US	1

1962

DATE	AIRLINE	AIRCRAFT AND LOCATION	KILLED
Jan 13	Ethiopian Air (Ethio.)	DC-3, on takeoff, Tippi, Ethio.	5
Jan 22	CBF (Bol.)	C-46, on takeoff, La Paz, Bol.	6
Feb 4	Faucett (Peru)	DC-3, northeast Peru	18
Feb 25	AVENSA (Venz.)	F-27, Margarita Is., Venz.	23
Mar 1	American (US)	B-707, on takeoff, New York City, US	95
Mar 4	Caledonian (GB)	DC-7, on takeoff, Douala, Camer.	111
Mar 8	THY (Turk.)	F-27, near Incirli, Turk.	11
Mar 15	Flying Tiger (US)	L-Super Constellation, disappeared between Guam Is., US–Phil.	107
Mar 27	Cubana (Cuba)	IL-14, hijack attempt, Cuba	2
Apr 22	AVISPA (Colmb.)	DC-3, Baudo Mountains, Colmb.	40
May 6	Channel (GB)	DC-3, near Ventnor, Isle of Wight, GB	12
May 9	Cruzeiro (Braz.)	CV-240, on landing, Vitoria, Braz.	23
May 12	EPA (Can.)	Con.–PBY-5A, on landing, Godthab, Greenland	15
May 22	Continental (US)	B-707, bomb explosion, near Unionville, Missouri, US	45
Jun 3	Air France (Fr.)	B-707, on takeoff, Paris, Fr.	130
Jun 22	Air France (Fr.)	B-707, Grande Terre, Guadeloupe, WI	112

DATE	AIRLINE	AIRCRAFT AND LOCATION	KILLED
Jul 6	Aeroflot (USSR)	IL-14, near Tashkent, USSR	14
Jul 6	Alitalia (It.)	DC-8, near Junnar, Ind.	94
Jul 16	Kalinga (Ind.)	DC-3, Assam state, Ind.	9
Jul 19	Egyptair (Egy.)	DH-Comet IV, near Bangkok, Thai.	26
Jul 22	C P Air (Can.)	BR-Britannia, on takeoff, Honolulu, Hawaii, US	27
Jul 23	BOAC (GB)	B-707, between Antigua–Barbados, WI	1
Aug 1	Nepal Air (Nepal)	DC-3, hit mountain, near Dhuli, Nepal	10
Aug 20	PAB (Braz.)	DC-8, on takeoff, Rio de Janeiro, Braz.	15
Aug 23	Taxader (Colmb.)	DC-3, on takeoff, Barrancabermeja, Colmb.	19
Aug 24	Faucett (Peru)	DC-3, near Cerro Puena Paz, Peru	9
Sep 10	Abaroa (Bol.)	DC-3, near La Paz, Bol.	4
Sep 21	Kalinga (Ind.)	DC-3, hit mountain, near the Se La Pass, Tibet	8
Sep 23	Flying Tiger (US)	L-Super Constellation, ditched in North Atlantic Ocean	28
Oct 10	Air Canada (Can.)	VC-Viscount, ground collision, Bagotville, Quebec, Can.	2
Oct 10	CSA (Czech.)	IL-14, near Brno, Czech.	11
Oct 12	Iberia (Sp.)	CV-440, near Carmona, Sp.	18
Nov 10	Air Vietnam (S. VN)	DC-3, near DaNang, S. VN	27
Nov 23	MALEV (Hung.)	IL-18, Roissy-en-France, near Paris, Fr.	21
Nov 23	United (US)	VC-Viscount, bird ingestion, near Ellicott City, Maryland, US	17
Nov 26	VASP (Braz.)	Saab-Scandia, midair collision, near Paraibuna, Braz.	23 / 4 *
Nov 26	Urraca (Colmb.)	C-46, near Port Henderson, Jam.	2
Nov 27	VARIG (Braz.)	B-707, near Lima, Peru	97
Nov 30	Eastern (US)	DC-7, on landing, New York City, US	25
Dec 6	Taxader (Colmb.)	DC-3, near Barrancabermeja, Colmb.	24
Dec 14	PAB (Braz.)	L-Constellation, near Manaus, Braz.	50

DATE	AIRLINE	AIRCRAFT AND LOCATION	KILLED
Dec 14	Flying Tiger (US)	L-Super Constellation, near Hollywood, California, US	5 3 *
Dec 19	LOT (Pol.)	VC-Viscount, on landing, Warsaw, Pol.	34
Dec 29	Air Nautic (Fr.)	B-307, hit Mt. Renoso, Corsica, Fr.	25

1963

DATE	AIRLINE	AIRCRAFT AND LOCATION	KILLED
Jan 15	Cruzeiro (Braz.)	CV-240, on takeoff, Sao Paulo, Braz.	8 7 *
Jan 29	Continental (US)	VC-Viscount, near Kansas City, Missouri, US	8
Feb 1	MEA (Leb.)	VC-Viscount, midair collision, near Ankara, Turk.	14 87 * 3 †
Feb 3	Slick (US)	L-Super Constellation, on landing, San Francisco, California, US	4
Feb 12	Northwest (US)	B-720, near Miami, Florida, US	43
Mar 2	PAL (Phil.)	DC-3, near Davao, Mindanao, Phil.	27
Mar 15	LAB (Bol.)	DC-6, hit mountain, southern Peru	39
Mar 20	King Saud (Libya)	DH-Comet IV, Maritime Alps, Fr./It. border	18
Mar 30	Itavia (It.)	DC-3, hit mountain, near Sora, It.	8
Apr 7	Syrian Arab (Syria)	DC-6, on takeoff, Hama, Syria	1
Apr 14	Icelandair (Ice.)	VC-Viscount, on landing, Oslo, Nor.	12
May 3	Cruzeiro (Braz.)	CV-340, on takeoff, Sao Paulo, Braz.	35
May 4	—	DC-6, hit Mt. Cameroun, Camer.	55
May 6	Air Canada (Can.)	VC-Vanguard, near Rocky Mountain House, Alberta, Can.	1
May 12	Egyptair (Egy.)	DC-3, near Ayayda, Egy.	34
Jun 3	Northwest (US)	DC-7, near Annette Is., Alaska, US	101
Jun 3	Indian Airlines (Ind.)	DC-3, near Pathankot, Ind.	29
Jun 10	UAB (Burma)	DC-3, Yunnan Province, China	20

DATE	AIRLINE	AIRCRAFT AND LOCATION	KILLED
Jun 16	TAROM (Rum.)	IL-14, southeast Hung.	31
Jul 1	VARIG (Braz.)	DC-3, near Bela Vista, Braz.	15
Jul 2	Mohawk (US)	M-404, near Rochester, New York, US	7
Jul 3	NAC (NZ)	DC-3, near Mt. Ngatamahinerua, NZ	23
Jul 13	Aeroflot (USSR)	TU-104, near Irkutsk, USSR	30
Jul 15	Air Madagascar (MR)	DC-3, on takeoff, Farafangana, MR	6
Jul 27	Egyptair (Egy.)	DH-Comet IV, near Bombay, Ind.	63
Aug 12	Air Inter (Fr.)	VC-Viscount, near Lyon, Fr.	20 1 *
Aug 17	FKKK (Jap.)	DH-Heron, near Hachyo Is., Jap.	19
Sep 2	—	—, refugee flight, northeast Laos	15
Sep 4	Swissair (Swit.)	SA-Caravelle, near Zurich, Swit.	80
Sep 11	Air Nautic (Fr.)	VC-Viking, hit mountain, near Perpignan, Fr.	40
Sep 11	Indian Airlines (Ind.)	VC-Viscount, near Patti, Ind.	18
Oct 4	SEA (Ecu.)	C-46, on takeoff, Cucuta, Colmb.	4
Nov 8	Finnair (Fin.)	DC-3, near Mariehamn, Aland Is., Fin.	22
Nov 29	Air Canada (Can.)	DC-8, on takeoff, Montreal, Quebec, Can.	118
Dec 7	Zantop (US)	C-46, near Nederland, Colorado, US	3
Dec 8	Pan Am (US)	B-707, near Elkton, Maryland, US	81

1964

Jan 9	ALA (Arg.)	DC-3, near Zarate, Arg.	30
Jan 15	LAB (Bol.)	DC-3, on takeoff, Potosi, Bol.	1
Feb 4	LAB (Bol.)	DC-3, near Yacuiba, Bol.	2
Feb 21	PAL (Phil.)	DC-3, near Marawi, Phil.	28
Feb 25	Eastern (US)	DC-8, on takeoff, New Orleans, Louisiana, US	58

DATE	AIRLINE	AIRCRAFT AND LOCATION	KILLED
Feb 27	FKKK (Jap.)	CV-240, on landing, Oita, Kyushu Is., Jap.	20
Feb 29	Eagle (GB)	BR-Britannia, near Innsbruck, Aus.	83
Mar 1	Paradise (US)	L-Constellation, near Lake Tahoe, California, US	85
Mar 8	Taxader (Colmb.)	DC-3, near Bogota, Colmb.	28
Mar 12	Frontier (US)	DC-3, near Miles City, Montana, US	5
Mar 28	—	DC-4, ditched in Pacific Ocean	9
Mar 28	Alitalia (It.)	VC-Viscount, near Mt. Vesuvius, It.	45
Apr 1	Kenting (Can.)	AC-680, near Thompson, Manitoba, Can.	4
Apr 17	MEA (Leb.)	SA-Caravelle, on landing, Dhahran, Saud.	49
May 7	Pacific (US)	F-27, pilot shot, near Concord, California, US	44
May 8	—	DC-4, ditched in sea, near Callao, Peru	49
May 20	PAL (Phil.)	DH(C)-3, near Zamboanga, Phil.	11
Jun 20	CAT B (Tai.)	C-46, near T'ai-chung, Tai.	57
Jun 21	TASSA (Sp.)	DC-3, ditched in sea, near Palma de Majorca, Sp.	1
Jul 9	United (US)	VC-Viscount, near Parrotsville, Tennessee, US	39
Aug 22	SAC (Bol.)	DC-3, hit mountain, near Tipuani, Bol.	4
Sep 4	VASP (Braz.)	VC-Viscount, hit mountain, near Nova Friburgo, Braz.	39
Oct 2	UTA (Fr.)	DC-6, hit mountain, near Granada, Sp.	80
Oct 9	TAROM (Rum.)	IL-14, near Sibiu, Rum.	32
Oct 10	TAROM (Rum.)	IL-18, near Sibiu, Rum.	43
Oct 19	Aeroflot (USSR)	IL-18, near Belgrade, Yug.	33
Nov 15	Bonanza (US)	F-27, near Las Vegas, Nevada, US	29
Nov 20	Linjeflyg (Swe.)	CV-340, near Angelholm, Swe.	31
Nov 23	TWA (US)	B-707, on takeoff, Leonardo da Vinci Airport, Rome, It.	49
Nov 29	BIAS (Belg.)	DC-4, on takeoff, Kisangani, Zaire	6

DATE	AIRLINE	AIRCRAFT AND LOCATION	KILLED
Dec 8	Abaroa (Bol.)	DC-3, bomb explosion, near Milluni, Bol.	17

1965

Jan 17	Norte (Mex.)	—, near Zapora, Mex.	10
Feb 3	Air Cameroun (Camer.)	C-46, near Garoua, Camer.	4
Feb 6	LAN (Ch.)	DC-6, near Santiago, Ch.	87
Feb 8	Eastern (US)	DC-7, near New York City, US	84
Mar 8	FOA (Phil.)	DC-3, near Mt. Tangcong, Phil.	10
Mar 17	EPA (Can.)	HP-Herald, near Upper Musquodoboit, Nova Scotia, Can.	8
Mar 22	AVIANCA (Colmb.)	DC-3, hit Pan de Azucar Peak, Colmb.	28
Mar 26	PIA (Pak.)	DC-3, near Lowarai Pass, Pak.	22
Mar 31	Iberia (Sp.)	CV-440, near Tangier, Morocco	50
Apr 10	ALIA (Jord.)	HP-Herald, near Damascus, Syria	54
Apr 14	BUCI (GB)	DC-3, near Jersey Airport, Channel Islands, GB	26
May 5	Iberia (Sp.)	L-Super Constellation, on landing, Tenerife, Canary Islands, Sp.	32
May 20	PIA (Pak.)	B-720, near Cairo, Egy.	121
Jul 8	C P Air (Can.)	DC-6, bomb explosion, British Columbia, Can.	52
Aug 4	Peruanas (Peru)	DC-4, near Panama City, Panama	7
Aug 12	Paraense (Braz.)	C-46, near Buraco, Braz.	13
Aug 16	United (US)	B-727, on landing, near Chicago Illinois, US	30
Sep 11	FOA (Phil.)	DC-3, hit mountain, near La Carlota, Phil.	11
Sep 16	TAO ᴬ (Ecu.)	DC-3, near Pastaza River, Ecu.	9
Sep 16	Air Vietnam (S. VN)	DC-3, on takeoff, Quang Ngai, S. VN	38

DATE	AIRLINE	AIRCRAFT AND LOCATION	KILLED
Sep 17	Pan Am (US)	B-707, on landing, Montserrat Is., WI	30
Oct 17	AVIANCA (Colmb.)	DC-3, midair collision, near Bucaramanga, Colmb.	18 1 *
Oct 20	PAL (Phil.)	DC-3, on takeoff, Manila, Phil.	1
Oct 27	BEA (GB)	VC-Vanguard, on landing, London, GB	36
Nov 4	Aeropesca (Colmb.)	Con.–PBY-5A, disappeared, southeast Colmb.	6
Nov 8	American (US)	B-727, on landing, Cincinnati, Ohio, US	58
Nov 11	United (US)	B-727, on landing, Salt Lake City, Utah, US	43
Nov 27	Edde (US)	DC-3, near Salt Lake City, Utah, US	13
Dec 4	Eastern (US)	L-Super Constellation, midair collision with TWA (US) B-707 near Danbury, Connecticut, US	4
Dec 7	Spantax (Sp.)	DC-3, near Tenerife, Canary Islands, Sp.	32

1966

DATE	AIRLINE	AIRCRAFT AND LOCATION	KILLED
Jan 1	Garuda (Indon.)	DC-3, midair collision, southern Sumatra, Indon.	17
Jan 2	—	G-21A, on landing, Vancouver, British Columbia, Can.	3
Jan 14	AVIANCA (Colmb.)	DC-4, on takeoff, Cartagena, Colmb.	56
Jan 22	COHATA (Haiti)	DC-3, near Duchity, Haiti	30
Jan 24	Air India (Ind.)	B-707, hit Mt. Blanc, Fr.	117
Jan 28	Lufthansa (W. Ger.)	CV-440, on landing, Bremen, W. Ger.	46
Feb 4	All Nippon (Jap.)	B-727, on landing, Tokyo, Jap.	133
Feb 7	Indian Airlines (Ind.)	F-27, near Banihal Pass, Kashmir	37
Feb 15	Indian Airlines (Ind.)	SA-Caravelle, near Palam, Ind.	2

DATE	AIRLINE	AIRCRAFT AND LOCATION	KILLED
Feb 17	Aeroflot (USSR)	TU-114, on takeoff, Moscow, USSR	21
Feb 18	BIAS (Belg.)	DC-6, near Milan, It.	4
Mar 4	C P Air (Can.)	DC-8, on landing, Tokyo, Jap.	64
Mar 5	BOAC (GB)	B-707, hit Mt. Fujiyama, Jap.	124
Mar 27	Cubana (Cuba)	IL-18, hijack attempt, Cuba	2
Apr 22	American Flyers (US)	L-Electra II, near Ardmore, Oklahoma, US	83
Apr 27	LANSA (Peru)	L-Constellation, hit mountain, near Lima, Peru	49
Jun 29	PAL (Phil.)	DC-3, near Sablayan, Mindoro, Phil.	26
Aug 6	Braniff (US)	BAC 111, near Falls City, Nebraska, US	42
Aug 11	TAROM (Rum.)	IL-14, near Sibiu, Rum.	24
Aug 31	Britannia (GB)	Br-Britannia, near Ljubljana, Yug.	98
Sep 16	Iberia (Sp.)	DC-3, on takeoff, Tenerife, Canary Islands, Sp.	1
Sep 22	Ansett (Austrl.)	VC-Viscount, near Winton, Queensland, Austrl.	24
Oct 1	West Coast (US)	DC-9, Cascade Mountains, Oregon, US	18
Nov 13	All Nippon (Jap.)	Nihon-YS-11, ditched in sea, near Matsuyama, Jap.	50
Nov 22	Aden Air (GB [1])	DC-3, bomb explosion, near Aden, S. Yemen	30
Nov 24	TABSO (Bul.)	IL-18, near Bratislava, Czech.	82
Nov 26	—	Con.–PBY-5A, near La Pedrera, Colmb.	9
Dec 18	Aerocondor (Colmb.)	L-Super Constellation, near Bogota, Colmb.	17
Dec 24	AVIANCA (Colmb.)	DC-3, near Cascubel River, Colmb.	29

1967

Feb 6	Syrian Arab (Syria)	DC-3, on landing, Aleppo, Syria	1

DATE	AIRLINE	AIRCRAFT AND LOCATION	KILLED
Feb 9	Cubana (Cuba)	AN-12, on landing, Mexico City, Mex.	10
Feb 16	Garuda (Indon.)	L-Electra II, on landing, Manado, Celebes, Indon.	22
Feb 20	SAHSA (Hond.)	DC-6, on landing, Tegucigalpa, Hond.	4
Feb 28	PAL (Phil.)	F-27, on landing, Cebu Is., Phil.	12
Mar 5	VARIG (Braz.)	DC-8, near Monrovia, Liberia	51 5 *
Mar 5	Lake Central (US)	CV-580, near Marseilles, Ohio, US	37
Mar 9	TWA (US)	DC-9, midair collision, near Dayton, Ohio, US	25 1 *
Mar 10	West Coast (US)	F-27, hit mountain, near Klamath Falls, Oregon, US	4
Mar 13	SAA (SA)	VC-Viscount, near East London, SA	25
Mar 30	Delta (US)	DC-8, training flight, New Orleans, Louisiana, US	6 13 *
Apr 2	Caribbean (US)	L-Lodestar, ditched in sea, near Lima, Peru	5
Apr 11	Air Algerie (Alg.)	DC-4, near Tamanrasset, Alg.	35
Apr 19	Globe Air (Swit.)	BR-Britannia, near Nicosia, Cyprus	126
Apr 27	AVIANCA (Colmb.)	DC-3, near Sogamoso, Colmb.	17
May 15	—	C-46, near Cape Dyer, Northwest Territories, Can.	4
Jun 3	Air Ferries (GB)	DC-6, hit Mt. Canigou, near Perpignan, Fr.	88
Jun 4	Midland (GB)	CL-Argonaut, near Stockport, Cheshire, GB	72
Jun 23	Mohawk (US)	BAC 111, near Blossburg, Pennsylvania, US	34
Jun 24	Saudi Arabian (Saud.)	DC-3, near Khalifnsier, Saud.	16
Jun 30	Thai (Thai.)	SA-Caravelle, ditched in sea on landing, Hong Kong, Tai.	24
Jul 6	PAL (Phil.)	F-27, near Bacolod, Negros Is., Phil.	21

DATE	AIRLINE	AIRCRAFT AND LOCATION	KILLED
Jul 19	Madagascar (MR)	DC-4, on takeoff, Tananarive, MR	42
Jul 19	Piedmont (US)	B-727, midair collision, near Hendersonville, North Carolina, US	79 3 *
Jul 25	Taiwan (Tai.)	DC-3, near Luang Prabang, Laos	18
Jul 25	Vehu Akat (Laos)	DC-3, near Luang Prabang, Laos	16
Aug 11	—	DC-3, near Marinilla, Colmb.	6
Aug 21	China (Tai.)	DC-3, ditched in sea, off S. VN	2
Sep 5	CSA (Czech.)	IL-18, near Gander, Newfoundland, Can.	37
Oct 9	LACSA (Costa Rica)	BAC 111, Caribbean Sea	2
Oct 12	BEA (GB)	DH-Comet IV, bomb explosion, near Rhodes Is., Gre.	66
Oct 24	China (Tai.)	DC-3, S. VN	16
Nov 3	Sadia (Braz.)	HP-Herald, hit hill, near Curitiba, Braz.	21
Nov 4	Iberia (Sp.)	SA-Caravelle, near Fernhurst, Sussex, GB	37
Nov 5	Cathay Pacific (HK)	CV-880, on takeoff, Hong Kong	1
Nov 6	TWA (US)	B-707, on takeoff, Cincinnati, Ohio, US	1
Nov 11	Aeroflot (USSR)	IL-18, hit mountain, near Sverdlovsk, USSR	130
Nov 20	TWA (US)	CV-880, near Constance, Kentucky, US	70
Dec 7	Lao Cathay (Laos)	DC-3, near Muong Soui, Laos	10
Dec 8	Faucett (Peru)	DC-4, Andes Mountains, near Juanuco, Peru	66

1968

Jan 27	Comores (Fr.)	DH-Heron, on landing, Comores Islands, Fr.	15

DATE	AIRLINE	AIRCRAFT AND LOCATION	KILLED
Feb 16	CAT B (Tai.)	B-727, near Taipei, Tai.	21 1 *
Feb 24	Air Lao (Laos)	DC-3, ditched in Mekong River, Laos	37
Mar 6	Air France (Fr.)	B-707, hit mountain, Guadeloupe Is., WI	43
Mar 8	Air Manila (Phil.)	F-27, near Ibajay, Phil.	14
Mar 24	Aer Lingus (S. Ire.)	VC-Viscount, ditched in Irish Sea, near Tusker Rock, S. Ire.	61
Apr 8	LADECO (Ch.)	DC-3, near Coihaique, Ch.	36
Apr 8	BOAC (GB)	B-707, on takeoff, London, GB	5
Apr 10	Rojas (Mex.)	DC-3, near Mexico City, Mex.	18
Apr 20	SAA (SA)	B-707, on takeoff, Windhoek, Namibia	123
May 3	Braniff (US)	L-Electra II, near Dawson, Texas, US	85
May 21	Garuda (Indon.)	CV-990, on takeoff, Bombay, Ind.	29 1 *
Jun 12	Pan Am (US)	B-707, on landing, Calcutta, Ind.	6
Jun 28	Purdue (US)	DC-3, passenger fell out open door, near Vichy, Missouri, US	1
Jul 1	—	L-Constellation, Nig.	4
Jul 3	BKS (GB)	DH-Ambassador, on landing, Heathrow Airport, London, GB	6
Jul 8	Saudi Arabian (Saud.)	CV-340, near Dhahran, Saud.	11
Jul 13	SABENA (Belg.)	B-707, near Lagos, Nig.	7
Aug 2	Alitalia (It.)	DC-8, near Malpensa. It.	12
Aug 9	Eagle (GB)	VC-Viscount, near Munich, W. Ger.	48
Aug 10	Piedmont (US)	FH-227, near Charleston, West Virginia, US	35
Aug 18	Egyptair (Egy.)	AN-24, ditched in Mediterranean Sea	40
Aug 20	LAV (Venz.)	HS-748, near Maturin, Venz.	4
Aug 23	TABSA (Bol.)	C-46, near Nieve, Bol.	4
Sep 3	Balkan (Bul.)	IL-18, near Karnobat, Bul.	47
Sep 11	Air France (Fr.)	SA-Caravelle, near Nice, Fr.	95

DATE	AIRLINE	AIRCRAFT AND LOCATION	KILLED
Sep 28	Pan African (US)	DC-4, near Port Harcourt, Nig.	57 1 *
Oct 11	CSA (Czech.)	IL-14, near Prague, Czech.	11
Oct 25	Northeast (US)	FH-227, near Hanover, New Hampshire, US	32
Nov 25	Air America (US)	C-46, near Savannakhet, Laos	26
Dec 2	Wien (US)	F-27, near Pedro Bay, Alaska, US	39
Dec 12	Pan Am (US)	B-707, ditched in sea, near Caracas, Venz.	51
Dec 24	Allegheny (US)	CV-580, near Bradford, Pennsylvania, US	20
Dec 26	—	DC-3, hit mountain, near Ensenada, Mex.	12
Dec 26	El Al (Israel)	B-707, terrorists aboard, near Athens, Gre.	1
Dec 27	North Central (US)	CV-580, on landing, Chicago, Illinois, US	27 1 *
Dec 31	SAESA (Mex.)	DC-3, hit mountain, northeast Mex.	26
Dec 31	McRobertson Millar (Austrl.)	VC-Viscount, near Port Hedland, Western Austrl.	26

1969

Jan 2	China (Tai.)	DC-3, near Nan-Pao Shan, China	24
Jan 5	Ariana Afghan (Afg.)	B-727, on landing, Gatwick Airport, London, GB	50 2 *
Jan 6	Continental A.S. (US)	DC-3, northeast Thai.	55
Jan 6	Allegheny (US)	CV-440, near Bradford, Pennsylvania, US	11
Jan 13	SAS (Swe.)	DC-8, ditched in Santa Monica Bay, California, US	15
Jan 18	Air America (US)	DC-3, hit mountain, near Da Nang, S. VN	9
Jan 18	United (US)	B-727, near Los Angeles, California, US	38

DATE	AIRLINE	AIRCRAFT AND LOCATION	KILLED
Feb 5	Angeles (US)	—, near Port Angeles, Washington, US	10
Feb 18	Hawthorne, Nevada (US)	DC-3, hit Mt. Whitney, California, US	35
Feb 24	FEAT (Tai.)	HP-Herald, near T'ai-nan, Tai.	36
Mar 5	PRINAIR (US)	DH-Heron, hit mountain, near San Juan, Puerto Rico	19
Mar 16	VIASA (Venz.)	DC-9, near Maracaibo, Venz.	84 70 *
Mar 20	Avion (US)	DC-3, near New Orleans, Louisiana, US	16
Mar 20	Egyptair (Egy.)	IL-18, on landing, Aswan, Egy.	100
Apr 2	LOT (Pol.)	AN-24, near Krakow, Pol.	53
Apr 7	Air Canada (Can.)	VC-Viscount, on takeoff, Sept. Iles, Quebec, Can.	1
Apr 16	—	C-46, ditched in River Congo	45
Apr 21	Indian Airlines (Ind.)	F-27, near Khulna, Pak.	44
Apr 24	Ghana Air (Ghana)	DC-3, near Takoradi, Ghana	1
May 23	UBA (Burma)	DC-3, on landing, near Lashio, Burma	6
May 24	Itavia (It.)	F-27, near Reggio di Calabria, It.	1
Jun 4	Mexicana (Mex.)	B-727, hit mountain, near Monterrey, Mex.	79
Jul 12	Nepal Airlines (Nepal)	DC-3, near Katmandu, Nepal	35
Jul 15	New York Airways (US)	DH(C)-6, JFK Airport, New York City, US	3
Jul 26	Air Algerie (Alg.)	SA-Caravelle, near Biskra, Alg.	33
Aug 26	Aeroflot (USSR)	IL-18, Vnukovo Airport, Moscow, USSR	16
Sep 6	PAL (Phil.)	HS-748, bomb explosion, near Zamboanga, Phil.	1
Sep 8	SATENA (Colmb).	DC-3, Meta Province, Colmb.	32
Sep 9	Allegheny (US)	DC-9, midair collision, near Fairland, Indiana, US	82 1 †

DATE	AIRLINE	AIRCRAFT AND LOCATION	KILLED
Sep 12	PAL (Phil.)	BAC 111, near Manila, Phil.	45
Sep 14	VASP (Braz.)	DC-3, on takeoff, Londrina, Braz.	20
Sep 17	Pacific Western (Can.)	CV-640, near Campbell River, British Columbia, Can.	4
Sep 20	Air Vietnam (S. VN)	DC-4, midair collision, near Da Nang, S. VN	74 2 *
Sep 21	Mexicana (Mex.)	B-727, on landing, Mexico City, Mex.	27
Sep 26	LAB (Bol.)	DC-6, hit mountain, near La Paz, Bol.	74
Nov 9	—	DC-3, near Timmins, Ontario, Can.	4
Nov 19	Mohawk (US)	FH-227, near Glens Falls, New York, US	14
Nov 20	Nigeria Airways (Nig.)	VC-10, near Lagos, Nig.	87
Dec 3	Air France (Fr.)	B-707, ditched in sea, near Caracas, Venz.	62
Dec 5	—	L-18, near Albuquerque, New Mexico, US	11
Dec 8	Olympic (Gre.)	DC-6, hit mountain, near Athens, Gre.	90
Dec 22	Air Vietnam (S. VN)	DC-6, near Nha Trang, S. VN	10 24 *
Dec 23	Lao Airlines (Laos)	DC-3, near Luang Prabang, Laos	6

1970

Jan 2	Air Lao (Laos)	—, hit mountain, near Vientiane, Laos	4
Jan 5	Spantax (Sp.)	CV-990, near Stockholm, Swe.	5
Jan 13	Faucett (Peru)	DC-4, Mt. Pozo Chuno, Peru	28
Jan 13	Polynesian (Samoa)	DC-3, ditched in sea, near Falula, Western Samoa	32
Jan 21	Rocky Mountain (US)	AC-680, near Aspen, Colorado, US	8
Jan 25	—	CV-340, near Poza Rica, Mex.	19
Jan 28	TAG (US)	DH-Dove, ditched in Lake Erie, US	9
Feb 4	TAROM (Rum.)	AN-24, Aprisini Mountains, Rum.	21

DATE	AIRLINE	AIRCRAFT AND LOCATION	KILLED
Feb 4	Aero. Argentinas (Arg.)	HS-748, near Loma Alta, Arg.	37
Feb 11	Pilgrim (US)	DH(C)-6, ditched in Long Island Sound, New York, US	5
Feb 12	Urraca (Colmb.)	DC-3, near Puerto Infrida, Colmb.	12 2 *
Feb 15	CDA (DR)	DC-9, ditched in sea, near Santo Domingo, DR	102
Feb 21	Swissair (Swit.)	CV-990, bomb explosion, near Wurenlingen, Swit.	47
Mar 6	Bavaria (W. Ger.)	HP-Jetstream, near Samedan, Swit.	11
Mar 14	Paraense (Braz.)	FH-227, ditched in sea, near Belem, Braz.	38
Mar 22	Commuter (US)	BE-Dumod, on takeoff, Binghamton, New York, US	3
Apr 1	Air Maroc (Morocco)	SA-Caravelle, near Casablanca, Morocco	61
Apr 21	PAL (Phil.)	HS-748, bomb explosion, near Manila, Phil.	36
May 2	Overseas National (US)	DC-9, ditched in sea, near St. Croix, Virgin Islands, WI	23
May 6	Somali Air (Somali Rep.)	VC-Viscount, on landing, Mogadiscio, Somali Republic	5
May 22	AVENSA (Venz.)	C-46, Costa Rica	6
May 26	Aeroservicios (Venz.)	DH-Heron, near Tocantin, Braz.	6
Jun 1	CSA (Czech.)	TU-104, near Tripoli, Libya	13
Jun 2	PAL (Phil.)	F-27, between Romblon-Roxas, Phil.	1
Jul 3	Dan Air (GB)	DH-Comet IV, Montseny Mountains, Sp.	112
Jul 5	Air Canada (Can.)	DC-8, on landing, Toronto, Ontario, Can.	109
Jul 18	Aeroflot (USSR)	AN-22, near Kflavik, Ice.	23
Aug 6	PIA (Pak.)	F-27, between Islamabad-Lahore, Pak.	30
Aug 9	LANSA (Peru)	L-Electra II, near Cuzco, Peru	99 2 *

DATE	AIRLINE	AIRCRAFT AND LOCATION	KILLED
Aug 12	China (Tai.)	Nihon-YS-11, near T'ai-pei, Tai.	14
Aug 29	Indian Airlines (Ind.)	F-27, near Silchar, Ind.	39
Sep 26	SAS (Swe.)	F-27, near Vagar, Faeroes, Dnk.	8
Oct 2	Richards (US)	M-404, near Loveland, Colorado, US	32
Oct 6	National Airways (SA)	DC-3, near Germiston, SA	3
Nov 14	Southern (US)	DC-9, near Huntington, West Virginia, US	75
Nov 27	Capitol (US)	DC-8, on takeoff, Anchorage, Alaska, US	47
Dec 5	Jamair (Ind.)	DC-3, near Delhi, Ind.	5
Dec 7	TAROM (Rum.)	BAC 111, near Constanta, Rum.	18
Dec 19	Aeroflot (USSR)	AN-22, near Calcutta, Ind.	17
Dec 28	Trans-Caribbean (US)	B-727, on landing, St. Thomas, Virgin Islands, WI	2
Dec 31	STA (Fr.)	Nord-262, disappeared, Alg.	31
Dec 31	PIA (Pak.)	F-27, on landing, Shamshernagar, Pak.	7
Dec 31	Aeroflot (USSR)	IL-18, on takeoff, Leningrad, USSR	93

1971

Jan 2	Egyptair (Egy.)	DH-Comet IV, near Tripoli, Libya	16
Jan 18	Balkan (Bul.)	IL-18, on landing, Zurich, Swit.	45
Jan 23	Korean (S. Kor.)	F-27, attempted hijack, near Kangnung, S. Kor.	2
Jan 25	LAV (Venz.)	VC-Viscount, near Merida, Venz.	13
Feb 19	TAM (Bol.)	C-46, Andes Mountains, near La Paz, Bol.	12
Mar 17	Air America (US)	—, collided with military aircraft, Can Tho, S. Vn	3
Mar 26	Jamair (Ind.)	DC-3, hit mountain, near Gauhati, Ind.	15
May 6	Apache (US)	DH-Dove, near Coolidge, Arizona, US	12

DATE	AIRLINE	AIRCRAFT AND LOCATION	KILLED
May 11	AVIATECA (Guat.)	C-46, near Guatemala City, Guat.	5
May 23	Aviogenex (Yug.)	TU-134, on landing, Belgrade, Yug.	78
May 25	Austral (Ch.)	C-46, on takeoff, Santiago, Ch.	4
Jun 6	Hughes Airwest (US)	DC-9, midair collision, near Duarte, California, US	49 1 *
Jun 7	Allegheny (US)	CV-580, on landing, New Haven, Connecticut, US	28
Jun 17	Copacabana (Braz.)	DC-6, hit Mt. Putre, Bol.	6
Jun 28	Shelter Cove (US)	DC-3, ditched in sea, on takeoff, Shelter Cove, California, US	17
Jul 3	TOA (Jap.)	Nihon-YS-11, hit mountain, near Hakodate, Jap.	68
Jul 24	Air Senegal (Senegal)	DC-3, near Bamako, Mali	6
Jul 30	All Nippon (Jap.)	B-727, midair collision, near Shizukuishi, Jap.	162
Aug 11	Aeroflot (USSR)	TU-104, near Irkutsk, USSR	100
Aug 28	MALEV (Hung.)	IL-18, near Copenhagen, Den.	32
Sep 4	Alaska (US)	B-727, hit mountain, near Juneau, Alaska, US	111
Sep 6	Pan International (W. Ger.)	BAC 111, near Hamburg, W. Ger.	22
Sep 16	MALEV (Hung.)	TU-134, near Kiev, USSR	49
Sep 16	Yemen (Yemen)	DC-3, near Presevo, Yug.	5
Sep 28	Cruzeiro (Braz.)	DC-3, near Madureira, Braz.	32
Oct 2	BEA (GB)	VC-Vanguard, near Ghent, Belg.	63
Oct 13	Aeroflot (USSR)	TU-104, on takeoff, Moscow, USSR	20
Oct 17	TAO ᴬ (Ecu.)	DC-3, near San Vicente del Caguan, Colmb.	19
Oct 21	Chicago & Southern (US)	BE-E18S, near Peoria, Illinois, US	16
Nov 10	Merpati Nusantera (Indon.)	VC-Viscount, ditched in sea, near Padang, Sumatra, Indon.	69

DATE	AIRLINE	AIRCRAFT AND LOCATION	KILLED
Nov 20	China (Tai.)	SA-Caravelle, ditched in Formosa Strait, Tai.	25
Dec 6	Sudan (Sudan)	F-27, Equatoria Province, Sudan	10
Dec 9	Indian Airlines (Ind.)	HS-748, near Chinnamanur, Ind.	21
Dec 21	Balkan (Bul.)	IL-18, near Sofia, Bul.	28
Dec 24	LANSA (Peru)	L-Electra II, near Puerto Inca, Peru	91

1972

DATE	AIRLINE	AIRCRAFT AND LOCATION	KILLED
Jan 6	SAESA (Mex.)	HS-748, near Chetumal, Mex.	23
Jan 7	Iberia (Sp.)	SA-Caravelle, near Ibiza, Balearic Islands, Sp.	104
Jan 17	TAO [B] (Colmb.)	DC-3, on landing, Caquetania, Colmb.	1
Jan 21	Urraca (Colmb.)	VC-Viscount, near Bogota, Colmb.	20
Jan 21	SATENA (Colmb.)	DC-3, near San Nicolas, Colmb.	39
Jan 26	JAT (Yug.)	DC-9, bomb explosion, near Hermsdorf, E. Ger.	27
Feb 3	Lao Air (Laos)	DC-6, near Tegal, Java, Indon.	6
Feb 5	TAC (Colmb.)	FH-227, near Valledupar, Colmb.	19
Feb 11	Lao Air (Laos)	DC-4, disappeared, Laos	23
Mar 3	Mohawk (US)	FH-227, hit house, near Albany, New York, US	16 1 *
Mar 14	Sterling (Den.)	SA-Caravelle, hit mountain, near Al Fujayrah, Trucial States	112
Mar 19	Egyptair (Egy.)	DC-9, hit mountain, near Aden, S. Yemen	30
Apr 13	VASP (Braz.)	Nihon-YS-11, hit mountain, near Rio de Janeiro, Braz.	25
Apr 16	Itavia (It.)	F-27, hit mountain, near Frosinone, It.	18
Apr 18	EAAC (Kenya)	VC-10, on takeoff, Addis Ababa, Ethio.	43
Apr 20	SATCO (Peru)	C-46, hit mountain, northern Peru	6
May 5	Alitalia (It.)	DC-8, hit mountain, near Palermo, Sicily, It.	115

DATE	AIRLINE	AIRCRAFT AND LOCATION	KILLED
May 8	Aerotecnica (Venz.)	DC-3, Venz.	7
May 18	Aeroflot (USSR)	AN-10, near Kharkov, Ukraine, USSR	108
May 21	DTA (Angola)	F-27, ditched in sea, near Lobito, Angola	22
May 29	Amazonense (Braz.)	L-Constellation, near Cruzeiro do Sul, Braz.	9
Jun 14	JAL (Jap.)	DC-8, near Delhi, Ind.	82
Jun 15	Cathay Pacific (HK)	CV-880, suspected bomb explosion, near Pleiku, S. VN	81
Jun 18	BEA (GB)	HS-Trident 1, Staines, Surrey, GB	118
Jun 24	PRINAIR (Puerto Rico)	DH-Heron, near Ponce, Puerto Rico	20
Jun 29	Air Wisconsin (US)	DH(C)-6 Midair collision, near Appleton, Wisconsin, US	8
Jun 29	North Central (US)	CV-580	5
Jun 29	Inter-city Flug (W. Ger.)	MMB-HFB-320, on takeoff, Blackpool, GB	7
Jul 29	AVIANCA (Colmb.)	DC-3 Midair collision, near Los Palamos, Colmb.	21
Jul 29	AVIANCA (Colmb.)	DC-3	17
Aug 11	Indian Airlines (Ind.)	F-27, on landing, Delhi, Ind.	18
Aug 14	Interflug (E. Ger.)	IL-62, near East Berlin, E. Ger.	156
Aug 16	Burma Airways (Burma)	DC-3, ditched in sea, near Sandoway, Burma	28
Aug 27	LAV (Venz.)	DC-3, near Canaima, Venz.	34
Sep 1	Ansett (Austrl.)	Short-Skyvan, hit Mt. Giluwe, Papua, NG	4
Sep 10	Ethiopian (Ethio.)	DC-3, near Gondar, Ethio.	11
Sep 13	Nepalese (Nepal)	DC-3, Katmandu Valley, Nepal	31

DATE	AIRLINE	AIRCRAFT AND LOCATION	KILLED
Sep 24	Air Vietnam (S. VN)	DC-4, near Saigon, S. VN	10
Oct 2	Cambodia Air (Cam.)	DC-3, hit by mortar, near Kampot, Cam.	9
Oct 2	Aeroflot (USSR)	IL-18, near Sochi, USSR	100
Oct 13	Aeroflot (USSR)	IL-62, near Moscow, USSR	174
Oct 21	Olympic (Gre.)	Nihon-YS-11, near Athens, Gre.	37
Oct 27	Air Inter (Fr.)	VC-Viscount, near Clermont-Ferrand, Fr.	60
Oct 30	Italiani (It.)	F-27, near Poggiorsini, It.	27
Nov 4	Balkan (Bul.)	IL-14, Rila Planina Mts., near Plovdiv, Bul.	35
Nov 18	—	DC-3, missing between Ice.– Newfoundland, Can.	3
Nov 28	JAL (Jap.)	DC-8, on takeoff, Moscow, USSR	62
Dec 3	Spantax (Sp.)	CV-990, on takeoff, Tenerife, Canary Islands, Sp.	155
Dec 8	PIA (Pak.)	F-27, near Jalkot, Pak.	26
Dec 8	United (US)	B-737, on landing, Chicago, Illinois, US	43 2 *
Dec 20	North Central (US)	DC-9, ground collision, Chicago, Illinois, US	10
Dec 23	Braathens SAFE (Nor.)	F-28, near Oslo, Nor.	40
Dec 29	Eastern (US)	L-1011 TriStar, Everglades Swamp, Florida, US	101

1973

Jan 22	ALIA [3] (Jord.)	B-707, on landing, Kano, Nig.	176
Jan 29	Egyptair (Egy.)	IL-18, near Nicosia, Cyprus	37
Feb 19	Aeroflot (USSR)	TU-154, near Prague, Czech.	66
Feb 21	Libyan Arab (Libya)	B-727, shot down, Sinai desert	106

DATE	AIRLINE	AIRCRAFT AND LOCATION	KILLED
Feb 21	Urraca (Colmb.)	DC-3, hit mountain, near Boquete, Panama	22
Mar 3	Balkan (Bul.)	IL-18, near Moscow, USSR	25
Mar 5	Iberia (Sp.)	DC-9, midair collision, near Nantes, Fr.	68
Mar 19	Air Vietnam (S. VN)	DC-4, near Ban Me Thuot, S. VN	57
Apr 10	Invicta (GB)	VC-Vanguard, near Basel, Swit.	108
May 5	Egyptair (Egy.)	B-707, in-flight turbulence, the Alps Mountains	1
May 19	Cambodia Air (Cam.)	DC-3, near Svay Rieng, Cam.	11
May 29	Air Gaspe (Can.)	DC-3, near Rimouski, Quebec, Can.	4
May 29	Amazonense (Braz.)	L-Constellation, near Lagoinha, Braz.	9
May 31	Indian Airlines (Ind.)	B-737, near Delhi, Ind.	48
Jun 1	Cruzeiro (Braz.)	SA-Caravelle, near Sao Luis, Braz.	23
Jun 20	Aeromexico (Mex.)	DC-9, near Puerto Vallarta, Mex.	27
Jun 30	Aeroflot (USSR)	TU-134, on takeoff, Amman, Jord.	2 7 *
Jun —	Catalina (Colmb.)	Con.–PBY-5A, near Villavicencio, Colmb.	1 *
Jul 11	VARIG (Braz.)	B-707, near Paris, Fr.	123
Jul 23	Ozark (US)	FH-227, near St. Louis, Missouri, US	39
Jul 23	Pan Am (US)	B-707, near Papeete, Tahiti Is.	79
Jul 31	Delta (US)	DC-9, on landing, Boston, Massachusetts, US	89
Aug 13	AVIACO (Sp.)	SA-Caravelle, on landing, La Coruna, Sp.	85 5 *
Aug 22	AVIANCA (Colmb.)	DC-3, central Colmb.	16
Aug 27	Condor (Colmb.)	L-Electra II, hit mountain, near Bogota, Colmb.	42
Aug 29	TWA (US)	B-707, turbulence, near Los Angeles, California, US	1

DATE	AIRLINE	AIRCRAFT AND LOCATION	KILLED
Sep 8	World (US)	DC-8, near King Cove, Alaska, US	6
Sep 11	JAT (Yug.)	SA-Caravelle, near Titograd, Yug.	41
Sep 28	Texas International (US)	CV-600, near Mena, Arkansas, US	11
Oct 1	—	DC-3, near Miritituba, Braz.	8
Oct 13	Aeroflot (USSR)	TU-104, near Domodedovo, Moscow, USSR	28
Oct 23	VASP (Braz.)	Nihon-YS-11, on takeoff, Rio de Janeiro, Braz.	5
Nov 2	Urraca (Colmb.)	HP-Herald, near Villavicencio, Colmb.	6
Nov 3	National (US)	DC-10, punctured fuselage decompression, near Albuquerque, New Mexico, US	1
Nov 17	Air Vietnam (S. VN)	DC-3, hit mountain, near Bato, S. VN	27
Dec 8	Aeroflot (USSR)	TU-104, on landing, Moscow, USSR	13
Dec 16	Aeroflot (USSR)	TU-124, near Moscow, USSR	65
Dec 21	TAM (Bol.)	CV-440, near Talara, Peru	6
Dec 22	Air Maroc [4] (Morocco)	SA-Caravelle, hit mountain, near Tangier, Morocco	106
Dec 28	National Airways (SA)	DC-3, near Durban, SA	1

1974

Jan 1	Itavia (It.)	F-28, near Turin, It.	38
Jan 9	SATENA (Colmb).	HS-748, hit mountain, near Florencia, Colmb.	31
Jan 10	TAM (Bol.)	DC-4, Bol.	24
Jan 17	Cessnyca (Colmb.)	DC-3, near Chigorodo, Colmb.	14
Jan 26	THY (Turk.)	F-28, near Izmir, Turk.	63
Jan 30	Pan Am (US)	B-707, on landing, Pago Pago, American Samoa	97

DATE	AIRLINE	AIRCRAFT AND LOCATION	KILLED
Feb 22	SAVCO (Bol.)	C-46, near San Francisco de Moxos, Bol.	7
Mar 3	THY (Turk.)	DC-10, explosive decompression, over Ermenonville, Paris, Fr.	346
Mar 13	Sierra Pacific (US)	CV-340, near Bishop, California, US	36
Mar 15	Sterling (Den.)	SA-Caravelle, on takeoff, Teheran, Iran	15
Apr 4	Wenela (Botswana)	DC-4, on takeoff, Francistown, Botswana	78
Apr 22	Pan Am (US)	B-707, near Denpasar, Bali Is., Indon.	107
Apr 27	Aeroflot (USSR)	IL-18, near Leningrad, USSR	118
May 2	ATESA (Ecu.)	DC-3, Andes Mountains, Ecu.	22
Jun 8	TAO B (Colmb.)	VC-Viscount, near Cucuta, Colmb.	44
Jun 27	Cambodia Air (Cam.)	B-Stratoliner, on takeoff, Battambang, Cam.	19
Aug 5	Laurentian (Can.)	DC-3, near Mt. Apica, Quebec, Can.	5
Aug 11	Air Mali (Mali)	IL-18, near Ouagadougou, Upper Volta	47
Aug 12	AVIANCA (Colmb.)	DC-3, missing, near Cali, Colmb.	24
Aug 14	LAV (Venz.)	VC-Viscount, hit mountain, near Porlamar, Venz.	47
Sep 7	Garuda (Indon.)	F-27, on landing, Tandjungkarang, Sumatra, Indon.	33
Sep 8	TWA (US)	B-707, bomb explosion, Ionian Sea	88
Sep 11	Eastern (US)	DC-9, on landing, Charlotte, North Carolina, US	70
Sep 15	Air Vietnam (S. VN)	B-727, hijack, near Phan Rang, S. VN	75
Oct 5	—	DC-3, Guat.	6
Oct 29	Panartic (Can.)	L-Electra II, near Rae Point, Northwest Territories, Can.	32
Nov 20	Lufthansa (W. Ger.)	B-747, on takeoff, Nairobi, Kenya	59
Dec 1	TWA (US)	B-727, near Washington, D.C., US	92

DATE	AIRLINE	AIRCRAFT AND LOCATION	KILLED
Dec 4	Martinair (Hol.)	DC-8, hit mountain, near Maskeliya, Sri Lanka	191
Dec 22	AVENSA (Venz.)	DC-9, near Maturin, Venz.	77
Dec 28	—	L-18, on takeoff, Tikal, Guat.	24
Dec 29	TAROM (Rum.)	AN-24, Lotru Mountains, Rum.	32

1975 [5]

Jan 16	Aeroflot (USSR)	AN-2, near Sam Neua, Laos	12
Jan 30	THY (Turk.)	F-28, ditched in Sea of Marmara, on approach, Istanbul, Turk.	42
Feb 3	PAL (Phil.)	HS-748, near Manila, Phil.	33
Feb 27	VASP (Braz.)	EMB-110, on takeoff, Sao Paulo, Braz.	15
Mar 12	Air Vietnam (S. VN)	DC-4, near Pleiku, S. VN	26
Apr 23	SAVCO (Bol.)	C-46, hit mountain, near Soyari, Bol.	4
May 7	Pan Am (US)	B-707, pressurization system malfunction causes passenger heart attack, near St. Johns, Newfoundland, Can.	1
Jun 3	PAL (Phil.)	BAC 111, in-flight bomb explosion, near Manila, Phil.	1
Jun 24	Eastern (US)	B-727, on landing, Kennedy Airport, New York City, US	113
Jul 31	FEAT (Tai.)	VC-Viscount, on landing, near Hua-lien, Tai.	27
Jul —	—	YAK-40, near Batumi, USSR	28
Aug 3	Air Maroc [6] (Morocco)	B-707, hit mountain, near Agadir, Morocco	188
Aug 19	CSA (Czech.)	IL-62, on approach, near Damascus, Syria	126
Aug 30	Wien Air Alaska (US)	F-27, hit mountain, Saint Lawrence Island, Alaska, US	10

DATE	AIRLINE	AIRCRAFT AND LOCATION	KILLED
Sep 1	Interflug E. (Ger.)	TU-134, on approach, near Leipzig, E. Ger.	26
Sep 24	Garuda (Indon.)	F-28, hit tree on approach, near Palembang, Sumatra, Indon.	26 1 *
Sep 27	Entre Rios (Arg.)	CL-44, on takeoff, Miami, Florida, US	5
Sep 30	MALEV (Hung.)	TU-154, ditched in sea on approach, near Beirut, Leb.	60
Oct 23	Connair (Austrl.)	DH-Heron, on approach, near Cairns, Austrl.	11
Oct 30	Inex Adria (Yug.)	DC-9, hit high ground on approach, near Prague, Czech.	74
Nov 18	AVIATECA (Guat.)	DC-3, en route, Peten State, Guat.	15
Nov 22	Balkan (Bul.)	AN-24, near Sofia, Bul.	2

FOOTNOTES TO THE CHRONOLOGY

* Bystander killed.

† Occupant of other aircraft killed.

—Unreported or untraceable.

1. This Japan Air Lines was founded in 1951 while Japan was still under Allied occupation following World War II. The present Japan Air Lines began operations in October 1953.

2. Aircraft on lease from BOAC.

3. Aircraft chartered by Nigeria Airways.

4. Aircraft owned by SABENA and chartered by Air Maroc.

5. 1975 accidents are included in the chronology for the sake of topicality. Because traffic statistics for 1975 were not available at the time of compilation all other tables in this Appendix only go up to 1974.

6. Aircraft owned by ALIA and chartered by Air Maroc.

Many of the airlines listed in the Chronological List of Air Disasters are no longer in business. In the United States, for example, Northeast and Lake Central have been swallowed up by Delta and Eastern respectively; British Eagle succumbed to bankruptcy and British United was taken over by Caledonian (now British Caledonian); and Deccan of India, LAI of Italy, and PAB of Brazil have all disappeared. The safety records of those airlines—and others that have suffered similar fates—are, obviously, of little relevance because the chief aim of our survey is to provide passengers with an approximate guide to the performance of airlines that are still flying.

The rest of the tables in this appendix, therefore, are devoted to airlines who (at the time of writing) still operate scheduled passenger services. The list of airlines we surveyed is not totally comprehensive but it does include most international "flag carriers" and all of the world's major airlines, measured in terms of the number of passengers carried annually. The first table below shows the number of fatal accidents suffered by each of seventy-four airlines between 1950 and 1974 and the number of people who died as a result. (Some carriers did not begin operations until after 1950. In those cases the commencement date is shown after the airline's name.)

TABLE ONE

AIRLINE	PASSEN-GER-FATAL ACCI-DENTS	PASSEN-GERS KILLED	CREW KILLED	OTHERS KILLED	TOTAL
Aer Lingus	2	77	7	0	84
Aeroflot [1]	28	1,375	157	7	1,539
Aerolineas Argentinas	12	282	58	2	342
Aeromexico	3	75	22	2	99
Air Afrique (1961)	0	0	0	0	0
Air Algerie (1962)	2	59	9	0	68
Air Canada [2]	8	305	24	3	332
Air Ceylon	0	0	0	0	0
Air France	19	829	110	0	939
Air India	3	157	27	0	184
ALIA (1963)	2	220	10	0	230
Alitalia	7	293	40	1	334
Allegheny	5	152	15	1	168
All Nippon (1953)	5	358	23	0	381
American	9	288	26	13	327
Ansett	3	34	9	0	43
Austrian Airlines (1958)	1	26	5	0	31
AVIACO	7	166	21	6	193
AVIANCA	16	310	56	3	369
BCAL (1961) [3]	1	101	10	0	111
Braniff	6	185	19	0	204
British Airways [4]	21	691	103	3	797
BWIA	0	0	0	0	0

AIRLINE	PASSEN-GER-FATAL ACCI-DENTS	PASSEN-GERS KILLED	CREW KILLED	OTHERS KILLED	TOTAL
Cathay Pacific	3	80	12	0	92
Continental	2	42	11	0	53
C P Air	6	163	37	1	201
Cruzeiro	11	146	34	7	187
CSA	10	212	49	0	261
Cubana	6	104	23	4	131
Cyprus Airways	0	0	0	0	0
Delta	2	99	9	13	121
EAAC	2	51	13	0	64
Eastern	9	389	33	4	426
Egyptair [5]	13	328	77	0	405
El Al	2	52	7	0	59
Ethiopian Airlines	3	14	10	0	24
Finnair	2	42	5	0	47
Garuda	7	130	37	1	168
Hughes Airwest (1968)	1	44	5	1	50
IAC	19	322	72	5	399
Iberia	9	317	42	0	359
Icelandair	2	25	7	0	32
Iran Air	3	29	14	0	43
Iraqi Airways	0	0	0	0	0
JAL (1953)	2	125	19	0	144
JAT	7	92	22	0	114
KLM	7	274	64	2	340
LAN	3	104	14	0	118
LOT	4	95	16	0	111
Lufthansa (1955)	3	126	15	0	141
MEA	2	53	10	90	153

AIRLINE	PASSEN-GER-FATAL ACCI-DENTS	PASSEN-GERS KILLED	CREW KILLED	OTHERS KILLED	TOTAL
Mexicana	5	124	16	2	142
NAC	2	22	3	0	25
National	6	139	20	4	163
Nigeria Airways	2	85	15	0	100
Northwest	13	357	49	2	408
Olympic (1957)	3	136	9	0	145
PAL	17	254	51	1	306
Pan Am	14	557	87	3	647
PIA (1955)	7	218	33	4	255
Qantas	1	6	1	0	7
SAA	4	158	30	0	188
SABENA	7	193	41	1	235
SAS	3	54	11	0	65
Swissair	3	115	15	1	131
TAA	1	25	4	0	29
TAP	0	0	0	0	0
TAROM	8	173	32	0	205
THY [6]	10	473	47	0	520
TWA	15	623	70	15	708
United	15	574	62	11	647
UTA (1963)	1	73	7	0	80
VARIG	5	294	33	5	332
VIASA (1961)	1	74	10	70	154

FOOTNOTES TO TABLE ONE

1. Until the USSR became a contracting member of ICAO in 1970, no records of Aeroflot's operations were made public. The figures in this table are therefore an estimate, based primarily on foreign press reports.

2. Until January 1965, Air Canada was called Trans Continental Airlines.

3. The accident occurred before Caledonian took over British United and changed its name to BCAL.

4. British Airways' figures refer to accidents suffered by both BEA and BOAC which were merged in April 1974.

5. Egyptair was formerly known as United Arab Airlines (when it ran a joint operation with Syrian Airways for ten months) and before that, Misrair.

6. THY was formerly known as DHY.

It would obviously be unrealistic to use only these death tolls in forming judgments about the safety standards of each airline. If that crude basis were employed, United Airlines (with 647 deaths) would emerge as practically the most dangerous carrier in the United States; in fact, United is one of the world's safest airlines.

In order to put the figures into perspective, it is necessary to find a factor common to all the carriers and against which they can each be judged. In the aviation industry, there is considerable disagreement over what that constant factor should be. But the most obvious is the number of people each airline has carried. That measuring stick provides, at least, a reasonable basis for comparison: we can compare the number of people each airline has carried with the number of people killed in its accidents.

Table Two gives the total number of passengers for each airline since 1950 or since the carrier began operations. In each case, that figure has then been divided into the number of people—excluding crew—who died. The result shows the number of people killed for every million passengers carried. (If it seems inhumane to refer to fractions of human beings—0.62, 0.71, 0.93 and so on—it can only be said that such apparent inhumanity is the inevitable result of using statistics to measure human affairs.)

The airlines are listed in order of merit—when judged by this single factor—beginning with those which have achieved the lowest average.

(Aeroflot has been omitted from this and all subsequent tables because of the lack of reliable data.)

TABLE TWO

AIRLINE	TOTAL KILLED	PASSENGERS CARRIED PASSENGERS	DEATHS PER * MILLION PASSENGERS
TAP	0	11.00	0
BWIA	0	5.62	0
Air Afrique	0	4.08	0
Iraqi Airways	0	2.51	0
Cyprus Airways	0	1.95	0
Air Ceylon	0	1.63	0
Delta	112	194.47	0.58
TAA	25	41.12	0.61
Qantas	6	9.86	0.61
Continental	42	56.78	0.74
SAS	54	69.42	0.78
Ansett	34	38.17	0.89
NAC	22	22.43	0.98
American	301	287.78	1.05
Eastern	393	323.75	1.21
United	585	346.56	1.64
Lufthansa	126	67.48	1.87
JAL	125	65.66	1.90
National	143	72.18	1.98
Braniff	185	87.39	2.12
Hughes Airwest	45	21.08	2.13
Swissair	116	52.07	2.23
Allegheny	153	66.76	2.29
Finnair	42	16.74	2.51
Air Canada	308	110.50	2.79
TWA	638	202.69	3.15
Iran Air	29	8.43	3.44

* Adjusted to the nearest decimal point.

AIRLINE	TOTAL KILLED	PASSENGERS CARRIED (IN MILLIONS)	DEATHS PER * MILLION PASSENGERS
Austrian Airlines	26	6.90	3.77
Aer Lingus	77	20.29	3.79
Ethiopian Airlines	14	3.69	3.79
Aeromexico	77	20.18	3.82
Northwest	359	89.97	3.99
Pan Am	560	139.00	4.03
British Airways	694	154.99	4.48
All Nippon	358	79.22	4.52
Mexicana	126	26.76	4.71
Iberia	317	65.63	4.83
BCAL	101	20.93	4.83
Alitalia	294	59.90	4.91
Olympic	136	21.48	6.33
JAT	92	13.92	6.61
KLM	276	37.87	7.29
EAAC	51	6.50	7.85
LOT	95	11.99	7.92
SAA	158	19.50	8.10
El Al	52	6.35	8.19
SABENA	194	23.40	8.29
AVIANCA	313	34.95	8.96
Icelandair	25	2.79	8.96
Air Algerie	59	6.53	9.04
Air France	829	91.38	9.07
C P Air	164	16.79	9.77
Cubana	108	10.91	9.90
Cruzeiro	153	15.32	9.99
Garuda	131	13.00	10.08
PAL	255	23.96	10.64

* Adjusted to the nearest decimal point.

AIRLINE	TOTAL KILLED	PASSENGERS CARRIED (IN MILLIONS)	DEATHS PER * MILLION PASSENGERS
CSA	212	18.56	11.42
IAC	327	28.58	11.44
Cathay Pacific	80	6.99	11.44
LAN	104	8.97	11.59
VARIG	299	24.01	12.45
AVIACO	172	11.22	15.33
MEA	143	8.65	16.53
PIA	222	13.29	16.70
Aerolineas Argentinas	284	16.52	17.19
UTA	73	3.63	20.11
TAROM	173	6.63	26.09
Nigeria Airways	85	3.23	26.32
THY	473	16.50	28.67
Air India	157	4.97	31.59
Egyptair	328	7.34	44.69
VIASA	144	2.76	52.17
ALIA	220	1.41	156.03

At the very least Table Two underlines the extraordinary safety record of the major U.S. domestic airlines—United, American, Eastern, and Delta—who among them carried more than 1,152 million passengers during the twenty-five-year period and suffered 1,390 casualties—just about *one ten-thousandth of one percent.* For comparison, Europe's leading airlines—British Airways, Air France, SAS, and Lufthansa—carried only one-third of the total of their U.S. counterparts, yet 17 percent more people were killed. South America's four biggest airlines did even worse: AVIANCA, VARIG, Aerolineas Argentinas, and Cruzeiro carried only 8 percent of the U.S. carriers' total and yet 1,049 people died, which translates into a death rate about nine times greater.

However, Table Two has one shortcoming in that it discriminates against airlines that carry relatively few people but transport them over much greater dis-

* Adjusted to the nearest decimal point.

tances: for example, the average Eastern customer flies less than six hundred miles on each trip, whereas every Air India passenger undertakes, on average, a three-thousand-mile journey.

It could be argued that Eastern's task is infinitely harder because the airline has to accomplish five takeoffs and five landings for each one of Air India and, as accident statistics prove conclusively, getting off and back on the ground are by far the most hazardous phases of each flight. But, obviously, each Air India passenger is airborne—and, in theory, therefore, at risk—for a much longer period than each Eastern customer.

Table Three measures the safety record of each airline by comparing the casualties to passenger-kilometers—which is the number of passengers carried, multiplied by the total distance flown. (For example, a flight which carried one hundred people five hundred kilometers would have traveled fifty thousand passenger-kilometers.) *

The death rate listed in the righthand column is the number—or fraction—of people who died for every one billion passenger-kilometers flown. Once again the airlines are listed in order of merit, beginning with those which have the best record.

* Kilometers have been used rather than miles because of the increasing international use of the metric scale.

TABLE THREE

AIRLINE	TOTAL KILLED	PASSENGER-KILOMETERS (in billions)	DEATHS PER * BILLION
TAP	0	23.64	0
Air Afrique	0	9.17	0
BWIA	0	6.71	0
Iraqi Airways	0	2.31	0
Air Ceylon	0	2.29	0
Cyprus Airways	0	2.15	0
Qantas	6	62.77	0.10
Delta	112	196.20	0.57
Continental	42	72.19	0.58
SAS	54	76.88	0.70
TAA	25	33.92	0.74
American	301	377.79	0.80
Ansett	34	30.36	1.12
JAL	125	109.10	1.15
United	585	487.67	1.20
Eastern	393	307.82	1.28
Lufthansa	126	89.94	1.40
National	143	89.00	1.61
Pan Am	560	337.48	1.66
TWA	638	380.68	1.68
El Al	52	29.08	1.79
Swissair	116	58.48	1.98
Air Canada	308	149.03	2.07
Braniff	185	84.63	2.19
NAC	22	9.46	2.33
BCAL	101	41.69	2.42
Northwest	359	125.48	2.86
British Airways	694	228.64	3.04
KLM	276	88.20	3.13

AIRLINE	TOTAL KILLED	PASSENGER-KILOMETERS (in billions)	DEATHS PER * BILLION
Alitalia	294	86.48	3.40
Aeromexico	77	21.82	3.53
UTA	73	20.12	3.63
C P Air	164	42.41	3.87
Iran Air	29	7.31	3.97
Finnair	42	9.28	4.53
SAA	158	34.04	4.64
Aer Lingus	77	15.76	4.89
Iberia	317	63.92	4.96
SABENA	194	38.75	5.01
Allegheny	153	29.69	5.15
Hughes Airwest	45	8.65	5.20
Ethiopian Airlines	14	2.56	5.46
Air France	829	149.73	5.54
EAAC	51	9.12	5.59
Olympic	136	23.47	5.79
Mexicana	126	20.48	6.15
Cathay Pacific	80	12.72	6.29
Air India	157	23.91	6.57
Austrian Airlines	26	3.62	7.17
All Nippon	358	48.80	7.34
VARIG	299	36.29	8.24
LAN	104	11.94	8.71
Air Algerie	59	5.84	10.11
Garuda	131	12.48	10.50
JAT	92	8.71	10.56
Aerolineas Argentinas	284	24.65	11.52
MEA	143	12.15	11.77
PIA	222	18.80	11.81
Cruzeiro	153	12.63	12.12

AIRLINE	TOTAL KILLED	PASSENGER-KILOMETERS (in billions)	DEATHS PER * BILLION
LOT	95	7.64	12.43
Icelandair	25	2.00	12.53
AVIANCA	313	22.68	13.80
IAC	327	21.58	15.15
Cubana	108	7.02	15.39
PAL	255	16.41	15.54
VIASA	144	8.82	16.34
CSA	212	12.45	17.03
Nigeria Airways	85	3.54	23.98
Egyptair	328	10.85	30.24
TAROM	173	4.59	37.72
AVIACO	172	4.13	41.61
THY	473	9.83	48.11
ALIA	220	1.85	118.85

One of the objections raised by some airlines against our survey was that accidents which occurred years ago are hardly relevant if the object of the exercise is to provide passengers with an approximate guide to *today's* standards. As Air France pointed out, the days when it was known as Air Chance are long gone and the airline has—at the time of writing—been accident-free for six years. That argument seems a little disingenuous because as far as we know Air France was not laboring under any peculiar difficulty when between 1950 and 1969 it notched up—in terms of numbers killed—the worst record of any airline in the world, save only for Aeroflot: a flawless six-year record does not altogether balance a dubious nineteen-year one.

But, of course, it is true that during the 1950s—and, in some cases, right through the 1960s—airlines were flying much less sophisticated aircraft. And technological advances, as well as increased experience, have rendered harmless many of the perils that threatened airliners and their passengers twenty and even ten years ago.

It is also true that airline managements and, hopefully, attitudes have changed for the better, regulations have been tightened, and the air-traffic control facilities provided at airports are generally speaking less primitive.

Table Four attempts to show which airlines have improved—or got worse—by analyzing their safety records for the last three consecutive five-year periods, 1960 to 1964, 1965 to 1969, and 1970 to 1974. The table gives the death rate for each period measured both in terms of passengers carried and passenger-kilometers flown. The airlines are listed in order of merit.

TABLE FOUR: DEATH RATE DURING PERIOD 1960–64

	PER MILLION PASSENGERS CARRIED				PER BILLION PASSENGER-KILOMETERS PERFORMED		
Airline	Total Killed	Passengers Carried (in Millions)	Deaths per Million *	Airline	Total Killed	Passenger-Kilometers (in Billions)	Deaths per Billion *
Delta	0	22.58	0	Delta	0	21.98	0
Braniff	0	12.57	0	Lufthansa	0	10.81	0
Lufthansa	0	8.89	0	JAL	0	10.20	0
JAL	0	7.41	0	Braniff	0	9.47	0
Allegheny	0	4.73	0	Qantas	0	8.12	0
Aer Lingus	0	3.74	0	Air India	0	3.87	0
Olympic	0	2.78	0	El Al	0	3.72	0
Aeromexico	0	2.57	0	SAA	0	3.70	0
PIA	0	2.56	0	PIA	0	3.04	0
SAA	0	2.34	0	Aeromexico	0	2.48	0
AVIACO	0	2.04	0	Olympic	0	1.85	0
JAT.	0	1.51	0	TAP	0	1.74	0
BWIA	0	1.44	0	EAAC	0	1.67	0
Qantas	0	1.40	0	Allegheny	0	1.55	0
EAAC	0	0.99	0	Aer Lingus	0	1.53	0
El Al	0	0.93	0	BWIA	0	1.31	0

* Adjusted to the nearest decimal point.

	PER MILLION PASSENGERS CARRIED				PER BILLION PASSENGER-KILOMETERS PERFORMED		
Airline	Total Killed	Passengers Carried (in Millions)	Deaths per Million [*]	Airline	Total Killed	Passenger-Kilometers (in Billions)	Deaths per Billion [*]
Iran Air	0	0.87	0	Cathay Pacific	0	1.27	0
TAP	0	0.87	0	AVIACO	0	0.87	0
Air India	0	0.75	0	Nigeria Airways	0	0.79	0
Cathay Pacific	0	0.70	0	JAT	0	0.79	0
Nigeria Airways	0	0.51	0	Iran Air	0	0.46	0
Iraqi Airways	0	0.44	0	Iraqi Airways	0	0.29	0
Air Ceylon	0	0.28	0	Air Ceylon	0	0.24	0
Cyprus Airways	0	0.24	0	Cyprus Airways	0	0.19	0
All Nippon	2	6.59	0.30	All Nippon	2	3.59	0.56
British Airways	21	28.72	0.73	British Airways	21	36.94	0.57
Mexicana	5	3.99	1.25	Pan Am	73	51.05	1.43
American	87	45.01	1.61	American	87	55.58	1.57
Ansett	11	5.42	2.03	Mexicana	5	2.94	1.70
Iberia	14	6.17	2.27	United	148	67.23	2.20
Eastern	132	50.21	2.63	National	29	11.85	2.45
United	148	54.28	2.73	Iberia	14	5.59	2.50
National	29	10.61	2.73	Ansett	11	4.31	2.55

* Adjusted to the nearest decimal point.

	PER MILLION PASSENGERS CARRIED				PER BILLION PASSENGER-KILOMETERS PERFORMED		
Airline	Total Killed	Passengers Carried (in Millions)	Deaths per Million *	Airline	Total Killed	Passenger-Kilometers (in Billions)	Deaths per Billion *
SAS	35	11.11	3.15	SAS	35	12.42	2.82
Pan Am	73	22.11	3.30	TWA	156	52.39	2.98
TAA	25	6.39	3.91	Eastern	132	40.83	3.23
TWA	156	31.25	4.99	C P Air	20	5.60	3.57
NAC	20	3.86	5.18	TAA	25	5.87	4.26
AVIANCA	35	6.56	5.34	KLM	64	13.80	4.64
Continental	42	7.53	5.58	Continental	42	8.42	4.99
Air Canada	113	19.08	5.92	Air Canada	113	21.33	5.30
Swissair	74	8.78	8.43	SABENA	62	7.06	8.78
IAC	40	4.59	8.71	Swissair	74	8.22	9.00
KLM	64	6.97	9.18	AVIANCA	35	3.81	9.19
C P Air	20	2.13	9.39	LAN	20	1.99	10.05
Ethiopian Airlines	6	0.62	9.68	IAC	40	3.56	11.24
LAN	20	1.90	10.53	Alitalia	148	12.30	12.03
SABENA	62	4.58	13.54	Northwest	228	16.39	13.91
Cubana	18	1.24	14.52	NAC	20	1.43	13.99
Alitalia	148	9.72	15.23	Air France	419	24.03	17.44
Finnair	42	2.68	15.67	Cubana	18	1.00	18.00

* Adjusted to the nearest decimal point.

PER MILLION PASSENGERS CARRIED			
Airline	Total Killed	Passengers Carried (in Millions)	Deaths per Million*
Icelandair	8	0.43	18.60
VARIG	93	4.89	19.02
Garuda	37	1.81	20.44
Northwest	228	10.82	21.07
PAL	114	5.14	22.18
THY	32	1.42	22.54
LOT	28	1.10	25.45
Air France	419	16.36	25.61
Cruzeiro	73	2.67	27.34
Austrian Airlines	26	0.89	29.21
CSA	116	3.91	29.67
Aerolineas Argentinas	125	3.03	41.25
BCAL	101	1.96	51.53
TAROM	92	1.18	77.97
Egyptair	120	1.48	81.08
MEA	143	1.38	103.62

PER BILLION PASSENGER-KILOMETERS PERFORMED			
Airline	Total Killed	Passenger-Kilometers (in Billions)	Deaths per Billion*
VARIG	93	4.25	21.88
Garuda	37	1.63	22.70
Icelandair	8	0.31	25.81
BCAL	101	3.90	25.90
Finnair	42	1.27	33.07
Ethiopian Airlines	6	0.17	35.29
Aerolineas Argentinas	125	3.47	36.02
Cruzeiro	73	1.95	37.44
LOT	28	0.69	40.58
THY	32	0.70	45.71
CSA	116	2.38	48.74
PAL	114	2.15	53.02
Austrian Airlines	26	0.40	65.00
Egyptair	120	1.75	68.57
MEA	143	1.70	84.12
TAROM	92	0.64	143.75

* Adjusted to the nearest decimal point.

DEATH RATE DURING PERIOD 1965–69

PER MILLION PASSENGERS CARRIED				PER BILLION PASSENGER-KILOMETERS PERFORMED			
Airline	Total Killed	Passengers Carried (in Millions)	Deaths per Million *	Airline	Total Killed	Passenger-Kilometers (in Billions)	Deaths per Billion *
Northwest	0	28.96	0	Northwest	0	37.94	0
National	0	21.82	0	National	0	26.92	0
JAL	0	15.89	0	JAL	0	26.02	0
Continental	0	14.41	0	KLM	0	20.50	0
Swissair	0	13.98	0	Continental	0	17.94	0
TAA	0	10.64	0	Qantas	0	16.88	0
KLM	0	9.35	0	BCAL	0	16.66	0
BCAL	0	8.87	0	Swissair	0	15.13	0
NAC	0	5.79	0	TAA	0	8.40	0
Finnair	0	4.17	0	UTA	0	7.09	0
Aerolineas Argentinas	0	4.16	0	Aerolineas Argentinas	0	6.08	0
AVIACO	0	3.86	0	TAP	0	5.62	0
Cruzeiro	0	3.23	0	MEA	0	3.28	0
THY	0	3.07	0	Air Afrique	0	3.12	0
JAT	0	2.91	0	Cruzeiro	0	2.83	0
TAP	0	2.80	0	EAAC	0	2.63	0
Qantas	0	2.56	0	NAC	0	2.50	0

* Adjusted to the nearest decimal point.

PER MILLION PASSENGERS CARRIED				PER BILLION PASSENGER-KILOMETERS PERFORMED			
Airline	Total Killed	Passengers Carried (in Millions)°	Deaths per Million°	Airline	Total Killed	Passenger-Kilometers (in Billions)	Deaths per Billion°
Iran Air	0	2.27	0	Finnair	0	2.23	0
Austrian Airlines	0	2.25	0	JAT	0	2.03	0
MEA	0	2.25	0	Iran Air	0	1.90	0
BWIA	0	1.77	0	BWIA	0	1.86	0
EAAC	0	1.77	0	THY	0	1.66	0
Air Afrique	0	1.48	0	AVIACO	0	1.27	0
UTA	0	1.28	0	Austrian Airlines	0	1.21	0
Ethiopian Airlines	0	1.02	0	Ethiopian Airlines	0	0.63	0
Icelandair	0	0.79	0	Icelandair	0	0.57	0
Iraqi Airways	0	0.62	0	Iraqi Airways	0	0.49	0
Cyprus Airways	0	0.43	0	Air Ceylon	0	0.47	0
Air Ceylon	0	0.41	0	Cyprus Airways	0	0.28	0
Air Canada	1	29.44	0.03	Air Canada	1	41.36	0.02
Delta	13	49.10	0.26	El Al	1	8.37	0.12
SABENA	2	5.92	0.34	SABENA	2	9.09	0.22
Aeromexico	2	4.61	0.43	Delta	13	53.66	0.24
El Al	1	1.79	0.56	Cathay Pacific	1	2.97	0.34
Cathay Pacific	1	1.65	0.61	Aeromexico	2	5.46	0.37

* Adjusted to the nearest decimal point.

PER MILLION PASSENGERS CARRIED				PER BILLION PASSENGER-KILOMETERS PERFORMED			
Airline	Total Killed	Passengers Carried (in Millions)	Deaths per Million *	Airline	Total Killed	Passenger-Kilometers (in Billions)	Deaths per Billion *
SAS	12	19.05	0.63	Alitalia	12	25.40	0.47
Alitalia	12	18.92	0.63	American	53	105.81	0.50
American	53	78.65	0.67	SAS	12	19.66	0.61
United	99	109.71	0.90	United	99	149.17	0.66
Eastern	82	91.09	0.90	Pan Am	68	96.97	0.70
TWA	88	60.02	1.47	TWA	88	109.99	0.80
Pan Am	68	39.74	1.71	Eastern	82	83.59	0.98
Lufthansa	42	20.39	2.06	Lufthansa	42	26.96	1.56
Ansett	20	9.40	2.13	Ansett	20	7.70	2.60
Cubana	7	2.62	2.67	British Airways	206	62.19	3.31
British Airways	206	44.08	4.67	Cubana	7	1.67	4.19
Braniff	118	24.98	4.72	Braniff	118	24.29	4.86
CSA	41	6.00	6.83	Air France	190	37.33	5.09
Iberia	102	14.72	6.93	VARIG	55	10.22	5.38
Allegheny	106	14.47	7.33	Iberia	102	15.50	6.58
Air France	190	23.05	8.24	C P Air	101	11.81	8.55
VARIG	55	6.27	8.77	CSA	41	3.81	10.76
IAC	75	8.00	9.38	IAC	75	6.15	12.20

* Adjusted to the nearest decimal point.

PER MILLION PASSENGERS CARRIED

Airline	Total Killed	Passengers Carried (in Millions)	Deaths per Million*
All Nippon	171	18.03	9.48
Aer Lingus	57	5.60	10.18
PAL	90	7.95	11.32
TAROM	20	1.67	11.98
AVIANCA	134	9.09	14.74
Olympic	85	5.49	15.48
LOT	48	3.06	15.69
PIA	126	7.50	16.80
Mexicana	94	5.05	18.61
C P Air	101	4.44	22.75
SAA	131	4.81	27.23
Garuda	64	2.06	31.07
LAN	80	2.27	35.24
Air Algerie	59	1.67	35.33
Egyptair	126	1.91	65.97
ALIA	50	0.60	83.33
Air India	106	1.27	83.46
Nigeria Airways	76	0.69	110.14
VIASA	144	0.79	182.28

PER BILLION PASSENGER-KILOMETERS PERFORMED

Airline	Total Killed	Passenger-Kilometers (in Billions)	Deaths per Billion*
Aer Lingus	57	4.04	14.11
Olympic	85	5.52	15.40
SAA	131	8.26	15.86
TAROM	20	1.20	16.66
All Nippon	171	9.79	17.45
PAL	90	4.95	18.18
Allegheny	106	5.74	18.47
Air India	106	5.71	18.56
PIA	126	6.46	19.50
Mexicana	94	4.81	19.54
AVIANCA	134	6.02	22.26
Garuda	64	2.59	24.71
LOT	48	1.91	25.13
LAN	80	2.82	28.37
Air Algerie	59	1.59	37.11
Egyptair	126	3.06	41.18
VIASA	144	2.63	54.75
Nigeria Airways	76	0.91	83.52
ALIA	50	0.57	87.72

* Adjusted to the nearest decimal point.

DEATH RATE DURING PERIOD 1970–74

PER MILLION PASSENGERS CARRIED

Airline	Total Killed	Passengers Carried (in Millions)	Deaths per Million*
Air France	0	35.29	0
Braniff	0	34.69	0
Northwest	0	34.41	0
Continental	0	29.54	0
TAA	0	16.83	0
KLM	0	14.54	0
BCAL	0	12.37	0
SAA	0	10.00	0
Mexicana	0	9.32	0
NAC	0	8.71	0
SABENA	0	7.99	0
C P Air	0	7.95	0
Finnair	0	7.44	0
Aer Lingus	0	7.43	0
TAP	0	6.96	0
LOT	0	6.41	0
Iran Air	0	4.81	0
Qantas	0	4.80	0

PER BILLION PASSENGER-KILOMETERS PERFORMED

Airline	Total Killed	Passenger-Kilometers (in Billions)	Deaths per Billion*
Air France	0	100.42	0
Northwest	0	54.85	0
Continental	0	42.05	0
Braniff	0	40.06	0
KLM	0	39.20	0
Qantas	0	32.99	0
C P Air	0	21.65	0
BCAL	0	20.29	0
SAA	0	18.90	0
SABENA	0	16.21	0
TAP	0	15.68	0
El Al	0	15.51	0
TAA	0	14.05	0
UTA	0	11.98	0
Air India	0	11.85	0
Mexicana	0	9.89	0
Aer Lingus	0	8.83	0
MEA	0	6.63	0

* Adjusted to the nearest decimal point.

PER MILLION PASSENGERS CARRIED				PER BILLION PASSENGER-KILOMETERS PERFORMED			
Airline	Total Killed	Passengers Carried (in Millions)	Deaths per Million *	Airline	Total Killed	Passenger-Kilometers (in Billions)	Deaths per Billion *
Cubana	0	4.43	0	LAN	0	5.57	0
MEA	0	4.15	0	VIASA	0	5.31	0
Air Algerie	0	4.04	0	Finnair	0	4.76	0
Austrian Airlines	0	3.66	0	Iran Air	0	4.69	0
El Al	0	3.25	0	Air Afrique	0	4.65	0
LAN	0	2.85	0	LOT	0	4.43	0
Air India	0	2.48	0	NAC	0	3.84	0
UTA	0	2.14	0	Air Algerie	0	3.42	0
VIASA	0	1.72	0	BWIA	0	2.98	0
Air Afrique	0	1.65	0	Cubana	0	2.64	0
BWIA	0	1.47	0	Austrian Airlines	0	1.85	0
Nigeria Airways	0	1.44	0	Nigeria Airways	0	1.58	0
Icelandair	0	1.17	0	Cyprus Airways	0	1.43	0
Iraqi Airways	0	0.99	0	Air Ceylon	0	1.28	0
Cyprus Airways	0	0.96	0	Iraqi Airways	0	1.26	0
Air Ceylon	0	0.55	0	Icelandair	0	0.86	0
National	1	28.45	0.04	National	1	38.70	0.03
Ansett	3	16.15	0.19	American	2	154.69	0.13

* Adjusted to the nearest decimal point.

	PER MILLION PASSENGERS CARRIED				PER BILLION PASSENGER-KILOMETERS PERFORMED		
Airline	Total Killed	Passengers Carried (in Millions)	Deaths per Million *	Airline	Total Killed	Passenger-Kilometers (in Billions)	Deaths per Billion *
American	2	101.26	0.20	United	42	209.69	0.20
SAS	7	29.48	0.24	SAS	7	33.46	0.21
United	42	137.70	0.31	Ansett	3	13.74	0.22
CSA	3	6.20	0.48	CSA	3	5.21	0.58
Allegheny	26	44.57	0.58	Delta	83	104.26	0.80
Delta	83	103.43	0.80	TWA	167	164.98	1.01
Eastern	164	123.12	1.33	Lufthansa	55	50.18	1.10
Lufthansa	55	36.57	1.50	Allegheny	26	21.60	1.20
Swissair	39	22.79	1.71	Eastern	164	129.08	1.27
British Airways	164	83.16	1.97	Swissair	39	29.60	1.32
TWA	167	75.09	2.22	Air Canada	101	69.76	1.45
Air Canada	101	44.23	2.28	British Airways	164	102.21	1.60
Aeromexico	22	9.38	2.35	Pan Am	252	147.50	1.71
Hughes Airwest	45	16.12	2.79	JAL	125	69.12	1.81
All Nippon	155	53.79	2.88	Aeromexico	22	11.22	1.96
Olympic	36	12.30	2.93	Olympic	36	15.66	2.30
JAL	125	39.28	3.18	Alitalia	109	45.61	2.39
Alitalia	109	29.03	3.75	Aerolineas Argentinas	33	11.46	2.88
Iberia	159	39.14	4.06				

* Adjusted to the nearest decimal point.

PER MILLION PASSENGERS CARRIED				PER BILLION PASSENGER-KILOMETERS PERFORMED			
Airline	Total Killed	Passengers Carried (in Millions)	Deaths per Million*	Airline	Total Killed	Passenger-Kilometers (in Billions)	Deaths per Billion*
PAL	33	7.61	4.34	Iberia	159	39.17	4.06
Pan Am	252	56.47	4.46	PAL	33	7.59	4.35
Garuda	30	5.75	5.22	All Nippon	155	34.98	4.43
Aerolineas Argentinas	33	5.97	5.53	Garuda	30	6.20	4.84
Ethiopian Airlines	8	1.42	5.63	Ethiopian Airlines	8	1.56	5.13
AVIANCA	66	10.32	6.40	VARIG	116	19.41	5.98
JAT	58	8.55	6.78	Hughes Airwest	45	7.15	6.29
Cruzeiro	44	5.56	7.91	PIA	55	8.29	6.63
IAC	108	12.77	8.46	AVIANCA	66	8.83	7.47
PIA	55	4.84	11.36	EAAC	35	4.25	8.24
VARIG	116	9.69	11.97	Cruzeiro	44	5.30	8.30
EAAC	35	2.78	12.59	Cathay Pacific	71	7.90	8.99
Cathay Pacific	71	4.33	16.40	JAT	58	5.49	10.56
Egyptair	60	3.03	19.80	Egyptair	60	5.44	11.03
AVIACO	84	3.69	22.76	IAC	108	9.50	11.37
TAROM	60	2.38	25.21	TAROM	60	1.89	31.75
THY	393	9.20	42.72	THY	393	6.55	60.00
ALIA	170	0.73	232.88	AVIACO	84	1.40	60.00
				ALIA	170	1.22	139.34

* Adjusted to the nearest decimal point.

From the confluence of figures it would obviously be possible to draw various and contradictory conclusions. But at least one conclusion is inescapable whatever measuring stick is used: some airlines have infinitely better records than others.

Table Five represents our judgment on the overall safety record of each airline. We have taken into account both the twenty-five-year records and the five-year analyses given in Table Four.

The airlines have been divided into six groups. Roughly speaking, each group has established a long-term safety record approximately twice as good as the group below it.

Inevitably, there are anomalies and some carriers have been categorized more harshly than they deserve if judged solely on their more recent records. Those airlines which had, as of December 31, 1975, been accident-free for at least ten consecutive years are indicated by an ↑.

Five of the seventy-four airlines we surveyed are not included in any of the six groups. They are, in alphabetical order: Air Afrique, Air Ceylon, British West Indian Airways, Cyprus Airways, and Iraqui Airways.

All five are small-league airlines, both in terms of how many people they have carried and how far they have flown them, and it would, possibly, be statistically invalid to compare them with their peers. It is also impossible to calculate their "death rates": all five share the distinction of never having killed a single passenger.

TABLE FIVE

Group One

1. TAP ↑	6. Trans-Australian	11. National
2. Qantas ↑	7. Japan Air Lines	12. Lufthansa
3. Delta	8. Continental ↑	13. Eastern
4. American	9. United	14. Iran Air ↑
5. SAS	10. Ansett (Australia)	15. Braniff

Group Two

1. Aeromexico	5. Pan Am	9. KLM ↑
2. TWA	6. El Al	10. Hughes Air West
3. New Zealand National ↑	7. Swissair	11. Allegheny
4. Air Canada	8. British Airways	12. Aer Lingus Irish

Group Three

1. Northwest ↑	7. JAT (Yugoslavia)	13. South African
2. Iberia	8. Finnair ↑	14. Ethiopian
3. Alitalia	9. All Nippon	15. Canadian Pacific
4. LAN Chile	10. Olympic	16. Air France
5. East African	11. Mexicana	17. British Caledonian
6. SABENA	12. Cathay Pacific	18. Icelandair ↑

Group Four

1. Cubana	6. Indian Airlines	11. Garuda
2. Austrian ↑	7. UTA ↑	12. Aero. Argentinas
3. AVIANCA	8. Cruzeiro	13. CSA
4. VARIG	9. Air Algerie	14. Philippine
5. Pakistan Internat'l	10. LOT (Poland)	15. Air India

Group Five

1. AVIACO	3. Middle East Airlines ↑
2. Nigeria	4. Turkish Airlines

Group Six

1. TAROM (Rumania)	3. VIASA (Venezuela)
2. Egyptair	4. ALIA-Royal Jordanian

The Basnight Memorandum

THIS IS THE full text of the "memorandum to file" written by the Western Region Director of the FAA, Arvin Basnight, on June 20, 1972, eight days after the Windsor incident:

On Friday, 16 June 1972, at 8:50 AM, I received a phone call from Mr. Jack McGowan [sic], President, Douglas Aircraft Company, who indicated that late on Thursday, 15 June, he had received a call from Mr. Shaffer asking what the company had found out about the problem about the cargo door that caused American Airlines to have an explosive decompression.

Mr. McGowan [sic] said he had reviewed with the Administrator the facts developed which included the need to beef up the electrical wiring and related factors that had been developed by the Douglas Company working with FAA.

He indicated that Mr. Schaffer had expressed pleasure in the finding of reasonable corrective actions and had told Mr. McGowan that the corrective measures could be undertaken as a product of a Gentleman's Agreement thereby not requiring the issuance of an FAA Airworthiness Directive.

In light of this data, I consulted with Dick Sliff * as we were already preparing an airworthiness directive and Mr. Sliff advised that several steps seemed advisable to prevent future explosive decompression on the DC-10 cargo doors and the Air Worthiness Directive Board was reviewing what we considered an appropriate airworthiness directive.

Mr. Sliff also indicated that earlier in the week when FAA engineers contacted the Douglas Company, the Company people had not made available any reports indicating problems encountered by the operating airlines with the cargo doors. Mr. Sliff stated that he had raised a fuss because they had not produced the information and on the following day (which was about Wednesday) the Company had produced data showing that approximately 100 complaints had been received by the Company indicating that the airlines using the DC-10 had noted and reported to the Company mechanical problems in locking the bulk cargo doors. Mr. Sliff was disturbed by the Company's attitude and felt they had not performed well in that their cooperation with the FAA in considering this data had been unresponsive.

With Mr. Sliff present, I called Mr. Rudolf, FS-1,† and reviewed with him what we were doing, the background data available to me and informed him our Airworthiness Board was in session and asked his guidance as to their continuing this effort. He suggested we continue what we were doing and wanted

---

* Richard Sliff, then Chief of the Aircraft Engineering Division, Western Region, FAA.
† James Rudolph, Director of Flight Standards Service, FAA, Washington, D.C.

a copy of our draft airworthiness directive, which Mr. Sliff had already furnished the Washington Office by telecopier.

Later in the day, I received a call from Ken Smith, DA-1,* and reviewed basically the same background information, the action of our Airworthiness Board, our judgment as to the appropriateness of issuing an airworthiness directive and was asked by Mr. Smith why our MRR's † had not made known the problem of the DC-10 bulk cargo doors so that we were as well informed as to their proper function as the Douglas Company. My response was the reporting system had not disclosed this type of data, the reason for which I would research and advise him.

Mr. Smith queried particularly why the Douglas Company's attitude might be one of not revealing the record of difficulty with the cargo doors to the FAA. Mr. Smith indicated that he concurred with our judgment that an airworthiness directive should be issued and that he would consult with the Administrator.

The Airworthiness Board continued to meet and refine the earlier drafted directive, and I continued to expect some advice from Washington as to a directive being issued. Quite late in the day, having received no advice, we called Mr. Rudolph again and were advised that they had had difficulty locating Mr. Shaffer, but Mr. Smith and Mr. Rudolph together discussed the matter with me. They were planning to have a conference by telephone with the Douglas Company and the three airlines using DC-10 equipment to assure that the objectives for effective operation of the DC-10 cargo doors sought by our proposed draft of an airworthiness directive were accomplished.

Additional time passed and we had no further advice so I again called Mr. Rudolph as we had learned through the Douglas Company that the Telecon had taken place and that steps were being taken to accomplish the objective of the proposed airworthiness directive.

Significance of this action includes the fact the drafted airworthiness directive was upgraded by the Airworthiness Directive Board based on data our engineering personnel had worked out with the Douglas Company to add an additional provision that would require drilling the fuselage near the lock bolts on the cargo door to allow a visual inspection of the locking mechanism after the doors were closed.

We so informed Mr. Rudolph at this time and asked him if it would be possible to include this provision in what we then understood to be the message transmitted by the Deputy Administrator to the Douglas Company and the three airlines involved.

Mr. Rudolph indicated that it was then after office hours in Washington. The message containing basically our draft phraseology had been released by Mr. Smith and since it was over the Deputy Administrator's signature, Mr. Rudolph could not modify the language and suggested we work the problem on Monday, but indicated he agreed with our proposed amendment.

Early on Monday, 19 June, I received a phone call from Joe Ferrarese, Acting FS-2,‡ who stated that present with him was Mr. Slaughter, Chief, Engineering and Manufacturing Division, FS-100, and that they were calling

---

* Deputy Administrator, FAA, Washington, D.C.
† MRR—Mechanical Reliability Reports, regularly sent to the FAA by airlines.
‡ Mr. Rudolph's deputy.

to advise that the teletype message signed by Mr. Smith and transmitted by Mr. Rudolph relating to the DC-10 cargo doors had been distributed to the three regions where the DC-10's are in primary use and that he was acting on instructions to ask that we destroy all but one copy of this message.

I then told Mr. Ferrarese that I had not seen the message, but would conform with his instruction. Shortly thereafter, the message came in a sealed envelope through our Duty Officer in three copies. I discussed the matter with Messrs. Blanchard and Sliff. We then called Mr. Ferrarese back, told him we had the message, were destroying all but one copy and reviewed with him Mr. Smith's inquiry regarding the MRR's.

I explained I had not engaged in trying to gain Mr. Smith's understanding of the scope of MRR's on Friday, but that I was sure he [Ferrarese] appreciated that the MRR's were designed to cover what was then considered to be significant safety factors involving maintenance and reliability and had not included items such as cargo doors which in earlier forms of aircraft were not judged to be critical to the equipment safety. Therefore, the MRR's available to us could not reasonably be expected to include data in this subject area.

There are other reports called Maintenance Information Summaries which include data of this category in abbreviated form which had surfaced some ten entries indicating such matters as AAL's N105 Trip 96/11 * being delayed 18 minutes at Los Angeles on account of difficulty in locking the bulk cargo door. This data normally is processed with a 30-day time delay and our FAA processes would not normally have surfaced a significant problem with the cargo door from this source of information, but that in retrospect, our personnel concentrating on scanning numerous entries of this nature had disclosed these ten entries.

I gave the original of the message signed by Mr. Smith and transmitted by Mr. Rudolph to Mr. Sliff for his Aircraft Engineering records and destroyed the other two copies in the presence of Messrs. Blanchard and Sliff.

Signed
ARVIN O. BASNIGHT
Director, Western Region

20 June 1972

* An American DC-10 flight from Los Angeles to New York, via Chicago.

Appendix E

The Trip Reports

THE REPORTS WRITTEN by L. F. Hazell and J. M. McCabe appear here exactly as they were written—and include their spelling and punctuation errors. Where the meaning of an abbreviation is not self-evident, an explanation in brackets follows the first use of that abbreviation. MDC is McDonnell Douglas Corporation.

MDC SENSITIVE
TRIP REPORT

9 February 1973

TO: C. L. Stout, C1-270
FROM: L. F. Hazell, C1-279
SUBJECT: TURKISH AIRLINES VISIT
COPIES TO: Distribution List

I. ITINERARY

On 2 December 1972, Turkish Airlines and DAC [Douglas Aircraft Company] crews, including the writer, ferried Ship No. 33 DC-10-10 to Yuma, Arizona for flight training. The airplane returned to Long Beach the evening of 7 December 1972. On the morning of 10 December 1972, the aircraft was ferried from Long Beach to Istanbul via Bangor, Maine and Frankfurt, Germany.

Flight Time

Long Beach–Bangor 4-40
Bangor–Frankfurt   6-05
Frankfurt–Istanbul   2-35

The DAC instructor flight crew onboard the delivery aircraft were: D. L. Mullin, pilot, and L. F. Hazell, flight engineer. Additional DAC flight crews were added to the training program as required by the delivery of the second aircraft, Ship No. 29. The additional and/or relieving personnel were messrs Milligan, Quinn (pilots) and C. E. Clark and S. Schlieper (flight engineers). The writer returned to Long Beach on 19 January 1973. The first aircraft, Ship 33, was used for trial flight training on 14 December. The following day, 15 December, an Istanbul demonstration flight was conducted in the morning, which was followed by an Ankara VIP demonstration flight in the afternoon. On 16 December 1972, the first revenue flight was made from Istanbul to Frankfurt and return. Following the inaugural flight to Frankfurt, the aircraft was put into schedule and charter service to the following cities: Munich, Dusseldorf, Hamburg, Hanover, Cologne, Stuttgart, London and Jedda. Initially, a Douglas pilot and flight engineer flew on every flight. As the THY flight crew proficiency improved, one pilot or engineer covered the flight. In order that safety was not compromised, the schedule was arranged so the DAC instructor pilot would have one of the better THY F/E's [flight engineers] on the panel and, inversely, the DAC instructor flight engineer would have two of the better THY pilots flying. This system worked out rather well and also relieved some of the cockpit congestion. As of 15 December 1972, eight of the original ten THY flight engineers were signed off for solo flight.

During the training program, several serious problems concerning the THY flight engineers were encountered. Specifically, they are as follows:

(a) F/E's training background inadequate for the job they are trying to perform.

F/E's training background is not commensurate with other world air carriers. The typical THY F/E is a recently retired or resigned Turkish Air Force single-engine pilot. In most cases, he has never served as a crewmember on a multi-engine transport aircraft either as pilot or engineer. His first exposure to a multi-engine jet transport is the DC-10 ground school at Douglas Long Beach. He has had no training in F/E theory similar to that which is necessary to pass U.S. license requirements. His background would even disqualify him from taking the U.S. F/E exam.

(b) English comprehension inadequate.

The majority of the THY F/E's could not understand or speak the English language at a proficiency required to accept DC-10 training in the time allocated. In many cases, an interpreter is required and this, of course, wastes one-half the training time.

(c) Turkish air regulations for flight engineer licensing almost non-existent.

Turkish air regulations, pursuant to F/E licensing, have been recently written to qualify a single-engine pilot for an F/E license. THY is the official examiner for the Turkish Government in the issuance of pilot and F/E licenses in that country. There is no check and balance or safety consideration with this practice since, in effect, the company polices itself.

(d) Flight engineers have no direct supervision (Chief Flight Engineer).

The THY F/E's have no direct supervisor or Chief Flight Engineer. They receive their assignments from a schedule published by the Chief DC-10 Training Pilot. There are no THY designated check flight engineers. With this policy, the F/E's are left to drift without direction or leadership. A typical case involving the lack of supervision resulted in a non-qualified THY F/E being sent solo on a DC-10 to London and back to Istanbul without a DAC or THY check F/E onboard the aircraft. Fortunately, the aircraft got back to Istanbul without disaster, but the number one hydraulic system was contaminated. This aircraft was out of service when this writer departed Istanbul.

(e) Company (THY) has no method of disseminating technical data to their flight engineers.

The company has no method of disseminating technical data such as AOL's [All Operators Letters], Know Your DC-10 Letters, etc., to their F/E's. This is a matter of serious concern since items like cold weather operation letters, revised hydraulic procedures, etc. will never get into the hands of, or be explained to, the THY F/E's after the DAC training crews depart Istanbul.

(f) Flight engineer flight kits inadequate.

The THY F/E's until recently had no idea of the type of books and equipment that is required in a flight bag: some of them do not even possess a flight bag. Most of them do not even have computers, and if they did they would have to be taught how to use them. They have no tools with them and in many cases do not carry a flashlight, night or day.

Based on the aforementioned problems outlined in paragraphs (a) through (f), the writer would recommend the following:

THY should employ contract flight engineers on a temporary basis until such time as they can be replaced by Turkish National F/E's. The replacement of the temporary F/E's by THY personnel should take place when the competency of the THY personnel is equal. This is probably a futile recommendation since Turkish law does not permit the hiring of aliens. As futile as the recommendation may be, it would make THY a safe airline now and preclude other problems which will be cited later in this report.

The problems outlined in paragraphs (a), (b), and (c) could be precluded by the following alternate recommendations. All present and future THY DC-10 F/E's should be enrolled in, and successfully complete, a course in technical English. After successful completion of the course, they should be enrolled in and successfully complete a basic flight engineer course similar to DAC or other scheduled airline requirements. With this basic minimum training, the THY F/E personnel would then be ready to advance to the Douglas DC-10 training school in Long Beach. If this recommendation is not followed, THY can never have a safe flight engineering program.

To eliminate the problems outlined in paragraph (d), the writer would recommend that THY appoint a Chief Flight Engineer immediately to direct their flight engineering program. The person selected to fill the position should be selected from outside the airline, since there is no one qualified in THY to fill the position competently.

The problem cited in paragraph (e) could be precluded by establishing a crew-read book which would contain the latest operational procedure changes and recommendations from DAC as they are published. A sign-off sheet should be included in the book, which would show that the F/E has read and understands the information in the book. If the person does not understand what he has read in the book, there should be a name and phone number on each document of a person to call (THY or DAC) to explain the material to the THY flight engineer.

To preclude the problems outlined in paragraph (f) it is recommended that each THY F/E carry a complete flight kit. A list was prepared which outlined the books, equipment, and tools that should be in a properly equipped DC-10 F/E's flight bag. The list was given to Captain D. Orhun, THY DC-10 Training Pilot. The latest class of F/E's to attend the DAC DC-10 school in Long Beach did not have flight kits. Apparently, the recommendation was not accepted for reasons unknown.

II. OPERATIONS

The following items are the most common type of operational mistakes committed by the THY F/E's. The writer will list them by number and expound on them in greater detail.

1. Takeoff with N_1 5% low, high, or staggered.

    The most common of all F/E mistakes, and one of the most dangerous, is the lack of attention to N_1 RPM, especially during takeoff. First of all, the THY flight engineers were taught to check the TRC N_1 [engine power setting] limit indicated with the chart value in FCOM, Volume 1. This is not done, although they are repeatedly reminded by DAC personnel. Secondly, after the final power adjustment, about 80 knots, neither the pilot or the co-pilot monitor N_1 RPM any longer. The attention of each pilot is devoted to flight instruments. The F/E's stare at the N_1's but do nothing regardless of how low, high, or staggered the N_1 RPM may be during the critical phase of flight. It was explained to the THY F/E's that they should immediately announce in a loud clear voice the N_1 condition so the co-pilot can correct it. It was explained that if the F/E was not sure of his responsibility pursuant to inaccurate power settings, or was afraid of embarrassing the co-pilot by announcing his bad power setting in a voice that could be heard by the captain, he should have this point clarified during the crew takeoff briefing. The THY crews were taught during simulator training that the co-pilot will set power. This is fine if the co-pilot would set and monitor power, but he sets it at 80 knots and then directs his attention to flight instruments, flight guidance, radio, and outside traffic (sometimes). Consequently, the throttles are on their own until the autothrottles are engaged. The DAC throttle policy whereby the co-pilot sets the power was re-emphasized by Mr. McCabe during his training assignment in Istanbul. The THY airplanes and engines will be in jeopardy until the co-pilot sets and devotes full attention to his task or the task is delegated to the F/E. European air carriers generally have the F/E set and monitor power, thus relieving the co-pilot of this task so he can assist the pilot in safely flying the airplane.

2. Use of the checklist for a work list and failure of the F/E to check to see if the pilot items were accomplished.

The F/E's use the checklist as a work list as they accomplish each item. This habit has developed because the F/E's are not sure of themselves in the airplane and use the checklist as a crutch not as a check. The F/E's also fail to crosscheck the pilot's response to see if the item has been accomplished as responded. Also, the F/E's do not remind the pilot if he forgets to call for a particular checklist. The checklist situation should improve as the F/E's gain much needed experience.

3. Failure to recognize weight and balance errors as high as 40,000 pounds in the paperwork completed by station agents.

During the early stages of line operation, the THY Load and Trim sheets showed errors as high as 40,000 pounds. The weight and balance forms are prepared by the individual station agents. The THY F/E's had no training in weight and balance theory prior to their enrollement in the DC-10 school in Long Beach. They received DC-10 weight and balance training, but they did not understand the subject. Initially, when the station agent gave the weight and balance form to the crew, just prior to aircraft departure, the crew just made a cursory check of the form and did not detect the errors. The errors were noted and called to their attention by the DAC flight crew onboard the aircraft at the time. One of the factors that contribute to the weight and balance errors and confusion is the fact the weight and balance form is calculated in kg. and the aircraft instrumentation (including, fuel flow, fuel used, fuel quantity, fuel catalizer quantity, and gross weight counter) are recorded in pounds. The writer would strongly recommend that THY convert their DC-10 load sheets to pounds. If THY does not want to do this, as an alternate they should convert the aircraft instrumentation to Kilograms. In addition, the writer recommends that the DC-10 F/E's receive additional weight and balance training beginning with the fundamentals.

4. F/E's do not recognize a malfunctioning airplane, therefore, do not write up squawks unless assisted by a DAC F/E.

The THY F/E's are inattentive and complacent pursuant to the observation and logging by FEFI [Flight Engineers Flight Instructions] code, aircraft squawks as they occur. In one case, the DAC F/E had recorded 15 squawks during a flight and the THY F/E was ready to leave the airplane at Istanbul without recording any of the 15 squawks. The DAC flight crew or the DC-10

Crew Chief onboard the aircraft have to consistently direct in this area. This is also due to inexperience, that the F/E's cannot differentiate between a normal and an abnormal functioning airplane. In the early stages of line operation the THY F/E's did not even check or understand the previous squawks on the airplane. They were ready to take off in an airplane completely ignorant of the aircraft's maintenance status. It would appear that the THY F/E's now understand this responsibility.

5. Shut off airconditioning packs in descent.

During line operation many of the F/E's were observed shutting down an airconditioning pack as the aircraft descent was initiated. This practice developed as a misunderstanding of a previously recommended practice. During low altitude training flights at Yuma, it was recommended that the F/E shut down a pack to reduce airconditioning noise caused by pneumatic switching. In the training situation this works out rather well since the DAC instructors can speak in normal tones and the radio volume can also be operated moderately. It was recommended that Pack 2 be shut down in order to keep the bleed extraction symmetrically on engines 1 and 3. In line operation when this practice is followed the number 2 pneumatic temperature drops from 200° C to 90° C. This results in a corresponding drop in the center and aft cargo compartment temperatures to about 68° F. In addition to the cargo temperature problem created, the cabin air change time is increased above the normal permitted. With 345 heavy smoking passengers onboard this creates a smoke removal problem. These points were explained to the THY F/E's so in the future they should be operating with 3 packs in lieu of 2 during descent.

6. Close cockpit air outlets.

It has been observed that the THY crews operate the DC-10 with the pilot and co-pilot air curtains closed, pilot's eyeball outlets closed, and in some cases, with the overhead outlets closed. THY explained that this was done for two reasons, i.e., to reduce cockpit air noise and to reduce cold drafts on the crewmembers. It was pointed out that normal air change in the cockpit should occur every 1½ minutes, based on all cockpit outlets open and the cockpit door closed. With the outlets closed, the cockpit temperature is always higher than selected and the cockpit duct temperature and pack one temperature drop to as low as 20° F. This, of course, is normal since the system is trying to cool the cockpit to the selected temperature. Avionic and instrument cooling is also affected by the reduced flow. The abnormal cockpit outlet configuration also in the writer's opinion created another problem, i.e., rain in the cockpit during descent. This opinion is based on the fact that with air curtains closed, after 2½ hours of cruise flight, frost can be felt in the cavity above each clearview. During descent, it melts and turns to water and drips from the overhead and the droplets fall on the pilot's knees. On subsequent flights the crews were requested to leave all cockpit air outlets open. The result was very good. In a short period of time, approximately 20 minutes, the cockpit temperature stabilized to the temp selected (auto dot) and duct temperature stabilized to approximately 78° F. The air coming from the cockpit air outlets, including the air curtains, was at a comfort level acceptable to the crew. In this configuration no frost formed above the clearviews as previously described, and no condensation was observed during descent. If the crews allow the cockpit temperature to stabilize, they should not have any further complaints concerning cold drafts and cockpit condensation, provided the cockpit temp knob is left in the auto set position. The two THY airplanes, 29 and 33, have very noisy cockpits, but when the latest flow control valves are installed, the noise level should be satisfactory. Until the new valves are installed, the crews will be tempted to close outlets and shut off packs.

7. F/E's leave and return to cockpit without advising the captain.

The THY F/E's have a habit of getting up and leaving the cockpit without advising the pilot. Once again, this is something that the F/E should already know, but due to their limited training and background, they do not know they are doing anything wrong. If an emergency should arise in the cockpit during the F/E's absence, obviously the safety of the DC-10 is in jeopardy to a greater degree with one less crewmember in the cockpit to handle the emergency.

8. They operate the DC-10 with the cockpit door unlocked.

The F/E's operate the aircraft with the cockpit door circuit breaker C/B pulled during line operation. They do this to accommodate the cabin crew that serve them food and beverage during the flight. Another reason for not locking the cockpit door is because in the past F/E's have locked themselves out of the cockpit because they did not have a key and were unaware that the door is self-locking if the C/B is engaged. Since skyjacking has been prevalent throughout the world, the writer has suggested that the door be locked and that all flight crews be issued keys. This suggestion was ignored in part in that some crewmembers were issued keys and THY still operates with the cockpit doors unlocked (C/B pulled).

9. Attempt to fly trans-Atlantic flight without a fuel log.

On the flight from Bangor to Frankfurt, the THY F/E had no idea that a fuel log is required for a trans-oceanic flight. This is understandable since it was his first trip over the ocean as a F/E and he was never trained for the task. The writer had some Northwest fuel and performance logs, so these were used for the flight. The writer recommends that the THY F/E's receive cruise control and performance training and that the airline develop and use their own performance logs. It is recommended that personnel from Operational Engineering be made aware of this deficiency.

10. Lack of attention to F/E panel and pilot's panels and failure to develop a systematic scan pattern.

The last of the serious F/E operational mistakes is their failure to develop a systematic scan pattern of aircraft instruments. When they do decide to look, they don't see. Once again, perhaps experience will correct this deficiency.

To summarize the THY flight engineer program, one could say conservatively that it has a long way to go. One of the most baffling aspects of the program is the fact that the THY F/E's ask few or no questions. Whether this is due to shyness or indifference, the writer cannot tell. Prior to the arrival of the second aircraft, it was decided to give the THY F/E's a review DC-10 course, using the slides and tapes used in the course at Long Beach. A duplicate set of slides and tapes were loaded on the second aircraft at Long Beach and have not been seen since that time. The general consensus is that the tapes and slides are lost in Turkish customs in Istanbul. One ground school session was given in an attempt to correct some of the things listed in this report. Five F/E's showed up and they had to leave early. A class was scheduled for the next day, but no THY F/E's showed up.

THY aircraft experienced two cases where the flaps were damaged during landings at Istanbul. Contrary to reported rumors the aircraft in each cases did not land short. The old runway at Istanbul is in a very bad state of dis-repair. The new runway cannot be used at night because it has no lights. THY has on occasion sent a jeep out on the runway to check for debris before a DC-10 is scheduled to land. It is the writers opinion that a new, clean runway with lights would preclude future flap damage on DC-10's at Istanbul.

THY has received very bad press during the flap damage incidents. The reports were very inaccurate and written to make the airline and the DC-10 look bad.

On the subject of maintenance, the major concern of THY is the poor performance and reliability of the APU [Auxiliary Power Unit]. When the APU doors were removed the APU reliability improved greatly. Frequent pneumatic abnormal lights and high pack flows are also quite chronic. The galley personnel lift has failed on occasion; however, the DAC crew chief managed to get it operable each time it failed. The galley cart lift sustained damage due to the cart door being left open.

The aircraft interior is very dirty due to lack of maintenance. The rugs and seats are stained very badly. The toilets are serviced in Istanbul without chemicals. The exterior of the aircraft is getting very dirty particularly in the area of the outflow valve. Tobacco stains are evident on the fuselage from the outflow valve all the way back to the wing. THY does not have a hanger large enough to accommodate the DC-10; consequently, all maintenance is performed under adverse conditions. The stabilizer is set to 0° after the last flight of the day to close the exposed opening above the stabilizer and to preclude the injestion of snow, rain, dirt, etc. The cabin outflow valve is closed after the last flight for the same reason; however, this procedure can cause a problem. Sometimes in the morning the ground crews will connect external power to the aircraft and leave the outflow valve closed. This procedure can lead to avionic cooling problems since the normal exit of cooling air is blocked. The thrust reversers have been troublefree; however, the flight crews are not using No. 2 engine reversing when landing away from Istanbul unless they really need it. Since maintenance support and equipment that would be required to repair the No. 2 reverser is not readily available at those stations, the policy was adopted. The first DC-10 tire change occurred 13 January 1973, after the passengers were loaded for a flight to Frankfurt. The THY mechanics could not find the adapter that is required to remove the wheel retaining nut. The nut was removed with a hammer and a brass punch. The new wheel/tire assembly was installed and "torqued" in the same manner.

In Germany, the ground handling and servicing of the THY DC-10 has been outstanding. Fueling, toilet servicing, baggage handling, potable water, and galley servicing has been accomplished in record time and without incident. This is due greatly to the assistance and instruction given to the German personnel by DAC crew chiefs that cover each THY trip. The German station agents also prepare accurate weight and balance forms. The only potential problem encountered in Germany was the failure of the Germans to understand THY radio request for a pneumatic cart, in the event the APU did not start. The APU started so there was no problem. However, it should be noted that the Germans call a pneumatic cart an air start. They are available in abundance at each airport in Germany travelled by THY.

In summary, the pilots love the airplane, and it is performing better than advertised. Airports in Germany have extended the night takeoff curfew for the DC-10 based on satisfactory sound recordings of the DC-10 on previous takeoffs at these airports. It is rumored among THY personnel that THY may buy more DC-10's and that they may fly charter flights this summer to New York. The flight engineer situation at THY should be resolved immediately. As insignificant as this job may seem to some people in the THY organization, it should be noted that one incompetent DC-10 flight engineer could trigger an incident and possibly jeopardize future sales by Douglas.

L. F. Hazell
Production & Delivery Pilot
Flight Operations

LFH:sr

MDC SENSITIVE
TRIP REPORT

6 April 1973

TO: C. L. Stout, C1-270
FROM: J. M. McCabe, C1-270
SUBJECT: TURKISH AIRLINES
COPIES TO: Distribution List

I. ITINERARY

This Trip Report covers the period of 26 November 1972 through 14 March 1973. During this time 15 crews from THY received simulator, flight transition and line training on the DC-10 Series 10. Fourteen different Douglas pilots and 12 different Douglas flight engineers participated in this program involving a grand total of 507 instructor man days and which represents an expenditure of 235 man days over and above that identified in their contract. Simulator training was conducted in the Douglas Flight Crew Training Center in Long Beach, flight training at Yuma for the first 10 crews and at Istanbul for the last 5 crews, line training was conducted on every revenue flight that THY performed in and through Europe and the Middle East until a satisfactory degree of proficiency was evidenced.

II. OPERATIONS

THY's introduction of the DC-10 into revenue service was accomplished in 11½ weeks from the signing of the Purchase Agreement. From the standpoint of flight crew training, this was a record and presented many unusual problems. To set the stage for this trip report, a brief review of this carrier is in order.

THY is the flag carrier of Turkey and, as is in many government-controlled airlines, management is frequently changed as the political scene shifts with each election. The Flight Operations management wants very much to run a safe, efficient and profitable airline but they are dealing with bureaucracy at every level, unwillingness to accept responsibility, and personnel that are technologically deficient. It was my observation that basic management techniques involving staffing, communications, and training are lacking when compared to U.S. or European standards.

General aviation, as such, is practically nil in Turkey. 100% of the pilots and flight engineers employed by THY come from the Turkish Air Force. Again, the political influence is felt. It has often been said that pilots run the airline and this is true. Four of the top operations management personnel flew together in the Turkish Air Force Acrobatic Team.

THY's first exposure to commercial jet operations was when the DC-9 was delivered in August 1967. The same situation as described above existed then to a greater degree. In the past five years there have been five different Chief Pilots.

Some progress has been evident in that the performance department has been established and substantial efforts made in the area of standardization. THY, since that time, has leaned heavily upon Douglas and uses the DAC DC-9 Flight Crew Operating Manual. From this first giant step to jet operations I am sure THY management felt confident that, with DAC assistance, the second step into wide-bodied tri-jet was feasible.

December 16, 1972 was established for entry into revenue service and this necessi-tated working backwards from this date to establish a flight training program. There was no time to spare in this accelerated program. It became evident as the ground school training progressed that a definite language barrier did exist with certain students. As is our custom, a progress report was made to THY Flight Operations management of the crew members that were weak and below average. Approximately ten hours of additional cockpit procedures trainer time was given these individuals. This group was mainly comprised of flight engineers. THY is a very unusual airline in that out of the 160 pilots employes, 153 are Captains and the 7 so-called First Officers are due to be upgraded in the near future.

They have few flight engineers since the previous requirement was only to crew three (3) B-707's. A management decision was made to remove only a few of these B-707 engineers into the DC-10 program. The balance of the flight engineers that were assigned to the DC-10 were new THY pilots coming from the Air Force where they flew fighter airplanes and with perhaps 200 to 300 hours on the B-707. This flight engineer knew that the DC-10 flight engineer's task was an interim stepping stone and eventually he would become an airline captain. As a result, he neither had the interest nor inclination to learn his job, nor could he fall back upon a solid background of flight engineer experience. The result was inevitable.

With the additional instruction, all THY crew members passed their final exam, however, it is suspected there was collaboration and their chief training pilot acknowledged this as a fact. This lack of proficiency was carried over to the simulator training program and was further aggravated by the fact that the airline was using a 24-hour per crew syllabus versus the Douglas recommended 32-hour per crew syllabus. Again, in certain individuals, a lack of the required level of proficiency was apparent and has been documented. As Harry Truman's desk sign said, "The buck stops here," and the last phase of flight crew training is airplane flight transition. In this phase, we were forced to disqualify 3 pilots and 2 flight engineers.

On 10 December 1972, the THY airplane departed for Istanbul via Bangor, Maine and Frankfurt, Germany. A Litton INS [inspection] system was installed for the trans-Atlantic crossing. It functioned admirably and the ferry flight was routine. Upon arrival in Istanbul, the next few days were spent doing local train-ing and V.I.P. demonstrations at Istanbul and Ankara. The DC-10 entered revenue service on 16 December 1972, with scheduled service to Munich and Frankfurt and charter flights to Dusseldorf, Hamburg, Hannover (Hanover), Cologne (Koln), Stuttgart and Juddah (Jidda). All line operations were covered by at least one Douglas flight crew member. THY's second DC-10 arrived from Long Beach on 19 December and it too was pressed into immediate service.

The consensus of the DAC flight instructors was that the THY pilots, with the exception of those that had failed their flight checks, were basically competent in their ability to fly the airplane and function as a line pilot on their route structure. This was not the case with the flight engineers, the majority of whom required several flights before they were released to conduct a trip on their own. This has been expounded upon in depth by Leo Hazell (Douglas Flight Engineer) in a prior trip report.

There are many operational items that were revealed, deficiencies in crew co-ordination, procedures, techniques, etc., as a result of our observation of THY's route flying. These will be broken down in various categories and elaborated upon.

A. *Government Control*

In Turkey there is no strong governmental regulatory body in aviation, such as our F.A.A. THY sort of writes their own rule book using our own Federal Aviation Regulations as a guide. At times there have been infractions due to lack of knowledge, lack of communications, and perhaps even deliberate waivers when, in the eyes of THY, safety is not a factor. The point is, there is no one to monitor and police other than their own check pilots. Flying in such an environment one can see the absolute necessity for standardization. This was one of the major areas of this writers' concentration during his visit.

B. *THY Flight Crew Publications*

Each crew member had his own idea of what required flight crew publications should be carried and the airplanes themselves had inadequate documentation aboard. A list was compiled for the airplane documentation that included a complete set of Maintenance Manuals, FEFI, TAFI, Minimum Equipment List with Operational and Maintenance Procedures, FAA Manual, Flight Crew Operating Manual (3 volumes), Flight Engineers Reference Manual, airport analysis, takeoff landing data cards, hard-card checklist. One or several of these items was repeatedly missing. This caused many delays and inconveniences. The problem was eventually solved by making airplane documentation a checklist item.

The Flight Engineers particularly were very casual in carrying the required manuals. While most of the pilots would not carry anything but their Jeppesen route manuals and the Company operations manual, which had not been updated to include the DC-10. They felt that their carrying the Flight Crew Operating Manual was redundant and what they really wanted was a condensation of their manual; its pertinent aspects, operational information that they would use on a day-by-day basis—without carrying a heavy volume. At the request of THY management such a compilation was made, approved by their Training and Operations Departments, and was met with wide approval by the line pilots. The THY Operations Manual was updated, actually it was re-written in a new format and given a new name, "Turkish Airlines Operational Handbook." It is composed of six sections:

1. Standardization and Flight Techniques
2. Performance
3. Limitations
4. Minimum Equipment List
5. Technical Bulletins
6. Training Bulletins

This handbook was written in Turkish by a Turkish Captain, under the tutelage of myself. I was given the prerogative of approval or disapproval. The first section (Standardization and Flight Techniques), which comprises 37 typewritten pages, required the greatest amount of editorializing.

C. *Standardization*

Some areas where standardization was achieved are listed below:

1. Standard callouts. The Turkish crews had already drifted away from using standard terminology on instrument approaches.
2. Lack of a standardized noise abatement profile for takeoff. They have now adopted the ATA [Air Transport Association] procedure.
3. Communications between the cockpit and ground lacked standardization in phraseology, procedures, and techniques. Use of the Flight Guidance and Control System particularly in the descent profile.
4. Use of autothrottles in cruise. A decision was made to use their autothrottles in cruise provided that excessive hunting was not evidenced.
5. Numerous checklist changes were made that more closely reflect the THY operations rather than the Laker (from which their checklist originated). Their normal checklist was modified and an intermediate stop checklist and a training checklist was recommended and adopted.
6. The deletion of the delayed engine start procedure and checklist. Their current route structure is such that it obviates any need for a delayed engine start.
7. Flight Engineer responsibility and preflight walk-around. They use the Douglas prepared form and, in addition, a procedure was established for a postflight walk-around immediately subsequent to engine shut down. This policy tends to minimize intermediate departure delays.
8. Crew coordination and responsibility in setting takeoff power and the use of reverse thrust. Initially many, many instances were observed of over and under boosting the engines that went either unnoticed or was not brought to the attention of the Captain in command of the airplane. The "First Officer" would either be still adjusting power at rotation speed or would make such a gross adjustment that it amounted to the equivalent of a 15% deration in thrust with subsequent performance loss.
9. Engine starting procedures. Maximum gas temperature EGT [exhaust gas temperature] was rarely observed as the pilot starting the engines would continue on with another engine after the starter button popped out. This condition was successfully solved by requiring a mandatory call-out and visual observance of EGT subsiding prior to starting the remaining engines.
10. The arming of cabin doors was not in accordance with FAA regulations. This was brought to their attention and corrected.
11. It was recommended that the cabin doors be opened from the outside by

station personnel rather than the cabin attendant. Many times these doors would be opened when there were no off-loading ramps in position. The typical Turkish passenger stampedes toward the open exit creating a potentially hazardous condition.

12. To prolong the service life of electronics equipment, a list of suggested avionics equipment circuit breakers was supplied to THY so that applicable C.B.'s could be painted and the flight crews could disable them at the termination of a flight. This, too, was made a checklist item.

13. THY had no standard operating procedure on their DC-10 in case of air piracy. A suggestion was made that the Pan Am procedure used on the B-707 be adopted in its entirety, which it was. This information was disseminated to all DC-10 flight crews.

14. A decision was made to use published minimums for landing and takeoff.

D. Weight and Balance

Weight and balance computation. What would normally be a relatively simple task becomes anything but, due to the fact that while the aircraft instrumentation is completely read out in pounds, the weight and balance form must be calculated in kilometers. It was a toss-up as to who made the most mistakes; the station agents or the flight engineers. The flight engineers did not have conversion tables nor computers to quickly determine the conversion. The weight and balance form is generally the last piece of paper to enter the cockpit prior to engine start. It is a time of pressure to get the airplane dispatched, but it is a time that requires a careful assessment of the airplane gross weight and C.G. [center of gravity]. At the outset there were mistakes on more than 50% of their weight and balance computations. Errors were made on the magnitude of 10,000 pounds, 22,000 pounds, and 40,000 pounds. Many of these were not caught by the THY crew, but only by the DAC flight instructor. Needless to say, a concentrated effort was made to educate all parties concerned. Even the Captains were given a quick rule-of-thumb to help identify a major discrepancy. This kind of problem is not new to Turkish Airlines. There have been incidents in the past involving other equipment as illustrated on a DC-9 flight from Istanbul to Frankfurt on 20 January 1971.

Turkish Airlines has recently adopted a policy of allowing their passengers to carry on all of the hand baggage they desire. There is a current advertising campaign to this effect which is very popular with the Turkish workers going to and returning from employment in Germany (similar to our Bracero program). Captain Bilgi indicated that he intended to run a sample of these passengers to determine a representative amount of baggage carried on by them. It is not known whether this has actually been accomplished. THY had made no provisions to supply their flight engineers with a weight and balance template which facilitates the execution of this form. This was brought to their attention, but no action was taken, so in desperation, a telex was sent to Long Beach where they were specially made and shipped to THY.

There were some instances reported of overloading the baggage containers. Too much weight in individual containers. There was at least one instance where a cargo container came loose in the forward baggage container in the airplane during flight. Knowing the in-service problems with baggage handling and operation of cargo doors, R. H. LaCombe had a flight cargo chief supervising all cargo handling which, for the most part, eliminated such instances.

E. Performance

Although THY was cognizant that other airlines were using reduced thrust on takeoff, they were reluctant to instigate this procedure. The crew members did not understand the presentation in their operations manual. When this became apparent, a suggestion was made to conduct a supplemental ground school on performance and reduced thrust in particular. This class could have been conducted by Mr. Gumusel, Captain Orhun or myself. It never materialized; however, before my departure, I tutored Mr. Gumusel and Captain Orhun in the use of de-rated thrust and left with them sample problems which could be used in a ground school. Another factor which made them hesitate to use de-rated thrust was the difficulty they were experiencing in takeoff gross weight determination.

None of the route segments that THY is currently flying are critical from a performance standpoint. Nevertheless, when a flight is dispatched it should be done knowing what the climb and descent schedule will be, as well as the Mach number, in order to accurately determine fuel requirements. Douglas flight instructors observed the entire spectrum in this regard—long range to high speed in climb, cruise and descent. This was eventually standardized to climb at 300 knots, .82 Mach number, cruise at Mach .82, descent at .82 to 340 or 250 knots below 10,000 feet.

The airplane was initially being dispatched with no company reserve. A flight planned for an .82 Mach cruise would have additional fuel aboard to fly to the alternate and hold for 30 minutes. However, flying at the higher speeds ate into the reserve fuel which would have created a very grave situation if it had proved necessary for the airplane to depart for its alternate. A company reserve of 5,000 pounds was recommended and accepted. It was brought to THY management's attention the importance of strict adherence to the cruise Mach as filed. The airplane was never flown at takeoff weights greater than 390,000 pounds and was generally much lighter; however, the pilots do not like the "slow" climb rate, especially at the higher altitudes. Frequent spot-checks were made and excellent correlation with book values was determined.

A standard flap setting of 15° was eventually decided upon. This would be used in all cases unless the airport analysis for second-segment climb could not be met with 15°. In this case, a 10° flap setting would be used. THY used the takeoff and landing data speed cards not only in training but in revenue operations. They were shown that these speed cards had *no correction* for wind, slope, or altitude and could not be used for de-rated thrust.

THY elected to use the Douglas prepared airport analysis. Takeoff performance for eleven (11) different airports was prepared by P. H. Patten's Operational Engineering Department. Three (3) additional airports were hand computed by Mr. Gumusel. They like the thoroughness and additional information which the Douglas format offers, but dislike the bulk of the manuals. They have this airport analysis in three (3) volumes. It is THY's Performance Department's intent to develop a simplified presentation for certain high frequency airports similar to Continental's format and reduced to Jeppesen size.

Two-engine ferry flight training was given to Captain Davut Orhun, Chief Training Pilot. This involved a page-by-page explanation of our FAA Manual Appendix, as well as a simulated two-engine ferry takeoff. Subsequent training of THY management pilots will be accomplished by Captain Orhun.

A personal search was made in THY's Operations Department for FAA Approved Flight Manuals. The Chief Pilot had a copy (locked up), Mr. Gumusel, Performance Engineer, had a copy, and that was all—other than the manuals on the airplane and in the Maintenance Library, which was more than ¼ mile away. Investigation with our FAA Liaison showed that THY received fewer manuals than any other airline and steps were taken to provide them with additional copies.

Considerable interest was expressed in using the autothrottles for takeoff. They had the erroneous impression that they could use the autothrottles in takeoff and were disappointed in training when they found out they could not. When this was relayed by telex to the Program Office with a request for a cost quotation, THY was told that once Service Change 790 was incorporated, they would be able to use autothrottles at no charge.

Excerpts from DAC Operational Engineering's Performance Handbook relating to estimated stopping distances due to abnormal configuration and conditions were disseminated verbally and included in their new DC-10 Operations Handbook.

THY requested a chart for determination of brake energy capability following a rejected takeoff. The chart that they would like is patterned after a Boeing chart. The Douglas Flight Crew Operating Manual at present does not have such a chart.

They also felt that our Flight Crew Operating Manual performance should have an optimum altitude chart two two-engine cruise.

The $V_2 + 10$ [takeoff speed plus ten knots] noise abatement takeoff profile to 3,000 feet on occasion involves a turn and a requirement for restricting the bank angle to 15°. This was discussed repeatedly and eventually became standard operating procedure.

Weather in Europe this past December and January was particularly bad and THY had made no provisions to compensate for runway surface conditions. Data extracted from the DAC Performance Handbook again was used to determine takeoff and landing performance degradation. This was put in each airplane and disseminated to each crew member.

F. Flight Crew Operating Manual

This airline is one of the many carriers that made the decision to use the Douglas Flight Crew Operational Manual. Although, in the broad sense, they like the manual, they submitted some criticisms of it which are being considered by this department. They are as follows:

1. Retrievability and understanding of emergency and abnormal procedures. They prefer the ATLAS plastic insert presentation over our Volume 1 and once the procedure has been found they prefer a DC-9 "black, line, follow the arrow" presentation which makes it practically impossible to go wrong. They are currently thinking about adopting the ATLAS Volume 1 concept; they will modify the inserts themselves.
2. A good tabulated conversion chart from pounds to kilos was needed. These were eventually furnished from Long Beach.
3. They are critical of the fact that the cold weather operations section is an appendix to the manual and not an integral part of it.
4. They felt that the Flight Crew Operating Manual should contain a buffet onset chart patterned after the FAA manual.
5. The Douglas Flight Crew Operating Manual does not show fuel tank capacity or undumpable fuel.

G. Training

With the delivery of the first THY DC-10, there went aboard a complete training program, approximately 3,500 35 MM slides, programmed instruction in the form of cassette tapes, movies, etc. This material somehow was never received by the THY Training Department. It has been documented to the fact that it was onboard the airplane. In addition, 60 sets of Flight Crew Operating Manuals, Flight Engineer Reference Manuals, Flight Study Guides, etc., were also sent onboard the first airplane. Only Volume III of the Flight Crew Operating Manual was ever found. These are the tools of the Training Department. Without them, recurrent training will be sadly deficient. This material could certainly have been used to advantage in doing remedial training for the THY flight engineers. A class for new flight engineers was started using just the Flight Crew Operating Manual with no training aids whatsoever. This was taught by Mr. Aykurt. This gentleman was appointed as Chief Flight Engineer for THY, a position that heretofore was non-existent, as inconceivable as that may be. The suggestion to establish a Chief Flight Engineer position was tactfully presented to THY management at every opportunity. His name still does not appear on any organization chart, although it is recognized by his colleagues that he does truly have this position.

The THY Training Department, under Captain Celebi, is in downtown Istanbul, approximately 45-minutes away from the airport where flight operations and maintenance personnel have their offices. Captain Celebi has the responsibility of doing all training, maintenance, operations, cabin attendant, etc., for the airline; but has not started yet on the DC-10. Product Support Maintenance Training representatives had been on-site for several weeks before their first class was scheduled. This could be attributed to the fact that their training materials had also been lost. When the class did convene, it convened at the airport, not in the training center. In defense of the separate location, there really is no adequate facility at the airport to do training and although the terminal and complex is in a continual state of construction, there are no plans to build a training center there.

Three standardization meetings were held with THY management and at this time it was strongly recommended that the Douglas flight instructors provide ground school on the material discussed in these meetings. We agreed that we would have repeated meetings as long as requested to cover all of the personnel. Two meetings were held for flight engineers, but only four of the ten flight engineers ever attended. The problem of the lack of basic experience of the flight engineers was discussed at length with Captain Celebi. It was suggested that basic training in the fundamentals involving weight and balance, wiring diagrams, avionics, etc. be taught these flight engineers. It was proposed that they attend an FAA Approved school in the United States or adopt their curriculum. The best suggestion was to send Mr. Aykurt and possibly one other flight engineer to IATA's basic flight engineer school that is taught in Beruit, Lebanon. This was in their route structure and involved a minimum amount of expense. Mr. Aykurt, once trained thoroughly, could gradually transmit this knowledge to other THY flight engineers. This has yet to be accomplished.

One of the gravest problems encountered in training was THY's lack of formalized training in emergency evacuation and ditching procedures. This is the responsibility of the carrier; however, when it became apparent that they had not taken any steps in this direction, this writer took the initiative and had the entire emergency evacuation and ditching program that was approved by our FAA (as demonstrated by CAL [Continental Air Lines] and AAL [American Airlines]) mailed to Istanbul. This slide package identified crew responsibility and was very explicit in its instructions. I am happy to relate that this was adopted in its entirety.

While searching THY's storeroom for the missing training materials with Captain Celebi, we stumbled upon the emergency evacuation instructions for passengers printed on hard-card forms. I realized then that the airplane had been flying in revenue service for three weeks without these important instructions covering the use of emergency equipment.

Many flight engineers had very limited knowledge of cabin equipment. A list was compiled for THY training department that covered the more salient items that a Flight Engineer should be knowledgeable in. Items such as how to restore oxygen mask, door operation, abnormal operation of the galley lift, water shutoffs in the lavatories and many others. We were promised that a program would be initiated to train the flight engineers in these areas.

There appeared to be a definite lack of communication between the flight deck and the cabin attendants, other than to ask for coffee, tea, or milk. A suggestion was made to conduct entire crew briefings involving all flight personnel in the ready-room prior to departure for the airplane. The advantages of this entire crew briefing were pointed out as practiced by many other airlines. Another area that was weak in communications involved alerting the cabin attendants when to "arm" and "de-arm" the doors.

There were three Captains that had failed the flight checks in the United States, and a THY decision was made that they would receive three months "on the line training" prior to being checked out again. Since it was in the area of procedures that these gentlemen failed their flight check, it was felt that this three month special time is a good solution. While waiting for a second and third airplane, the flight engineers undergoing training in Turkey were also scheduled for frequent line observation flights. THY felt, and this writer agrees, that this kind of training is invaluable. The Turkish Airlines plan to accomplish the flight training for the third group of 5 crews was altered due to the alleged high cost that was quoted to them. Douglas management felt we should still play a part in their flight training to assure ourselves that at least the same minimum level of proficiency was attained. Therefore, the services of a pilot and flight engineer were offered to THY to assist in the ferry flight and two weeks of flight training in Turkey. This was naturally accepted—it was free. The flight training was accomplished by THY Captains Bilgi, Celebi, and Orhun and Douglas flight instructor O. T. Quinn. In the 5 to 6-hour flight syllabus, the same pattern was apparent in the flying ability of the pilots and the lack of experience of the flight engineers.

Mr. Quinn reports they have increased their requirement to 50 hours of line training under supervision prior to release as pilot in command. One of the areas brought to the attention of THY in the standardization meetings involved cockpit crew discipline. As has already been reported by Mr. Hazell, Douglas Flight Engineer, the flight engineers were prone to leave the cockpit for prolonged periods without permission. In a cockpit environment where two pilots, both captains, are flying, there was no clear-cut crew member responsibilities defined. As the saying goes, "more than one cook spoils the broth." These responsibilities were identified, spelled-out in a 37-page Standards Procedures Techniques section of the THY Operations Handbook. The same lack of cockpit discipline was noticed by General Leuman, USAF retired, who rode as a guest of THY on the third aircraft delivery flight. His comment was that he intended to broach the subject with General Ihsan Goksaran. The latest report (from Douglas flying crew chiefs) smacks of rebellion in the cockpit with flight engineers refusing to obey the Captain's commands. Such an infraction in most airlines is grounds for dismissal, or strong disciplinary action.

Turkish Airlines is looking to the U.S. to do its annual and semi-annual recurrent familiarization training. National Airlines and American Airlines are contenders for this training along with ourselves. Knowing the Turks, they will probably choose the cheapest program. It is suggested that our own management seriously consider a major reduction in our simulator prices to capture this airline, so that we can continue to exercise a level of control over their operation.

Yesilkoy Airport has had a new 10,000 foot runway for approximately two years, but still does not have any lights—runway lights, taxi lights or approach lights, it is not serviced by an ILS [Instrument Landing System], and its use therefore is restricted to daylight operations. The weather is frequently below that required for a non-precision approach, thus further minimizing its utilization.

It is difficult for an American to appreciate the problems of disseminating information. At Yesilkoy Airport, where the Flight Operations department is located, there is one Gafax machine for reproduction which was broken down a good percentage of the time.

III. MAINTENANCE AND DESIGN

Due to the short time period between the signing of the contract and delivery of the first DC-10, it would have been impossible for THY to support their own airplane initially. A contract was signed with Douglas Product Support organization to provide the necessary technical assistance in troubleshooting, on-the-job training, setting-up logistic support, etc. Mr. R. H. LaCombe, Director, Product Support, headed up this task force of approximately 30 people. Mr. LaCombe literally did the impossible, battling against customs, spare parts shortages, language barrier, lack of technical equipment. He was able to achieve a mechanical delay rate substantially below the overall DC-10 fleet average. Although the statistical delay rate appears low, during my 24 flights with the Turkish Airlines I *never* departed on time. Still, during my stay there, I saw progress; where a two-hour flight delay was typical, I saw it visibly diminish. Most of the delays involved passenger handling, i.e., baggage loading, connecting flights from Ankara, etc.

On the plus side of the ledger, THY has just received FAA certification for their maintenance shop. The presence of the Douglas crew chief onboard is extremely valuable and has prevented many delays by a timely bit of on-the-spot instruction. When the Douglas flight crews were doing route checking with THY, the DAC crew chiefs resided in the cabin unless their services were specifically requested on the flight deck. However, it was discussed with THY management and Mr. LaCombe and it was mutually agreed that, space permitting, it would be advantageous for them to be in the cockpit after our departure. Many of the THY flight engineers lack fluency in their ability to express themselves in English and cannot spell correctly. By promoting a closer relationship with the crew chief and the THY flight engineers, it was hoped the overall level of systems knowledge of the flight engineer would improve—allowing better reporting practices. All discrepancies were to be reported with a FEFI code. One of the difficulties that allegedly did arise with non-flight crew members in the cockpit is that they occasionally gave advice on use of the flight guidance and control system that involved its operational use (procedures and techniques).

A fly-away kit was carried on the airplane, however, a list of its contents was not carried aboard. Fortunately, this was rectified before the need arose to use items in the kit.

THY had established a practice of not removing the landing gear pins until after all engines were started. They were then brought aboard and laid on the floor between the left and right forward cabin doors. These were the four-foot long gear pins and were frequently wet and dirty and presented a very bad appearance to the travelling public. We were eventually able to convince the Turks that a set of small clean gear pins should be carried in the cockpit for emergency use and that the other pins, both at Istanbul and route stations, could be pulled prior to engine start as is the practice by airlines throughout the world. A screwdriver was frequently used in lieu of the nose gear bypass pin which was left in on occasion. A suggestion was made to fabricate a bypass pin with a long flag and make this an integral part with the tow-bar so that when the tow-bar was disconnected, the mechanic would remember the bypass pin. This was adopted.

There is considerable congestion at Yesilkoy Airport and it is a wonder there are not more accidents. The ramp area is not stripped nor are there stanchions, or any other means of determining adequate wing clearance. Airplanes do not always park in the same place, therefore, the flight crew must be continually alert while pushing back or taxiing in these close quarters. I saw a Pan American Boeing 747 wait for 45 minutes for ground control to give them directions for parking on the ramp. Finally, the Captain of the 747 shut down his engines on the taxiway and disgorged all of his passengers—which added another one-hour and 15-minute delay to the mass confusion. One of our recommendations to THY management was to immediately take steps to paint wing clearance lines along the ramp taxiways, similar to what we have on our own Douglas flight ramp. They agreed but this has not been accomplished as yet.

Yesilkoy Airport has no snow removing equipment to clear the taxiways or runways but must rely on natural means. They have one Gylcol truck but the first day that we had accumulated a fair amount of ice and snow on the wings they ran out of Gylcol after doing one wing—and there was no more Gylcol in Istanbul! Again, all flights were cancelled until the weather melted the ice. This shows a definite lack of planning and foresight. As the ice started to melt, I was approached and asked if we could take a fire hose and wash the airplane down. Such ignorance really surprised me because Istanbul does get snow and ice conditions every winter. Gradually we were able to institute some practices, that of closing the outflow valve and repositioning the horizontal stabilizer in inclement weather; and were able to eliminate the practice of parking the airplane in a turn. The taxi man would frequently turn the airplane in too tight an arc.

IV. PERSONNEL

The following personnel were contacted during this assignment:

> Gen. Yellman, General Manager
> Nezihi Unsal, Technical Manager
> Captain Aytekin Bilgi, Chief Pilot
> Captain Davut Orhun, Chief Training Pilot
> Captain Atila Celebi, Chief Training Manager
> Captain Erhan Suar, Assistant Chief Pilot
> Hikmer Gumusel, Chief of Performance & Operational Engineering

V. SALES

Every year there is a mass exodus of non-skilled Turkish workers to European countries primarily Germany. These Turkish workers save their money and either send it home or bring it home. This improves the Turkish balance of payments. The Turkish government pays 40% of their air fare home. THY, prior to the DC-10, was unable to handle but a small portion of this traffic. The Turkish workers are nationalistic and recognize that if they fly on THY, which is a government airline, they will save the government money. The first two weeks that the DC-10 was in service, they carried 15,000 passengers from Europe to Istanbul.

During our stay in Istanbul, the F-28 arrived from Holland and was introduced into revenue service. THY has expressed interest in the DC-10 Twin but will need an interim airplane. I was asked about the possibility of leasing stretched DC-8 airplanes which would be supported by Douglas Product Support in a similar fashion as the DC-10. There are mixed feelings on this subject. The Boeing sales-men were in town early in March making a strong push to sell THY the Boeing 727-200. One thing is certain, some aircraft will be needed as their DC-9 Series 10 is due to be returned to Douglas this month and their three Boeing 707's lease expires the end of May and December, this year.

Unless THY demonstrates a greater concern for their passengers, I would antici-pate a decrease in their load factor. For example, when a delay is incurred, the passengers are seldom notified. I do not mean a 5 or 10 minute delay, I am refer-ring to delays of from 45 minutes to 1-hour and 45-minutes. There are no in-flight announcements from the flight deck as to weather, points of interest, etc. When a flight of the Turkish workers was cancelled due to weather, they were bussed from one airport to another, with their baggage following. I witnessed an overnight delay where no provisions were made for the ticketed passengers and they had to sleep in the terminal in some cases, and in another case the airplane was made available to them and they slept on the unheated airplane at the ramp. I witnessed decisions wherein luggage for 50 passengers was left in Germany (on such a transfer of personnel) because it involved an additional 30-minute delay. The flight had been delayed 30 minutes but it meant that the passengers baggage now would not arrive until more than 24 hours later and upon their arrival at Istanbul, they would be confronted with another 24-hour wait for their baggage before they could depart for Ankara and other eastern cities. The Turkish workers accept this kind of treatment because of their anxiousness to get home and because they know nothing better. The worst illustration of this occurred in Frankfurt where the workers had been waiting all night for the airplane (many more workers than space was available). When it became apparent to them that they could not all have seats they threw their hand-luggage into the engines in-takes, sat down in front of the landing gear, and refused to move. The Germans had to use a fire hose to clear the ramp so the airplane could be dispatched. The Turks advertised the DC-10 as the "First in Europe" which was not really true but it was good adver-tising. It did improve the image of the Turks throughout Europe.

VI. MISCELLANEOUS AND RECOMMENDATIONS

Perhaps the picture of Turkish Airlines painted in this trip report is too severe, but it's important that Douglas management understand reality in that part of the world. The Turks are a proud people and a curious mixture of Eastern and West-ern cultures. They know they need help and yet are reluctant to admit this to someone that might look down upon them as an inferior people. However, if they sense a genuine interest in the individual or company attempting to assist them they are most receptive and accept advice willingly.

Prior to my departure on 2 February, I met at length separately with Captain Bilgi, Chief Pilot, and Mr. Uncil, Technical Manager, at which time I summarized the observations of myself and of the Douglas flight personnel. The subject matter was not solely related to flight training by any means, but were pertinent to the airline operation as a whole. Most of the recommendations have already been enumerated in the context of this report. On 17 March, Mr. O. T. Quinn, also confirmed the fact that the airline has adopted the majority of these recommendations.

However, three sensitive areas have not as yet been covered. I tactfully pointed out the need for better communications between their respective departments and facilities recommending an approach similar to our Smart Meetings. Another area involved THY's organizational structure. They appear to be over-staffed in the menial tasks and under-staffed in middle management positions. Mr. Uncil himself, although very competent, appeared to have attempted to do too much himself; and should delegate more, but there is no one to delegate to. In the Operations and Training Department, there is a need for more technical assistance by non-flying personnel, employes that could perhaps be ex-crew members, so that they understand the problems. Importance of additional education for flight engineers in basic knowledge of DC-10 systems also was emphasized.

These meetings went extremely well and I returned home confident that our recommendations would be either incorporated or given serious consideration. In a prior trip report by Leo Hazell, he pin-pointed many operational mistakes which were reiterated prior to my departure.

The airline is smoothing out, significant progress can be noted especially by one that is intimate with the whole situation from its beginning. Nevertheless, it is this writer's opinion that the Douglas influence and support either contractually or non-contractually should continue. We have been told that the Flight Operations Department has an open door for our Commercial Operations Revisit Program.

J. M. McCabe
Manager, Flight Crew Training
Flight Operations

The Victims of Ship 29

IDENTIFYING ALL THOSE who died in the Ermenonville crash was an enormous task for the French police. Because of the general confusion at Orly Airport on that Sunday, and the last-minute rush for seats just before the DC-10's departure, the Turkish Airlines passenger manifest contained dozens of inaccuracies: many of the victims' names were spelled incorrectly and some didn't appear on the list at all. For at least six weeks the police were not even certain of how many people had died and it required painstaking inquiries in twenty-one countries to produce the full catalog of victims.

The list below is based on the records of the French police. The victims are listed according to their nationality.

ARGENTINE:

Mr. Pedro Pernias, 52, married, two children, writer and his wife Ilda Lina Elisa, 53, and their daughter Alejandra Maria, 15, student

AUSTRALIAN:

Mrs. Christine Carmen Bland, 46, married, librarian
Miss Rachel Margaret Cowie, 47, accountant

BELGIAN:

Mr. Jean-Jacques Coussens, 42, married, two children, driver

BRAZILIAN:

Mr. Nelson Lins Bahia, 32, married, two children, civil engineer and his wife Nailka, 29
Mr. Amadeu Pumar Bergamini, 22, student
Mr. Mario Jorge de Souza Castanon, 26, architect
Mr. Roberto Antonio Campanella dos Santos, 22, student

BRITISH:

Mr. Harry Abrahamson, 67, married, two children, jeweller and his wife Florrie, 65
Mr. Thomas William Addis, 71, widower, four children, retired driver
Mr. Brian Lucius Arthur, 32, married, two children, real estate agent
Mr. John Steven Backhouse, 43, married, three children, insurance agent

Mr. Roy Bacon, 32, married, two children, secondhand dealer
Mrs. Theresa Gene Edith Barrow, 41, married, three children, sales manager
Mr. Ronald Charles Frederick Beasley, 56, married, one child, merchant and his
wife May Florence, 54, factory worker
Mr. Michael George Beckwith, 28, married, one child, doctor and his wife Patricia
Ann, 30, teacher
Mrs. Claire Black, 46, married, three children, artist
Mr. Charles Christopher Bowley, 53, married, two children, doctor
Mr. Ivor John Bradley, 46, married, two children, marketing director
Mr. Sydney Maurice Bressloff, 52, insurance agent
Mr. Robert Anthony Breton, 33, married, three children, accountant
Mr. Geoffrey Reginald William Brigstocke, 56, married, four children, civil servant
Mrs. Bessie Victoria Brown, 55, married, one child, traffic warden
Miss Susan Mary Burn, 23, secretary
Mrs. Jean Mary Burris, 48, married
Mr. Jack Burtonwood, 40, married, three children, engineer
Mrs. Georgina Evelyn Byatt, 46, married, one child, teacher
Mr. David John Steven Cain, 22, student
Miss Esther Winifred Collen, 20, computer programmer
Mrs. Diana Connelly, 43, married, one child, shorthand-typist
Mr. James Conway, 58, married, one child, trade union officer
Mr. John Hugh Cooper, 33, married, business representative
Mr. Arthur Lawrence Cornish, 39, married, one child, printer
Mr. Richard Tomlin Coult, 22, civil engineer
Mr. David Lewis Cowell, 36, married, one child, farm products representative
Mr. William Claude Dack, 39, married, three children, sales manager
Mr. Trevor Vincent Dangerfield, 37, married, company director
Mr. Hubert James Davies, 49, married, two children, lawyer
Mrs. Margaret Eileen Davies, 58, married, one child, social worker
Mr. William Harry Deane, 59, married, one child, company director and his wife
Brenda Winifred, 45
Mr. Ronald William Delaney, 47, married, three children, merchant
Mrs. Marjorie Magdalen Dickman, 73, widow, three children and her sister
Mrs. Gladys Mary Rowlandson, 77, widow, one child, company director and Mrs.
Rowlandson's daughter Joyce Fairburn, 44, and her husband Bryan Egerton
Fairburn, 45, architect
Mr. Kenneth Christopher Doran, 60, married, retired air force officer and his wife
Phyllis Gertrude, 52, two children (previous marriage)
Mr. Mark Dunsford Dottridge, 31, writer
Mr. Eric George Dullforce, 61, married, one child, shipping line manager
Mr. Bryan Graham Ellis, 35, married, two children, company director
Ms. Jean Constance Fillery, 35, divorced, secretary
Mr. Carlo Alberto Fontana, 38, married, one child, engineer
Mr. Joseph Jack Freer, 54, married, two children, engineer
Mr. Nicholas Jeremy Stewart Fripp, 25, married, one child, wine merchant
Mr. Ian Roger Fuller, 24, fashion model
Mr. Robert James Gillman, 21, trainee print buyer
Mr. Peter Jeremy Green, 27, married, two children, farmer
Mr. Ernest John Griffin, 42, married, five children, engineer

Mrs. Isobel Kennedy Griffin, 52, widow, two children, shop assistant

Mr. Michael Hannah, 24, actor-fashion model

Mr. William James Patrick Hayes, 50, married, six children, sales executive

Mrs. Francis Eustace Hill, 55, married, two children, company director

Mr. Paul Anthony Hillman, 25, real estate agency director

Mr. Francis Michael Hope, 35, married, one child, journalist

Ms. Joyce Marjorie Hope, 53, divorced, two children, film librarian and her son
   Mr. Keith John Madge, 25, film technician

Mr. George Desmond Hunt, 47, married, two children, court official

Ms. Sheelagh (Cave) Hurble, 45, divorced, four children

Mr. Patrick John Hutton, 43, married, physicist and his wife Anne Elizabeth, 40,
   secretary

Mr. Malcolm Harry Jackson, 29, married, three children, fashion buyer

Mr. Nicholas Paul Jones, 16, student

Mr. Eric Kelsey, 52, married, three children, café owner and his wife Catherine
   Esther, 50, teacher and their daughters Helen Esther Ann, 10, and Ann Mary
   Olga, 6

Mr. Christopher Mather Kendall, 31, married, two children, company director

Mr. Stuart Frank King, 19, salesman

Mr. Otto Sigmund Kohnstamm, 72, married, one child, doctor and his wife Winifred, 51, secretary

Mr. David Herman Kween, 30, married, two children, merchant (and his wife
   Phyllis, listed under U.S. victims)

Mr. John Anthony Law–Wright, 33, married, two children, engineer and his wife
   Susan Vanessa, 28

Mr. Graham Cecil Levet, 18, grain trading trainee

Mr. Morton Lloyd Lewis, 56, divorced, journalist

Mr. Elfryn Lloyd, 46, married, five children, aeronautical designer

Mrs. Jessica Emma Maney, 64, married, one child, merchant and her sister-in-law
   Mrs. Sarah Elizabeth Rosher, 71, widow, one child, merchant and Mrs. Rosher's
   daughter Mrs. Sarah Elizabeth Sykes, 52, widow, one child

Mr. Thomas Roger Marriage, 37, married, three children, farmer

Mr. John Patrick McClinton, 26, sales representative

Mr. Peter Graham McDonald, 35, married, two children, professor and his wife
   Wendy, 34

Mr. James Wilson McMeekin, 47, married, three children, civil servant

Mr. Jack Middleton, 54, married, one child, company director and his wife Joan
   Margareth, 51

Mrs. Anna Philomena Middleton, 47, married, personnel director

Mr. Robert Arthur Milne, 36, married, compositor and his wife Pauline, 35, design
   artist

Mrs. Hedwig Morley, 65, widow, one child

Mr. Rex Morley, 40, married, three children, farmer

Mr. Maurice Clive Morris, 40, numismatist

Mr. Peter Graham Morris, 29, married, lawyer

Miss Helen Victoria Rumsey Neale, 38, research economist

Mrs. Mable Doris Newton, 53, widow, civil servant

Mr. David John Russell Nicholls, 35, two children, photographer

Mr. Tadeusz Okuniewski, 46, married, three children, engineer

Mrs. Gaynor Mary Osborne, 26, married, teacher

Mr. James Allan Overton, 37, married, three children, engineer and his wife Susan Brenda, 31

Mr. Leslie George Paine, 47, married, four children, mining engineer

Mr. Julian Glynne Parish, 31, helicopter pilot

Mr. Terence Frederick Paul, 42, married, three children, company representative

Mr. Herbert Payman, 76, married, one child, electrical engineer

Mr. Martin Robert Perry, 26, married, insurance agent

Mr. Malcolm Ronald Portch, 47, married, one child, company director and his wife Enid Dilysea, 41, company director and their daughter Trudi Melinda, 6

Mr. Ronald Edward Powell, 46, married, two children, chartered accountant and his wife Ann, 45

Miss Prudence Hedley Pratt, 30, fashion model

Mr. Anthony Price, 39, businessman

Mr. Anders Sipo Qunta, 59, married, one child, doctor and his wife Margaret Mary, 60

Mr. Geoffrey Rawlinson, 52, company director

Mr. Paul Raymond, 28, lecturer

Mr. Desmond Armstrong Reay, 59, company director

Mrs. Hannah Marianne Richards, 59, married, nurse

Mr. Richard Anthony Riley, 35, married, three children, company director

Miss Lynne Barbara Roberts, 23, student

Mr. Nigel John Norcliffe Roberts, 24, wine merchant

Mrs. Sylvia Mary Beaumont Roch, 61, married, one child

Mr. George Thomas Rogers, 63, married, two children, retired company secretary and his wife Ivy Maud Letitia, 57

Miss Patricia Margaret Rogers, 46, company director

Mr. Anthony Colin Ryan, 21, student

Mr. Gregory Robert Anthony Rynsard, 18, student

Mrs. Rosalind Vivian Saeger, 29, married

Mr. Ronald Mark Sanders, 46, married, one child, jewelry representative and his wife Dorothy May, 51, company director

Mr. Frederick Ernest George Sanders, 42, married, one child, company director

Mr. Charles Henry Sargeant, 49, married, three children, farmer

Mr. Robert Briggs Savidge, 20, assistant farm manager

Mr. David George Scott, 27, bakery worker

Miss Pamela Sheel, 28, fashion model

Mr. John Frederick Slater, 33, married, one child, steward and his wife Gloria, 31, and their daughter Nadine Nicolas, 4

Mr. Laurence Congleton Smith, 46, managing director

Mr. Malcolm Sorkin, 44, married, lawyer and his wife Ruth Yvonne, 43

Mr. Edward William George Sturgess, 57, married, lithographer and his wife Edith Amelia, 57

Miss Erica Ann Sworder, 34, nurse

Mr. Michael Tilbrook, 36, married, agricultural merchant

Mr. David Alexander Tomlin, 43, married, two children, accountant

Mr. Michael Frederick Townshend, 43, bank manager

Mr. Robert William Wallis, 30, married, one child, student

Mr. Peter John Walsh, 23, lithographer

Mr. Edwin William Wardley, 46, married, five children, leather merchant
Mr. Lionel James Ware, 28, married, two children, helicopter engineer
Mr. Peter Frederick Warnett, 36, married, one child, technical services representative
Mr. Sidney Waterhouse, 48, married, four children, textile representative
Mr. John Victor Watkins, 49, married, two children, company director and his wife Mary, 44, nurse
Mrs. Georgina Weld-Forester, 22, married, fashion model
Mrs. Wendy Wheal, 24, married, fashion model
Mrs. Glenda White, 35, married, two children
Mr. Michael Rodney Whitehead, 32, married, two children, engineer
Mrs. Isabella Wilkinson, 51, widow
Mr. Owen James Williams, 28, merchant
Mr. John North Lewis Wilson, 25, married, research economist and his wife Joyce, 24, teacher
Mr. John Winston Winterton, 63, married, three children, industrialist and his wife Winifred, 61, company director
Mr. Peter John Withers, 38, married, three children, company director
Mr. Graham Milton Wood, 37, married, three children, insurance broker and his wife Margaret Rosemary, 33
Miss Linda Woods, 27, commercial artist
Mr. Peter William Woolvett, 25, fashion designer
Mr. Peter Walter Leslie Zuelchaur, 28, married, marketing development assistant

CYPRIOT:

Mr. Nebil Mustafa, 23, student

FRENCH:

Mr. François de Bellegar-Malhortie, 44, married, four children, engineer
Miss Françoise Bernard, 22, student
Mrs. Tamar Bouhadana, 31, married, two children, fashion designer
Mr. Abel Roger Donnat, 21, assistant personnel manager
Mr. François Druesne, 35, married, five children, chemical engineer
Mr. Robert André Jean Lerendu, 34, married, one child, bank director and his wife Josette, 30 and their daughter Corinne Elise Simone, 5 and Mr. Lerendu's brother Mr. Jacques Auguste Louis Lerendu, 38, married, two children, company secretary
Ms. Monique Lezay, 33, divorced, five children, clerk
Miss Lucie Ludmilla Loubenzow, 26, secretary
Mr. Ladislas Mayer, 42, married, two children, aeronautical engineer
Miss Françoise Pauly, 26, teacher
Miss Martine Poisson, 18, student
Mr. Robert G. Reynaud, 43, married, two children, sales supervisor
Miss Betty Françoise Wormser, 21, student

GERMAN (WEST):

Miss Christiane Muller, 26, chief librarian

INDIAN:

Mr. Rasiklal N. Kothary, 40, married, one child, textile industrialist
Mr. Ashok Ramanlal Shah, 37, married, three children, industrialist

IRISH:

Mr. James McDonald, 59, priest

JAPANESE:

Mr. Takanori Awakura, 23, bank management trainee
Mr. Fumihiro Fujita, 22, trading company management trainee
Mr. Junichi Goto, 23, bank management trainee
Mr. Masaki Hayashi, 24, trading company management trainee
Mr. Takehiro Higuchi, 26, married, architect and his wife Atsuko, 25, clerk
Mr. Masamichi Hiramatsu, 24, bank management trainee
Mr. Shigeo Hosoi, 23, bank management trainee
Mr. Masaki Imazeki, 23, bank management trainee
Mr. Hiroshi Ishibashi, 23, trading company management trainee
Mr. Tomo-o Ishikawa, 23, trading company management trainee
Mr. Ikuo Itoh, 23, businessman
Mr. Bobuo Iwanaga, 23, bank management trainee
Mr. Makoto Kameyama, 22, trading company management trainee
Mr. Takashi Kamiya, 24, bank management trainee
Miss Rieko Kanai, 23, student
Mr. Junzo Kato, 22, trading company management trainee
Mr. Kazumasa Kawabata, 26, textile company manager
Mr. Kenichi Kawaguchi, 22, bank management trainee
Mr. Kazuyuki Kembo, 23, trading company management trainee
Mr. Motohisa Kinugasa, 22, bank management trainee
Mr. Hiromasa Kobayashi, 22, bank management trainee
Mr. Yasushi Kohama, 23, bank management trainee
Mr. Yushitami Kojima, 22, bank management trainee
Mr. Yoshiro Kumezawa, 23, bank management trainee
Mr. Tadanobu Matsume, 28, married, one child, tour director
Mr. Yasuyuki Mohri, 23, city welfare employee
Mr. Norikazu Morikawa, 22, bank management trainee
Mrs. Fumi Nagao, 48, married, two children
Mr. Takashi Ogura, 22, trading company management trainee
Mr. Michiaki Ohhama, 23, bank management trainee
Mr. Yuji Okamura, 23, bank management trainee
Mr. Tetsuo Okura, 23, bank management trainee
Mr. Tadaharu Sakata, 22, bank management trainee
Mr. Kenichiro Sakurai, 26, student
Mr. Kakuichi Sato, 23, bank management trainee
Mr. Tetsuro Sato, 23, bank management trainee
Mr. Masami Shimizu, 23, bank management trainee
Mr. Takeo Shinmi, 22, bank management trainee
Mr. Hiroshi Shiraki, 23, trading company management trainee

Mr. Yasunori Takashima, 23, bank management trainee
Mr. Shoji Tani, 23, bank management trainee
Mr. Yasuhiro Tominaga, 44, married, two children, company manager
Mr. Hiroshi Tsuchiya, 23, bank management trainee
Mr. Yoshihiko Tsuji, 22, trading company management trainee
Mr. Kimio Tsukada, 24, married, trading company management trainee
Mr. Mitsuo Yano, 23, bank management trainee
Mr. Tsukuru Yoshitake, 22, trading company management trainee

MOROCCAN:

Mr. Mohamed Saad Chaouni, 22, student

NEW ZEALANDER:

Mr. James Wedgwood Leslie Berg, 37, married, two children, company director

PAKISTANI:

Mr. Abdul Razzak Lak-Hani, 34, married, two children, importer

SENEGALESE:

Mr. Boubacar Sy, 40, married, merchant

SPANISH:

Mr. Ernesto Justo, 25, head waiter

SWEDISH:

Mrs. Margot Vilma Lassagnes, 78, married, one child, film and television director

SWISS:

Mr. Rainer Pfister, 26, race car driver

TURKISH:

Mr. Omer Faruk Afir, 21, student
Mrs. Serife Tulga Akbulut, 34, married, two children, television producer and her
son Mehmet Yunus, 2
Mrs. Clarita Arditti, 27, married, two children, and her son Julian Edgar, 2
Mr. Zeki Dogan Beller, 37, married, two children, maintenance manager for THY
Mr. Halil Bengi, 26, married, one child, merchant and his wife Yildiz Saime, 24,
receptionist
Mr. Mehmet Bircan, 36, hotel waiter
Mr. Oktay Cebeci, 35, electronics engineer
Mr. Mustafa Ihsan Celebi, 20, student
Mrs. Selvinaz Cetin, 30, married, two children
Miss Meral Ciril, 17, student
Mrs. Fatma Gonul Cizravi, 25, married

Miss Sevim Emeksiz, 20, student
Mrs. Ilgun Erdin, 31, widow, civil servant
Mr. Fikri Gurel, 29, married, waiter and his wife Hasibe, 21
Miss Fani Ime, 21, student
Mr. Mehmet Engin Inan, 21, student and his sister Hatize Burcin, 25, student
Mrs. Aliye Filiz Kapur, 29, married, one child, engineer and her son Bazak Kerime, 2
Mr. Hayim Kohen, 51, married, two children, exporter and his wife Coya, 43, exporter
Mr. Engin Koray, 24, student
Mr. Ali Nihat Koz, 66, married, two children, engineer and his daughter Mevhibe, 17, student
Miss Sukran Mihlar, 17, student
Mr. Metim Oner, 29, married, one child, assistant professor and his wife Seniha Gonul, 24, receptionist
Mr. Ahmct Oral, 27, student
Mrs. Vilma Ortaagopyan, 37, married, two children
Mr. Mustafa Serafettin Ozturk, 41, married, three children, office worker
Mr. Mehmet Sadettin Pekerol, 55, divorced, three children, importer
Mr. Nazif Sen, 33, married, one child, waiter
Mr. Necmettin Senyurt, 24, engineer
Mr. Ahmed Simsek, 19, student
Mr. Islam Temizkan, 25, airline employee
Mr. Mose Uziyel, 24, married, one child, chemical engineer and his wife Rasel, 24, and their daughter Liat, 9 months
Mrs. Vildan Yardimci, 24, married, civil servant
Miss Gunseli Yuruker, 17, student

CREW OF SHIP 29:

Miss Fatima Rona Altinay, 29, stewardess
Miss Fatma Barka, 25, stewardess
Mr. Mejat Berkoz, 43, married, two children, captain
Miss Ayse Birgili, 21, stewardess
Miss Semra Cazibe Hidir, 19, stewardess
Mr. Huseyin Erhan Ozer, 36, married, three children, flight engineer
Miss Selim Sibel Sahin, 22, stewardess
Miss Gulay Naciye Sonmez, 21, stewardess
Mr. Hayri Tezcan, 29, steward
Mr. Hasan Engin Uzok, 44, married, three children, ground mechanic
Mr. Oral Mehmet Ulusman, 38, married, co-pilot
Miss Nilgun Yilmazer, 22, stewardess

UNITED STATES:

Mr. Penn Boyd Blair, 29, married, one child, machine company vice-president
Mr. Gary Ray Chard, 36, married, four children, airline sales manager
Mr. Roman Wojciech Filipkiewicz, 63, married, two children, petroleum engineer
Mr. John Hanessian, Jr., 49, married, four children, professor

Mr. Loren Mell Hart, 41, married, three children, geologist and his wife Alice
   Joan, 40, their son Robert Raymond, 12 and their daughters Nancy Joan, 14,
   and Terri Diane, 16
Miss Nancy Beth Kalinsky, 22, student
Mrs. Phyllis Kween, 29, married, two children (wife of David Kween, listed under
   British victims)
Mrs. Carol Lemmer, 28, married, one child, hospital administrator
Mrs. Clara Lux, 79, widow, one child, typing school director
Mr. Rehn Claus Peterson, 49, married, three children, company president
Mr. Milton Safran, 49, married, four children, antique dealer
Mr. Ronald Lee Smith, Jr., 12
Mr. Wayne Ayres Wilcox, 41, married, four children, cultural affairs officer and
   his wife Ouida Rae, 38, their son Clark, 12 and their daughter Kailan, 15
Mr. Thomas Prescott Wright, 40, married, three children, banker and his wife Fay,
   38, their son Jackson, 7 and their daughters Carlaine, 10 and Sherryl, 8

VIETNAMESE (NORTH):

Miss Danielle N Guyen Thi, 22, student

Index

A-300B Airbus, 73*n,* 163, 202, 282
aerodynamics, 22-3, 26-7, 29; "coffin corner," 40, 95; "deep stall," 91-2; "jet upset," 94-5; pressurization problems, 124-5, 135ff
aero-engines, 288-9;
    "compound engine," 41;
    fuselage mounting problems, 89-93;
    General Electric *CF-6,* 73, 86, 98, 112, 119, 281-2, *J-79,* 73, *TF39,* 28, 67, 73;
    jet propulsion:
        fan-jet, 27-9, 67, 73-4, 86, 92-3, 98, 281-2, bypass, 74-5, 91, 143, 291, 295, diameter fans, 75, turbine blades, 93-4, 143-5, 147
        gas turbine "pure jet," 25-8, 41, 75, 89, 290-1
        ramjet, 290
    metallurgy, 24, 25, 74, 101;
    military stimulus, 24, 26, 28, 29, 51;
    noise levels, 90-1;
    piston: high-compression horsepower development, 23-5, 37, Liberty engine, 24, fuel, 24;
    power to weight ratio, 23-4, 29, 30;
    Pratt and Whitney, 37, 41, 119; Double Wasp, 25, *JT3D-7,* 75, *JT-7D,* 295, *JT-9D,* 112, 143-5;
    propellers, 26-7, 290;
    rear-mounted, 90-1;
    Rolls-Royce, 24, 29, 41, 75, 103, Dart, 74, 103*n,* fan-jet bypass, 74, 75, *RB-211,* 29, 75-6, 78, 211, 295, costs and price, 100-2, 114-15, 116, technology, 101, 104, 114, 115, 201, *RB524,* 86*n;*
    "shrapnel," 93-4, 144, 197;
    supersonic, 28-9;
    turboprop, 27, 28, 74, 291;
    Wright designs, 35, 36, 37, 40, 41

Aeroflot, 38, 191
Aerospace International, 70
aerospace industry, *see* aircraft industry
Agnew, Spiro, 85
Air Canada DC-8 flight 621 crash (1970), 172-4
Air France: B747 engine explosion (1970), 144; safety record, 198, 382
Air Holdings Ltd., 78
air hostesses, 192-3, 198-9
air speed: Mach meter flight, 95; and stalling, 94-5;; transonic, 94-5
air speed records: transport aircraft, 36, world absolute speed record, 52
Air Transport Association, 169, 173, 194
Air Transport Group (Australia), 198
airbus, 65-8; "all-purpose airplane," 76, 86; engines, 73-6; financial lift, 85; orders competition, 71-3, 76-82, 107, 111, 117, 201-2, 281, 282; tri-jet concept, 68, 76, 89, 92; wide-body formula, 68, 77, *see also* Boeing 747, Douglas DC-10 and Lockheed L-1011 Tri-Star
aircraft accidents, 30, 92, 95, 122, 172-4, 188, 192, 194, 196, 305, 311-83; B707 (1963), 197; B727